T0339867

Energy-Growth Nexus in an Era of Globalization

Energy-Growth Nexus in an Era of Globalization

Edited by

Muhammad Shahbaz
Energy Economics, School of Management and Economics,
Beijing Institute of Technology, Beijing, China

Aviral Kumar Tiwari
Rajagiri Business School (RBS), Kochi, Kerala, India

Avik Sinha
Centre for Excellence in Sustainable Development,
Goa Institute of Management, Goa, India

ELSEVIER

Elsevier
Radarweg 29, PO Box 211, 1000 AE Amsterdam, Netherlands
The Boulevard, Langford Lane, Kidlington, Oxford OX5 1GB, United Kingdom
50 Hampshire Street, 5th Floor, Cambridge, MA 02139, United States

Notices
Knowledge and best practice in this field are constantly changing. As new research and
experience broaden our understanding, changes in research methods, professional
practices, or medical treatment may become necessary.

Practitioners and researchers must always rely on their own experience and knowledge in
evaluating and using any information, methods, compounds, or experiments described
herein. In using such information or methods they should be mindful of their own safety
and the safety of others, including parties for whom they have a professional responsibility.

To the fullest extent of the law, neither the Publisher nor the authors, contributors, or
editors, assume any liability for any injury and/or damage to persons or property as a matter
of products liability, negligence or otherwise, or from any use or operation of any methods,
products, instructions, or ideas contained in the material herein.

Library of Congress Cataloging-in-Publication Data
A catalog record for this book is available from the Library of Congress

British Library Cataloguing-in-Publication Data
A catalogue record for this book is available from the British Library

ISBN: 978-0-12-824440-1

For information on all Elsevier publications visit our website at
https://www.elsevier.com/books-and-journals

Publisher: Joseph P. Hayton
Acquisitions Editor: Graham Nisbet
Editorial Project Manager: Chris Hockaday
Production Project Manager: Maria Bernard
Cover Designer: Victoria Pearson

Working together
to grow libraries in
developing countries

www.elsevier.com • www.bookaid.org

Typeset by TNQ Technologies

Contents

3.   The energy consumption-growth nexus in Jamaica: does globalization matter?

Adian McFarlane, Anupam Das and Kaycea Campbell

7.   The effect of globalization on energy consumption: evidence from selected OECD countries

Burcu Ozcan, Ali Gokhan Yucel and Mehmet Temiz

8.   The electricity retail sales and economic policy uncertainty: the evidence from the electricity end-use, industrial sector, and transportation sector

Faik Bilgili, Pelin Gençoğlu, Sevda Kuşkaya and Fatma Ünlü

15. Making green finance work for the sustainable energy transition in emerging economies

Suborna Barua and Shakila Aziz

18. **Energy consumption, financial development, globalization, and economic growth in Poland: new evidence from an asymmetric analysis**

Yılmaz Toktaş and Agnieszka Parlinska

List of contributors

Alex O. Acheampong, Newcastle Business School, University of Newcastle, NSW, Australia; Centre for African Research, Engagement and Partnerships (CARE-P), University of Newcastle, NSW, Australia

Andrew Adewale Alola, Department of Economics and Finance, Istanbul Gelisim University, Istanbul, Turkey; Department of Economics, School of Accounting and Finance, University of Vaasa, Finland

Uju Violet Alola, Department of Tourism Guidance, Istanbul Gelisim University, Istanbul, Turkey

Erkan Alsu, Department of Business Administration, Gaziantep University, Gaziantep, Turkey

Rafael Alvarado, Carrera de Economía and Centro de Investigaciones Sociales y Económicas, Universidad Nacional de Loja, Loja, Ecuador

Mary Amponsah, Newcastle Business School, University of Newcastle, NSW, Australia; Centre for African Research, Engagement and Partnerships (CARE-P), University of Newcastle, NSW, Australia

Shakila Aziz, School of Business & Economics, United International University, Dhaka, Bangladesh

Eugenio Baita-Saavedra, Saitec, Leioa, Bilbao, Spain

Suborna Barua, Department of International Business, University of Dhaka, Dhaka, Bangladesh

Faik Bilgili, Erciyes University, Faculty of Economics and Administrative Sciences, Kayseri, Turkey

Elliot Boateng, Newcastle Business School, University of Newcastle, NSW, Australia; Centre for African Research, Engagement and Partnerships (CARE-P), University of Newcastle, NSW, Australia

Kaycea Campbell, Department of Economics, Los Angeles Pierce College, Woodland Hills, CA, United States

Laura Castro-Santos, Universidade da Coruña, Departamento de Enxeñaría Naval e Oceánica, Escola Politécnica Superior, Ferrol, Spain

David Cordal-Iglesias, Universidade da Coruña, Escola Politécnica Superior, Ferrol, Spain

Anupam Das, Department of Economics, Justice, and Policy Studies, Mount Royal University, Calgary, AB, Canada

Mehmet Akif Destek, Department of Economics, Gaziantep University, Gaziantep, Turkey

Almudena Filgueira-Vizoso, Departamento de Química, Escola Politécnica Superior, Universidade da Coruña, Ferrol, Spain

Pelin Gençoğlu, Erciyes University, Research and Application Center of Kayseri, Kayseri, Turkey

Akram Shavkatovich Hasanov, Department of Econometrics and Business Statistics, Monash University, Subang Jaya, Selangor, Malaysia

Aarushi Jain, Indian Institute of Management, Indore, Madhya Pradesh, India

Abhinav Jindal, NTPC Ltd., NTPC Bhawan, New Delhi, Delhi, India; Indian Institute of Management, Indore, Madhya Pradesh, India

Cengizhan Karaca, Department of Office Management and Executive Assistance, Gaziantep University, Gaziantep, Turkey

Abiral Khatri, Erasmus University, Rotterdam, Netherlands

Sevda Kuşkaya, Erciyes University, Justice Vocational College, Department of Law, Kayseri, Turkey

Patrícia Hipólito Leal, University of Beira Interior, Management and Economics Department, Covilhã, Portugal; NECE-UBI, University of Beira Interior, Covilhã, Portugal

António Cardoso Marques, University of Beira Interior, Management and Economics Department, Covilhã, Portugal; NECE-UBI, University of Beira Interior, Covilhã, Portugal

Adian McFarlane, School of Management Economics and Mathematics, King's University College at Western University Canada, London, ON, Canada

Walid Mensi, Department of Finance and Accounting, University of Tunis El Manar, Tunis, Tunisia; Department of Economics and Finance, College of Economics and Political Science, Sultan Qaboos University, Muscat, Oman

Muhammad Shujaat Mubarik, College of Business Management (CBM), Institute of Business Management (IoBM), Karachi, Pakistan

Yusuf Muratoglu, Department of Economics, FEAS, Hitit University, Corum, Turkey

Navaz Naghavi, School of Accounting & Finance, Faculty of Business & Law, Taylor's University, Lakeside Campus, Subang Jaya, Malaysia

Cristian Ortiz, Esai Business School, Universidad Espíritu Santo, Samborondon, Ecuador

Yessengali Oskenbayev, Business School, Suleyman Demirel University, Kaskelen City, Kazakhstan; Department of Finance and Accounting, University of International Business, Almaty, Kazakhstan

Burcu Ozcan, Faculty of Economics and Administrative Sciences, Department of Economics, Firat University, Elazig, Turkey

Nirash Paija, Tribhuwan University, Kirtipur, Kathmandu, Nepal

Agnieszka Parlinska, Institute of Economics and Finance, Warsaw University of Life Sciences WULS − SGGW, Warsaw, Poland

Pablo Ponce, Carrera de Economía and Centro de Investigaciones Sociales y Económicas, Universidad Nacional de Loja, Loja, Ecuador

Devran Sanli, Department of Economics, FEAS, Bartın University, Bartın, Turkey

Nida Shah, Business Administration, IQRA University, Karachi, Pakistan

Muhammad Shahbaz, School of Management and Economics, Beijing Institute of Technology, Beijing, China

Arshian Sharif, Othman Yeop Abdullah Graduate School of Business, University Utara Malaysia, Malaysia

Mehmet Songur, Department of Economics, FEAS, Dicle University, Diyarbakır, Turkey

Mehmet Temiz, Faculty of Economics and Administrative Sciences, Department of Economics, Firat University, Elazig, Turkey

Fatma Ünlü, Erciyes University, Faculty of Economics and Administrative Sciences, Kayseri, Turkey

Yılmaz Toktaş, Department of Economics, Merzifon Faculty of Economics and Administrative Sciences, Amasya University, Amasya, Turkey

Elisa Toledo, Departamento de Economía, Universidad Técnica Particular de Loja, Loja, Ecuador

Ali Gokhan Yucel, Faculty of Economics and Administrative Sciences, Department of Economics, Erciyes University, Kayseri, Turkey

Ali Syed Raza, Business Administration, IQRA University, Karachi, Pakistan

Chapter 1

Exploring the linkages between technological advancements and environmental degradation on the energy-growth nexus

Aarushi Jain[1], Abhinav Jindal[2,3]
[1]*Indian Institute of Management, Indore, Madhya Pradesh, India;* [2]*NTPC Ltd., NTPC Bhawan, New Delhi, Delhi, India;* [3]*Indian Institute of Management, Indore, Madhya Pradesh, India*

1. Introduction

1.1 A brief overview

Greta Thunberg, a popular Swedish environmentalist stated in her speech,

> *"We deserve a safe future. And we demand a safe future. Is that really too much to ask?"*

The rapid increase in technological advancements and economic development has resulted in the rise of the consumption of energy in the developing as well as in the developed countries due to globalization. Popular definitions of globalization include, "a state of the world involving networks of independence at multi continental distances." Globalization is a multidimensional process with one dimension being that of environment. Environmental globalization is becoming palatable due to the connectedness in the regular environmental management practices and increasing spatial uniformity (Grainger, 2005). Therefore, different nongovernment organizations, governments, and intergovernment organizations are increasingly getting concerned with the environmental aspects of globalization like and aims to control or regulating. Due to the rapid production and financial advancements, the problems related to energy have been persistent and increasing due to several reasons.

One of the major concerns for the rise in energy consumption is due to the population growth, there is huge demand for energy consumption due to which

Energy-Growth Nexus in an Era of Globalization. https://doi.org/10.1016/B978-0-12-824440-1.00012-6

the energy consumption and environmental degradation are increasing when the globalization and energy-growth nexus increases. If the energy-growth nexus has a negative impact such as decrease in usage of fossil fuels and carbon emissions, then it will lead to the reduction of environmental degradation and energy consumption which in turn will lead to preservation of the energy (Chen et al., 2019).

Although the consumption of energy leads to the growth in economy, it is now becoming a major cause for the degradation of the environment (Saidi et al., 2016). Environmental degradation such as pollution has become one of the major issues in the last few years due to the increase in the greenhouse gas (GHG) emissions (Dogan et al., 2016). The recent case of Australian bushfire and Amazon forest fire has already killed many animals and two dozen people. The forests are abnormally dry and need a significant amount of rainfall in order to thrive. The prediction by IPCC (2007) explains that there will be 40%−110% increase in the energy related CO_2 emissions by 2030. Therefore, to curb these, it is essential to explore the consumption of energy-related factors which effect the emission of CO_2 and to reduce the carbon emissions to achieve a carbon free economy.

The high CO_2 emission which is the dominant contributor of the GHGs is aggravating the problem of environmental degradation (Zhang and Cheng, 2009). In recent years, among both the social and physical scientists, environmental degradation is the most discussed topic. Economists with other professionals have developed a research group who work extensively on the energy and growth nexus and its repercussions on the environment (Ahmed et al., 2015). Among the various options which are open to the society for the reduction in environmental degradation, technology is best suited. Technology has been able to identify patterns (Chen et al., 2021). For instance, the use of electronic components in consumer products and equipment, the trend to use lighter materials, reductions in material equipment, etc. The developing economies like China, India, using technological manifestations have improved the global competitiveness and efficiency in the energy sector. The countries such as Malaysia, Korea, China, India, which initially had lower technological efficiency have now become technologically efficient and advanced and more capable of handling the futuristic challenges, hence validating the Porter's hypothesis (1991). Porter stated that "no severe environmental regulation may have a positive effect on firm's performance by stimulating innovations. Properly designed environmental regulations can trigger innovation partially or fully offset the costs of complying with them." (Zahra and Covin, 1994) suggested innovation as "Innovation is widely considered as the life blood of corporate survival and growth." The significance of innovation is not only restricted to business organizations, it also has wider implications in the face of globalization and environmental challenges.

1.2 Technological innovation

Technology means the ability of humans to create things with the help of machines or hands. It has been derived from two Greek words (Techne and logos) which mean the ability of humans to create things using hands or machines (Energypedia, 2018). Technology is a means to develop the new products, services, and process. Innovation in technology is a very crucial aspect in today's modern times. The products and services are getting smart and smarter day by day which no one has thought of. Technology and humans are inseparable from each other. There is a cyclical dependence on technology in society. Technology is capable of enhancing our physical and intellectual capabilities, emotions, and also guides our decision making.

However, Schumpeter (1939) pointed out that technological innovation is a cyclical process which follows a defined path. The S-curve theory of innovation posits the limits to technology in enhancing performance beyond a limit. The impact of innovations is rather limited and may not be sufficient alone to mitigate all energy-related challenges in the near future. It would also require efforts on energy conservation, behavioral measures, etc. Innovations are expected to reduce carbon footprint by fostering the use of renewable energy for mitigating CO_2 emissions. Further, in an era of globalization, technological spill-over and foreign investment can potentially cause a positive influence on energy consumption, green energy, and promoting growth. This is evidenced by the rapid growth of solar power in India as technology is borrowed from other countries namely China and the United States and increasing investment by foreign companies in India in this sector. There is a general opinion that an increase in the use of renewable energy decreases the CO_2 emissions in a country. However, the impact of deployment of renewable energy on economic growth remains unexplored and needs further investigation.

There are various literatures which examines the nexus between environmental degradation and economic growth using the environmental Kuznets curve (EKC) hypothesis empirically proving causal relationships between the variables environmental pollution and economic growth, income and energy usage, gross domestic product (GDP) and energy usage (Zhang et al., 2009; Soytas and Sari, 2006; Yuan et al., 2008). According to EKC, "environmental pollution increased with economic growth up to a certain income level and then decreased, it is stated that industry-induced environmental pollution was not encountered in preindustrial societies, who earn their living via agricultural activities." However, due to rapid urbanization and industrialization, consumption of natural resources and use of technologies resulting in the destruction of the environment is causing environmental pollution. Societies have become cognizant and now demand clean environmental qualities. Therefore, the use of cleaner technologies has started on the purpose to reduce environmental degradation so as to reduce the GHG's emissions. Accordingly, in the advanced industrialization phases, there is an increasing demand for cleaner technologies (Destek and Ozsoy, 2015).

The chapter discusses the significance of technological innovations in the energy-growth nexus and highlights the role of recent innovations like electric vehicles, energy storage, etc., on energy consumption as well as growth. Subsequently, we point out how environmental degradation is related to growth and analyze a few recent incidents of environmental degradation like Australian wildfires, Amazon fire, cyclones, and typhoons in India, etc. Finally, the chapter concludes with policy suggestions. In the next section, we will discuss about how technological innovations have helped in increasing the energy efficiency in different geographic regions of the world and helped in contributing to the energy sector.

1.2.1 Technological innovations: Contributions to energy sector

In every part of energy framework, energy efficiency plays a crucial role as it ensures a country's sustainable growth and energy security (Liu et al., 2020; Destek and Sinha, 2020; Sinha et al., 2020a). The energy efficiency is crucial as it results in the reduction of energy intensity at both the industrial and household level (Saudi et al., 2019). Therefore, in this section, we will discuss about the contributions made by technological innovations in the energy sector.

There is various literature which deals differently with technology and energy efficiency in various ways. The literature deals with energy efficiency either through energy intensity or energy consumption examining with the impact of technology directly or indirectly. For example, Jin et al., (2018) the authors utilized three stage approach to analyze the nexus between the energy consumption and technological innovation. The authors found out that the energy consumption is not reduced due to the usage of technological innovations. They also further suggested that the policies should focus on technological innovations to reduce the energy consumptions.

There have been various studies which have been done in the country specific context as different countries have different energy usage. For example, the authors in Saudi et al. (2019) in the Indonesian context assessed the impact of technological innovations on energy intensity. The results highlighted that the technological innovations negatively impacts the energy intensity in Indonesia. The literature has used patent as a proxy of techno-logical innovations (Owoeye et al., 2020; Sohag et al., 2015). The authors suggested that the technological innovations which are measured by trademark registered, and patents have nullified effect on the electricity consumption. Shahbaz et al. (2019) conducted a study on top 10 polluted Middle East and North Africa (MENA) countries from the period of 1990–2017 to analyze the impact on technological innovations and environmental degradation.

The authors developed technological innovation index using various different measures. The results gave several insights such as the economic growth has a positive impact on technological innovations on the countries which enable technological advancements.

There have also been studies conducted on the MENA countries as these regions have vast oil, petroleum, and natural gas reserves. Sinha et al. (2020a) analyzed the impact of technological innovations on the energy efficiency from the period of 1990–2016 for the MENA countries and found that structural transformation and technological innovations have positive impact on the energy efficiency. The authors also found that the development of shadow economy negatively impacts the energy efficiency. The authors (Shahbaz et al., 2019b) have used economic indicators to measure technological progress as the research and development (R&D) expenditures cannot alone be used to measure the growth thereby making it unadaptable to measure technological innovation. The study found that higher economic complexity leads to the improvement in reduction of carbon emissions.

The other studies are conducted in Asian regions which discusses the technological advancements being utilized to reduce carbon emissions and increase energy efficiency. Sinha et al. (2020b) the authors also analyzed the impact of technological advancements, population, and gross national income, consumption in renewable energy for the Asia–Pacific countries. They found that technological innovations have positive impact on the air pollution index; on the other hand, renewable energy is negatively impacted by the same.

The authors Sinha et al. (2020c) have also conducted the study on N11 countries. The countries have huge growing populations and are large in size. The countries are basically categorized as "emerging economies" of the world. The authors in this study analyzed the impact of technological innovations, renewable energy, gross national income, and population on air pollution in N11 countries found various insights. Technological innovation, renewable energy negatively impacts air pollution. Whereas environmental policies and technological innovations are impacting the air pollution level in these countries as the growth in national income and environmental deterioration are being caused by the technological innovations taken up by these nations. The authors suggested that in order to have sustainable development and higher growth trajectory, these countries need to be restructured for internalizing the negative externalities.

1.3 Scheme of study

The chapter has been conceptualized through the energy-growth nexus domain taking into account the aspects of technological innovations and environmental degradation. The chapter looks at the sources in the form of existing literature in the energy-growth nexus area. In this chapter, to find out the new perspectives and future directions from the energy-growth nexus, we have analyzed the energy-growth nexus, technological innovation, and environmental degradation areas, respectively, as mentioned in the figure above. The sources used to search through these areas are current literature, current technologies, and social media applications such as blogs, vlogs, social networking sites such as facebook, reddit, and many other sources. The objective is to find out new perspectives and directions from this chapter to which we have analyzed the aforementioned sources for this chapter (Fig. 1.1).

1.4 Related literature on energy-growth nexus

The technological advancements in the energy sector have recently gained a lot of interest within the economists. There have been many empirical studies which have focused on finding the causal relationship between economic growth, environmental pollution, renewable energy, CO_2 emissions, and technological advancements variables. Causal relationships are often complicated and complex (Rogers, 2008). Causal relationships or the statements explain the events, allow actual predictions about the future and based on the predictions makes it possible to take actions in the future. Therefore, this literature review is divided into three parts which talks about the causality in the energy-growth nexus and how different authors have used casualty to explain in the energy sector.

The causal relationship between the growth of the economy and the consumption of energy was initially introduced in the article published by Kraft and Kraft (1978). Causality has also been explored in a detailed literature survey conducted in the paper by Ozturk in 2010. The authors conducted the study to examine the relationship between these variables in the context of USA and suggested that the causality relationship has several policy

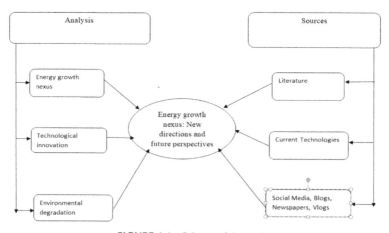

FIGURE 1.1 Scheme of the study.

implications which are (Shiu and Lam, 2004; Jumbe, 2004; Yoo, 2005; Mozumdar and Marathe, 2007; Chen et al., 2007; Squalli, 2007; Apergis and Payne, 2009; Ozturk, 2010):

1. No causality—It is called a "neutrality hypothesis" as there is no causality between energy consumption and GDP which means that neither expansive policies and conservative policies in relation to energy consumption have any effect on economic growth. Therefore, it is supported by the absence of a causal relationship between real GDP and the energy consumption.
2. The unidirectional causality—It is called "conservation hypothesis" which suggests that the policy conservation of energy can be implemented with little or no adverse effect on economic growth. This can be applied for the less energy-dependent economy.
3. The unidirectional causality—It is known as "growth hypothesis." It suggests that directly or indirectly the economic growth plays an important role in the process of production which in addition to labor and capital acts as a complement. This implies that the restriction in the usage of energy may negatively affect the growth of the economy and lead to a subsequent increase in the energy which then contributes to the economic growth.
4. Bidirectional causality—It is known as "feedback hypotheses." It is the causality between the economic growth and energy consumption which determines that both these variables are determined together and are affected at the same time.

However, the causal relationship studies among different variables but did not lead to any consensus and it is still the widely debated topic in the energy sector domain. The results produced are conflicting in nature, and there exists no consensus on whether the relationship between different variables for instance, economic growth, and energy consumption is bi-directional or uni-directional. This area of literature survey is important for policy implications on the appropriate implementation of economic and environmental policies as this issue needs further attention.

The table below provides a literature review of the major studies done by various authors using Granger causality to find a unidirectional, bidirectional, or no direction in the relationship among the energy growth, energy consumption and economic growth variables, and technological innovation. This table gives a comprehensive view that what are the variables which have been taken to find out the causal relationship between energy growth, energy consumption and economic growth variables, and technological innovation. It helps us understand which variables affect what and how. There are other studies also in this field which use these variables to calculate the nature of relationship among certain variables using the Granger causality method in different contexts.

The authors contributing to this debate to reach to a clear conclusion have ample scope of improvement (see Table 1.1). The authors need to identify what has been done and what needs to be done to provide a scope of improvement and contribute to this field.

TABLE 1.1 Salient studies using Granger causality investigating the relationship.

Studies	Year	Variables used	Findings
Murry and Nan (1996)	1996	Electricity consumption with economic growth	23 countries studies done from the period of 1970–93 using Granger causality test within vector autoregressive mode. Findings suggest that: Granger causality causes the growth of economy in Canada, whereas in Norway this relationship is not evident.
Mehrara	2007	Economic growth with electricity consumption	The study was conducted for the period of 1971–2002 for 11 oil exporting countries. Unidirectional strong granger causality between economic growth and electricity consumption.
Yoo and Kwak	2010	Electricity consumption and economic growth	Study conducted from the period of 1976–2006 in Ecuador. Granger causality is found between the consumption of electricity and the growth of economy.
Apergis and Payne	2011	Electricity consumption and economic growth	Unidirectional granger causality between the consumption of electricity and the growth of economy for low-middle income countries such as Ecuador.
Narayan et al.	2010	Electricity consumption and economic growth	There exists a bidirectional relationship between electricity consumption and economic growth.
Begum et al.	2015	Energy growth and carbon emissions	Established the causal relationship between carbon emissions, per capita energy, and growth in the Malaysian context.

2. Our study

2.1 Energy consumption

The total energy used and produced by the entire human civilization is what is defined as energy consumption. It does not need to come from a single source. It could totally be a different source which has a greatest impact on the certain process. Because of the globalization, the energy consumption around the world has increased rapidly. The table below shows the energy consumption in Mtoe (Megatonne of oil equivalent) of the top 10 countries around the world as to how much energy they have consumed since the year 2000−2019 Table 1.2.

China has overtaken the United States in energy consumption. China produced one third of the hydropower and half of the world's coal in the year 2019. Moreover, 50% of the nuclear gas was produced by France and USA, 40% of the natural gas was produced by Russia and the United States, and the production of crude oil came from the Middle Eastern country, Saudi Arabia. Canada replaced Qatar as the fourth largest producer of natural gas as a leading producer of energy in 2017 (Enerdata, 2020). Because of the decline in the climatic conditions and growth of economy in 2018, the global electricity consumption was increasing at a slower rate than previous years due to the lower economic growth which was partly also bolstered by strong demands from service and residential sectors as well. However, the slowdown in industrial sector was stable in Russia and India and in USA, the lower demand

TABLE 1.2 Top 10 countries for energy consumption.

Year ending March 31, 2000	Year ending March 31, 2010	Year ending March 31, 2019
USA	China	China
China	United States	United States
Russia	India	India
Japan	Russia	Russia
India	Japan	Japan
Germany	Germany	Germany
France	Brazil	South Korea
Canada	France	Canada
United Kingdom	Canada	Brazil
South Korea	South Korea	Iran

Source: Global energy statistical year book, 2020

from residential and industrial sectors contributed to the cut in electricity consumption by 2.2%. With the economic slowdown, the consumption of electricity also reduced in European Union by 1.4%, Japan, South Korea, and Africa. However, from the year 2019 the global electricity consumption in China grew by 4.5% since 2018 as China accounts for 28% of the world's global energy consumption.

2.2 Economic growth

It is defined as the increase in the production of economic services and goods from one time period to another. To simply put it, it is referred to as an increase in the aggregate production in the economy.

It is measured by GDP or gross national products. However, GDP is the most suitable way for measuring the growth of the economy. GDP measures the final production as it does not involve the parts that are constructed to make the products. It involves exports as they are produced within the country. From the growth of economy, imports are subtracted Table 1.3.

The tables below depict GDP growth in the years 2000, 2010, and 2019.

From the above table, we can see that from the Asian region, Japan, has sustained its GDP growth. The growth of Japan in terms of GDP has been indigenous as compared to United Kingdom and France to become the world's second largest economy after China in the Asian region. Japan has supplanted the title of Asian economic miracle. The other economic miracle title holder is

TABLE 1.3 Gross domestic product growth of top 10 countries.

Year ending March 31, 2000	Year ending March 31, 2010	Year ending March 31, 2019
United States	United States	United States
Japan	China	China
Germany	Japan	Japan
United Kingdom	Germany	Germany
France	France	United Kingdom
China	United Kingdom	India
Italy	Brazil	France
Canada	Italy	Brazil
Mexico	India	Canada
Brazil	Canada	United States

Source: World Bank Data.org, 2020

China. China faced extreme deprivation before the year 1993 and has lifted hundreds and thousands of people since then. By 1993, China became one of the largest economies squeezing into the list of countries with highest economic growth. From the year 2010, China had surpassed Japan, France, and United Kingdom to obtain the second rank on the list of countries with highest economic growth. By far, they have managed to retain their position till today. The United States has also retained the first spot since 2000. The US economy accounts for 20% of the total global economy output, and in terms of economy, it is much ahead of China.

India on the other hand has improved from 2010 to 2019 after nine years to gain the spot on number seven from number nine in the world GDP. India has surpassed France and has even managed to have a higher growth rate since 2010. India is set to move from sixth position to fifth position in the upcoming years as it has been credited as the fastest growing economy among the other large economies surpassing that of China, recently. Brazil was showing 3.4% growth every year. After the shrinking of the economy of Brazil for a brief time in 2009, it rebounded with a 7.5% growth in the following year. Meanwhile Canada has shown considerable growth as it is just one place ahead of Russia.

2.3 Globalization

Globalization was introduced by Dreher (2006) and is based on Clark (2000) & Norris (2000). It is defined as:

"Globalization describes the process of creating networks of connections among actors at intra- or multicontinental distances, mediated through a variety of flows including people, information and ideas, capital, and goods. Globalization is a process that erodes national boundaries, integrates national economies, cultures, technologies and governance, and produces complex relations of mutual interdependence".

An important issue to measure globalization as it is multifaceted concept including economic, social, and political aspects which go beyond indicators such as capital movements and trade indicators (Potraflake, 2015). For the measurement of globalization, KOF globalization index is the most popular one. The KOF globalization index measures the rate of globalization in the countries around the world. For the measurement of the KOF globalization index, there are a core set of indicators or dimensions which are economic, social, and political. The overall index of globalization tries to assess current economic flow, data on information flows, data on personal contact, data on cultural proximity and economic restrictions with the surveyed countries (Table 1.4).

TABLE 1.4 Top 10 countries with highest globalization index (KOF index).

Year ending March 31, 2000	Year ending March 31, 2010	Year ending March 31, 2019
Switzerland	Belgium	Switzerland
Sweden	Sweden	Netherlands
Belgium	Switzerland	Belgium
Netherlands	Netherlands	Sweden
Denmark	United Kingdom	United Kingdom
Austria	Denmark	Austria
United Kingdom	Austria	Germany
Finland	Luxembourg	Denmark
Germany	Germany	Finland
Norway	France	France

Source: Statista, 2020

The KOF globalization of Switzerland is the highest in the year 2019 with the index being 91.19 points. The pattern being seen here is that European Union has the highest number of countries with high KOF globalization index as compared to the rest of the countries in the world.

2.4 Energy-growth nexus

Energy is an essential output in all of the production and consumption activities. It is also a key source of industrialization, urbanization, and economic growth as they induce more energy, specifically induced energy (Paul and Bhattacharya, 2004). Most of the studies done in the area of energy-growth nexus are empirically tested studies. Several researches have focused on the variables such as electricity consumption, energy consumption, economic growth, and technological innovations. The studies have used Granger causality tests or Vector AutoRegressive (VAR) models to empirically determine the direction of relationship between the aforementioned variables. The causal relationship between the consumption of energy and the growth of economy has been examined in the context of various countries with the help of country specific data, different variables used, and different techniques have been applied (Ozturk, 2010; Fankhauser and Jotzo, 2018). The studies which have been conducted are empirical in nature and have provided uncertain results which are found sometimes inconclusive (Morimoto and Hope, 2004; Stern, 2004; Lau et al., 2014). However, the studies have indicated that there is a causal relationship which is classified as either short run or long run between energy consumption and economic growth (Acaravci and Ozturk, 2010; Shahbaz et al., 2013).

The empirical relationship of the variables such as unidirectional, bidirectional, or no direction relationship. The direction of causation between these variables has significant implications. For instance, if there exists a unidirectional causality in between the consumption of energy and the growth of economy. It implies that the conservation of energy policies may be implemented with a little or no adverse effects on the growth of the economy of the country. If there exists a unidirectional causality relationship between the consumption of energy to income, the reduced consumption of energy will lead to a fall in income. No causality in either direction implies that energy conservation policies do not affect economic growth which is also referred to as "neutrality hypothesis" (Asafu-Adjaye, 2000). Understanding the direction of causality is important for the policy makers for the formulation of the economic and energy growth and development policies to ensure sustainable economic development (Tang and Tan, 2013) and weight of the pros and cons of conservation of electricity policies (Fei et al., 2014).

Granger causality: It is a hypothetical test to investigate causal relationships between two variables. In economics, the word "cause" is referred to as "Granger cause," although a more appropriate word might be "precedence" (Leamer, 1985). The idea of cause and effect is known as causality; however, it is not the same. For example, a variable X is causal to variable Y, if X is the cause for Y or vice versa.

Vector AutoRegressive (VAR) model: It is used to predict multiple time series variables using a single model. Many authors who have researched upon the energy-growth nexus area have constructed a VAR model to predict the relationship between different variables.

Environmental Kuznets curve: The relationship has been examined between the environmental pollution and the economic growth in other literatures (Pao and Tsai, 2010). The research explores the validity of the EKC. The EKC (Grossman and Kreuger, 1995) curve suggests that is the existence of a U-shaped relationship between the economic growth and the level of pollution, i.e., income per capita and pollutant per capita increases at the same level until a threshold point of income is achieved in which the curve of pollutant growth gets flattened and reversed. This infers that when a certain threshold level of income is achieved, the growth can automatically be achieved without a relative rise in emissions (Irandoust, 2016). This concludes that the problem of environmental degradation can be resolved by economic growth. The nexus between economic growth and environmental degradation has already been studies by various authors. However, it does not test the validity of the EKC. This is because of the following assumptions:

1. Instead of aggregate energy, renewable energy consumption is taken.

2. The EKC method requires modeling of nonlinear relationships, but the methodology used to test the hypothesis is linear.

   1. **Technological advancements in the energy sector (Pack, 1994):** There are also some studies which have not taken into account the technological innovation aspect into the energy-growth nexus. The energy generating methods and the advancements in the production techniques in the energy sector are improving due to the technological advancements because of which the applicability of endogenous growth theory is important. The endogenous growth theory has a robust structure which explains the variations in the renewable energy resources (Pack, 1994). Innovations in technology play a crucial role to reveal the variations in the renewable energy usage and economic growth. Therefore, it is important for the policy makers in order to implement energy policy for sustainable economic growth in the country to find the causality between the energy consumption and the economic growth.

   2. **Endogenous growth theory:** The causal relationship has been established between renewable energy, CO_2 emissions, and economic growth. The relationship between environmental pollution, energy, and economic growth has been examined for different countries (Ang, 2007; Soytas and Sari, 2009; Hatzigeorgiou et al., 2011). The results were similar; however, there were some narrow differences due to the adoption of diverse methodologies, different countries, and different time periods. The other study done by the authors was to find the effective relationship between pollutant emission and energy consumption in the United States for the period of 1960–2000. They found that the energy usage and the growth of pollution are affected by the economic growth in the long run (Soytas et al., 2007). Fie et al. (2014) finds the causal relationship between CO_2 emissions, technological innovations, clean energy, and growth in Norway and New Zealand from 1971 to 2010. The findings of this study are based on the ARDL approach which says that innovation in technology plays a crucial part in energy-growth nexus. The other empirical study done in the context of Nordic countries (Denmark, Finland, Sweden, and Norway) to find out the causal relationship between economic growth, technological innovation, renewable energy, and CO_2 emissions by using the "Granger noncausality procedure" developed by Yamada and Toda (1998) and Toda and Yamamoto (1995). There is a unidirectional causality between renewable energy to CO_2 emissions in Finland and Denmark and also establishes a bidirectional causality between renewable energy to CO_2 emissions in Norway and Sweden and an unidirectional causality between renewable energy and technological innovations in Denmark, Sweden, Norway, and Finland (Irandoust, 2016).

2.5 Information and communications technology in energy sector

With modern technological equipment, the economic activities also expand with the passage of time (Danish & Ahmad, 2018). Information and communications technology (ICT) has revolutionized works, entertainment, and communications all around the world specifically in the industrialized and emerging economies. Moreover, ICT has also proven itself to be a powerful source in the structural transformation and economic growth. ICT adoption although has been challenging, but it also has been proven successful in almost every sector including energy. Shifting from renewable energy resources to conventional energy resources is a paradigm shift which however is costly but also is supported by the smart ICT usage (Mattern et al., 2018). Conventional ICT used in the preservation of energy are embedded web systems, Internet of Things (IOT) (low power sensors), networking and ubiquitous computing. This section discusses ICT as being able to manifest its role in the area of energy sector.

ICT is being able to make better decisions day by day with improving technologies related to resource and energy consumption, supply chain and optimization of production. ICT is being used as a tool to provide energy efficiency in the energy sector. For example, reducing traveling by teleworking and helping to save energy at home. Energy conservation is an important issue in emerging countries such as India, China. Due to colossal population in these countries, environmental sustainability has become an important issue with the rising energy costs, growing infrastructure, and therefore, optimization of processes has been initiated to consume minimal power (Mattern et al., 2010).

The IOT is a huge savior in saving the energy process as they communicate with the household meter for scanning for an alternative cheap source of energy that may be available on the power grid. For example, the energy produced by the intermittent renewable energy sources to cool itself down to below its normal operating temperature and thus store energy. Apart from IOT, ubiquitous computing technologies such as embedded web servers, wireless communications, and low power sensors are also being utilized nowadays in the electronic gadgets and day-to-day activities in order to save energy. These technologies do not need direct user involvement providing new opportunities for energy consumption. Moreover, autonomous vehicles, the concept driven by Tesla CEO, Elon Musk, are making it possible to drive an autonomous car without the usage of fuels such as petrol and diesel. If in future such cars can be utilized, the energy consumption will decrease hence maintaining the preservation of the natural resources in its purest form (Fig. 1.2).

The innovation in the power sector as explained in the figure above is an important illustration of the energy sector to illustrate the transformation from traditional power sector to the developed power generation infrastructure (Gielen et al., 2019). The figure above depicts that for the rapid change in renewable energy to take place, there is a dire need for the innovations in

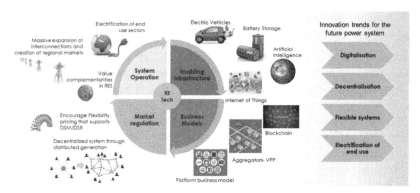

FIGURE 1.2 Power sector Innovation. *Source: IRENA, Abu Dhabi.*

technologies. One such example is the emergence of smart batteries. Much attention is needed for the creation of infrastructure of emerging technologies such as intelligent battery charging of the electric vehicle, smart grid reinforcement, and solar and wind power, and management strategies for biomass feedstock and district heating networks. There is a high demand for ICT in the high performance materials' areas, formulations for a new battery, and many others (Chen et al., 2018).

With the advent of new and smart technologies, the new business models are emerging in the domain of electricity including power aggregators and virtual power plants. There are 30 types of new revolutions which have been recognized which among them cover several 100 discrete cases (IRENA, 2019). The emergence of pollution and environmental degradation has evolved a lot due to which smart technologies are needed to resolve these ongoing problems.

2.6 Role of renewable energy in energy transformation

Renewable energy is the alternative source of energy production of fossil fuels. These forms of energy sources are preferred over fossil fuels because of their contribution to greenhouse gas reduction. Different forms of renewable energy sources are wind energy, solar thermal energy, photovoltaic energy, hydropower, and biomass. Renewable energy is particularly suited for the developing countries as the transmission and distribution of energy from fossil fuels can be expensive. Therefore, renewable energy is an economically viable alternative for the rural and remote areas energy source.

To match contemporary human development and requirements, energy is the crucial feature of human life. The energy usage per capita of the global average did not exceed 20 Giga Joules in the rural areas of the developing countries, which shows the historic status of the past and the present energy usage (Asif and Munner, 2005). Since the population increased six times, the energy consumption is also estimated to increase 20 times the present energy consumption. A transition from penury to abundance was seen when there was a far excess increase in the global population growth which is the first and major energy transition. The developed countries mimic the use of energy consumption to that of developed countries (Fig. 1.3).

The above figure helps to understand the global energy consumption patterns of the popular nations across the globe in order to achieve their quest of prosperity. The most important energy transformation is that of decarburization which has increased the quality of energy. The transformation from wood to coal, from coal to oil with each successive transition led to the shift to a combustible or a fuel which were not only transported and utilized more economically, but they also had much lower carbon content and much higher hydrogen. The third wave of decarbonization is natural gas which among fossil fuels is growing fastest. The fourth wave which is certainly on horizon is pure use of hydrogen and the production. For the advancements, the major drivers are technological advancements to combat the pressure of increasing climate change and air pollution.

With high speed economic development and tremendous energy consumption, the world is facing the ever increasing challenges of energy supply and demand (Zhang et al., 2017). High energy consumption is responsible for the greenhouse gas emissions which contribute to global warming subsequently. To protect the environment from global warming and climate change, there is a need to reduce the energy consumption from fossil fuels to renewable energy sources. Renewable energy is gaining popularity among the countries, and new efficiencies are also paving the way for the development of the nations. Renewable energy growth is crucial to understand the energy demand

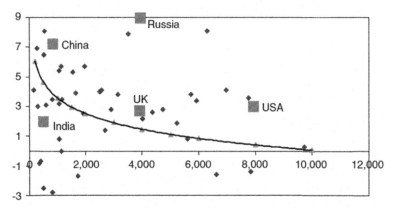

FIGURE 1.3 Percentage (%) increase in the consumption of energy versus amount of oil consumption/capita of 50 countries (in Kgs). *Source: Asif and Muneer, 2007.*

curve which will level off by 2050, and these energy sources have the prospective to deliver energy with zero chances of greenhouse gases and air pollutants (Asif and Muneer, 2005).

The intensity of energy is decreasing, renewable energy is rising, and new energy efficiencies are in the queue for the progress of the nations. Renewable energy growth is crucial to understand the energy demand curve which will be eventually leveled off by 2050. We imagine in terms of fuels needed as input when we talk about the amount of electricity that needs to come out of the socket to ensure that an electric appliance works, the amount of nuclear fuel that should be fed to the power plant for steam generation which enables the turbines. In the case of renewable energy, all these metrics are practically meaningless. The measurement of the energy which comes out of the solar panels or pushes the windmill, we do not measure fuel. There is a whole debate on energy demand and economic growth. As the economy grows, the demand for energy increases. If there is a constraint in energy, the GDP growth of the country plummets.

The developing countries such as China and India are facing colossal demands of energy consumption in order to sustain themselves. China is the primary facing country which faces the challenges of huge energy consumption with an annual increase rate of 5.58% (China Statistic Press, 2015). The role of China is critical in the energy transition globally being the global energy producer and the consumer. To reduce air pollution China is utilizing the renewable energy resources to fulfill its increasing energy demands. They have set the target to minimize the carbon emissions per unit of GDP by 60%−65% by 2030 (Gielen and Boshell, 2019). India also recognized the need to use renewable energy for sustainable development in the late 80s. The state of Maharashtra has rich potential for renewable energy resources such as biomass, wind energy, solar energy, biofuel, hydropower, and photovoltaic cells. These forms of renewable energies play a crucial role in the economic energy development of the country when compared to conventional energy resources (Bhoyar et al., 2013).

According to the McKinsey report of 2020, the growth of economy and the energy demand growth will be the function of four forces; these drivers have larger implications for a wide range of industries and will reform the world's energy growth story.

1. Increase in energy efficiency due to the technological improvements and behavioral changes
2. Increase in renewable energy usage
3. Advancement of electrification is an orderly way to meet the energy requirements in many applications
4. Decline in energy intensity of GDP is a primary concern due to the shift from industrial-based economy to service-based economy in developing countries mainly in China and India.

3. Findings

This chapter employs the various improvements in the technological innovation in the energy sector. These innovations are crucial in the development of the energy sector such as improvement of energy efficiency, energy consumption and conservation, and the overall growth of this sector. There is a need for change in the energy sector in order to conserve renewable energy, and there are a number of ways through which the innovations could help in improvising this sector. Therefore, for the policies to be implemented, there are different measures which should be adapted which we think might be useful for the policy makers. The important action issues are presented below:

3.1 Transformation of energy systems

For addressing the changes in the climatic conditions and broadening the access to affordable and secure energy, the transformation of the global energy system needs to be transformed which is enabled by the renewable power which is much cost effective, and is innovative in terms of market solutions and policies.

1. The cost of renewable energy technologies have fallen drastically in the recent years. It has fallen to such an extent that the commercially available renewable power generation would be more competitive with fossil fuels with respect to cost by 2020.
2. The considerable progress has been made by the power sector in the past 20 years; however, more acceleration is required in terms of progress. The entire renewable energy systems must be scales up to six times faster to meet the demands and needs across the entire energy systems. Emission trends are not included in the current and planned policies. It is estimated that the world would exhaust its budget under 20 years.
3. By 2050, the energy intensity's fallout of the global economy must be less than two third. Despite economic growth and significant population, it can be achieved by substantially improving energy efficiency by increasing the efficiency from 1.8% per year to 2.8%.
4. The most important challenge for the government bodies around the world is to find out solutions to build upon low cost renewable power and how to accelerate the pace of energy transition.

3.2 Technological innovations can accelerate progress

Technological innovations have played an important role in authorizing the early development of the world's energy transition as it has contributed in almost every field. However, in the energy sector, the innovation priorities need to be refreshed to address the new challenges.

1. The innovations must be broad ranging rather than just focusing on technological R&D. To address the gap in capability and to support the accelerated progress of the energy systems, more and faster innovation is needed. The improved technological innovations accompanied by innovations must be further utilized in the business models, markets and designs, and implementation of policies.
2. There is a need to integrate high shares of renewable power and the sectors such as transport, energy, and buildings need to be electrified.
3. The new and emerging technologies such as smart charging of electric vehicles, wider utilization of mini grids, local and grid scale energy storage, and many other technologies are needed to integrate high shares of variable renewable power. These technologies will be crucial in preserving the renewable form of energy. Apart from electrification, there is also need to affordably decarbonize industrial activities such as cement production, iron and steel production, petrochemical and chemical production including maritime transport and aviation and freight activities.

The aforementioned innovation requires cooperative and collaborative measures by multiple parties on multiple fronts.

3.3 Governmental support

Government support is needed to support these innovations as it has a crucial role at all the stages of the innovation journey.

1. Because of the policy driven transformations, governments play a crucial role in the innovation journey, therefore, the government schemes need to be carefully prioritized.
2. The private sector helps in commercialization and successful innovations.
3. For the innovations to grow significantly, both public and private sectors must invest to bring significant change to the world. The support of the innovation should be coordinated with the private sectors, national governments, and international initiatives. For instance, technologies such as solar photovoltaics and wind technologies are a path to for reduction of the cost competitiveness when the government and the private sector invest in innovation.
4. Public sector spending toward clean energy innovation must be boosted. Boosting spending on the private sector required coordinated actions and fresh thinking.
5. International collaborations are required for increased and stronger support to strengthen innovations and growth in:
    (i) The operation and integration of the innovations in energy system: to integrate and electrify end use sectors with more renewable energy sources.

  (ii) The industrial process innovation: to reduce carbon dioxide emissions, specifically iron, steel, and cement chemicals which together contribute toward 17% of carbon emissions.
  (iii) The transport innovation: to reduce carbon emissions, specifically freight and aviation which together accounts for 11% carbon emissions.

4. Conclusion and future scope

This chapter introduces the impact of technological innovations on the energy-growth nexus and the future directions in this field. The increase in usage of technological innovations in the energy sector has sparked a revolution which can be used to improve this sector making it carbon free and protecting the environment at the same time which is also the main purpose of this chapter. Technology can make or break an individual which is explained by Schumpeter in 1939 in the technology S-curve as it how much positive aspect technology has on an individual's social skills and life, it can be dangerous as well. It is the responsibility of the society or a sector to use technology in a productive way in view of sustainable development.

There have been many empirical studies done in lieu of how technological innovations impact the energy-growth nexus. The studies have determined the direction of Granger causality between the variables such as energy such as electricity consumption, environmental growth, technological innovations, environmental degradation, and carbon emissions. The results vary according to different countries. For example, the countries with high carbon dioxide emissions have more negative effect due to technological innovations instead of countries with less carbon emissions. This implies that in ameliorating the energy efficiency and reducing the consumption of energy, technological innovations are crucial.

To take mitigation measures for the reductions of the emissions of the carbon effectively and efficiently in the countries with low carbon emissions, it is important to focus on renewable energy resources. Therefore, to meet the energy needs, the proportion of renewable energy sources must be sufficient so that the countries can benefit from these renewable forms of energy sources.

The future directions could glance on the following questions:

1. To achieve the objective of emission reduction and low carbon economy, the usage of elevating the technological innovations is beneficial for the economy or the country?
2. To find out the innovation trends in the energy sector using cross country comparison.
3. How ICT can be used to save energy by inducing changes in the user behavior?
4. A study on the policy decisions to accelerate the energy transition: comparative study on developed versus developing economies in the world.

These are some of the roadmaps which we could identify from the literature. There are other areas also which can be pondered upon using this context.

References

Acaravci, A., Ozturk, I., 2010. On the relationship between energy consumption, CO_2 emissions and economic growth in Europe. Energy 35 (12), 5412−5420.

Ahmed, K., Shahbaz, M., Qasim, A., Long, W., 2015. The linkages between deforestation, energy and growth for environmental degradation in Pakistan. Ecol. Indicat. 49, 95−103.

Ang, J.B., 2007. CO_2 emissions, energy consumption, and output in France. Energy Pol. 35 (10), 4772−4778.

Apergis, N., Payne, J.E., 2009. CO_2 emissions, energy usage, and output in Central America. Energy Pol. 37 (8), 3282−3286.

Asafu-Adjaye, J., 2000. The relationship between energy consumption, energy prices and economic growth: timeseries evidence from Asian developing countries. Energy Econ. 22, 615−625.

Asif, M., Munner, T., Kelley, R., 2005. Lifecycle assessment: a case study in dwelling home in Scotland. Build. Environ. 42 (3), 1391−1394.

Asif, M., Muneer, T., 2007. Energy supply, its demand and security issues for developed and emerging economies. Renew. Sustain. Energy Rev. 11 (7), 1388−1413.

Bhoyar, R.R., Bharatkar, S.S., 2013. Renewable energy integration in to microgrid: powering rural Maharashtra State of India. In: 2013 Annual IEEE India Conference. (INDICON). https://doi.org/10.1109/indcon.2013.6725877.

Chen, M., Sinha, A., Hu, K., Shah, M.I., 2021. Impact of technological innovation on energy efficiency in industry 4.0 era: moderation of shadow economy in sustainable development. Technol. Forecast. Soc. Change 164, 120521.

Chen, S.-T., Kuo, H.-I., Chen, C.-C., 2007. The relationship between GDP and electricity consumption in 10 Asian countries. Energy Pol. 35, 2611−2621.

Chen, W., Lei, Y., 2018. The impacts of renewable energy and technological innovation on environment-energy-growth nexus: new evidence from a panel quantile regression. Renew. Energy 123, 1−14.

Chen, Y., Zhao, J., Lai, Z., Wang, Z., Xia, H., 2019. Exploring the effects of economic growth, and renewable and non-renewable energy consumption on China's CO_2 emissions: evidence from a regional panel analysis. Renew. Energy 140, 341−353.

China Statistical Press, 2015. https://www.chinayearbooks.com/tags/china-energy-statistical-yearbook. Accessed 7 July 2021.

Clark, W.C., 2000. Environmental globalization. In: Joseph, S.N., Donahue, J.D. (Eds.), Governance in a Globalizing World. Brookings Institution Press, Washington, D.C.

Danish, M., Ahmad, T., 2018. A review on utilization of wood biomass as a sustainable precursor for activated carbon production and application. Renew. Sustain. Energy Rev. 87, 1−21.

Destek, M.A., Ozsoy, F.N., 2015. Relationships between economic growth, energy consumption, globalization, urbanization and environmental degradation in Turkey. Int. J. Energy Res. 3 (4), 1550017.

Destek, M.A., Sinha, A., 2020. Renewable, non-renewable energy consumption, economic growth, trade openness and ecological footprint: evidence from organisation for economic co-operation and development countries. J. Clean. Prod. 242, 118537.

Dreher, A., 2006. Does globalization affect growth? Evidence from a new index of globalization. Appl. Econ. 38 (10), 1091−1110.

https://yearbook.enerdata.net/electricity/electricity-domestic-consumption-data.html Accessed on 10 July, 2020.

Energypedia. 2018. https://energypedia.info/wiki/Impact_of_Technology_and_Energy_Innovation_on_the_Development_of_Society. Accessed 15 May 2020.

Fankhauser, S., Jotzo, F., 2018. Economic growth and development with low-carbon energy. Wiley Interdiscipl. Rev. Clim. Change 9 (1), e495.

Fei, Q., Rasiah, R., Leow, J., 2014. The impacts of energy prices and technological innovation on the fossil fuel-related electricity-growth nexus: an assessment of four net energy exporting countries. J. Energy South Afr. 25 (3), 37−46.

Gielen, D., Boshell, F., Saygin, D., Bazilian, M.D., Wagner, N., Gorini, R., 2019. The role of renewable energy in the global energy transformation. Energy Strategy Rev 24, 38−50.

Grainger, A., 2005. Environmental globalization and tropical forests. Globalizations 2 (3), 335−348.

Grossman, G.M., Krueger, A.B., 1995. Economic growth and the environment. Q. J. Econ. 110 (2), 353−377.

Hatzigeorgiou, E., Polatidis, H., Haralambopoulos, D., 2011. CO_2 emissions, GDP and energy intensity: a multivariate cointegration and causality analysis for Greece, 1977−2007. Appl. Energy 88 (4), 1377−1385.

IPCC, 2007. Climate changes the physical science basis. In: Agu Fall Meeting Abstracts, Vol. 2007, pp. U43D-01

Irandoust, M., 2016. The renewable energy-growth nexus with carbon emissions and technological innovation: evidence from the Nordic countries. Ecol. Indicat. 69, 118−125.

IRENA, 2019. Innovation Landscape Report for the Power Sector. IRENA, Abu Dhabi.

Jin, L., Duan, K., Tang, X., 2018. What is the relationship between technological innovation and energy consumption? Empirical analysis based on provincial panel data from China. Sustainability 10 (1), 145.

Jumbe, C.B.L., 2004. Cointegration and causality between electricity consumption and GDP: empirical evidence from Malawi. Energy Econ. 26, 61−68.

Kraft, J., Kraft, A., 1978. On the relationship between energy and GNP. J. Energy Dev. 3, 401−403.

Lau, L.S., Choong, C.K., Eng, Y.K., 2014. Investigation of the environmental Kuznets curve for carbon emissions in Malaysia: do foreign direct investment and trade matter? Energy Policy 68, 490−497.

Leamer, E.E., 1985. Vector autoregressions for causal inference?. In: Carnegie-rochester Conference Series on Public Policy, vol. 22, pp. 255−304 (North-Holland).

Liu, H., Zhang, Z., Zhang, T., Wang, L., 2020. Revisiting China's provincial energy efficiency and its influencing factors. Energy 208, 118361.

Mattern, F., Staake, T., Weiss, M., 2010. April). ICT for green: how computers can help us to conserve energy. In: Proceedings of the 1st International Conference on Energy-Efficient Computing and Networking, pp. 1−10.

McKinsey Report, June 2020.

Morimoto, R., Hope, C., 2004. The impact of electricity supply on economic growth in Sri Lanka. Energy Econ. 26 (1), 77−85.

Mozumder, P., Marathe, A., 2007. Causality relationship between electricity consumption and GDP in Bangladesh. Energy Pol. 35, 395−402.

Murry, D.A., Nan, G.D., 1996. A definition of the gross domestic product−electrification inter-relationship. J. Energy Develop. 19 (2), 275−283.

Norris, P., 2000. Global governance and cosmopolitan citizens. In: Joseph, S.N., Donahue, J.D. (Eds.), Governance in a Globalizing World. Brookings Institution Press, Washington, D.C.

Owoeye, T., Olanipekun, D.B., Ogunsola, A.J., Kutu, A.A., 2020. Energy prices, income and electricity consumption in Africa: the role of technological innovation. Int. J. Energy Econ. Pol. 10 (5), 392.

Ozturk, I., 2010. A literature survey on energy−growth nexus. Energy Policy 38 (1), 340−349.

Pack, H., 1994. Endogenous growth theory: intellectual appeal and empirical shortcomings. J. Econ. Perspect. 8 (1), 55−72.

Pao, H.T., Tsai, C.M., 2010. CO_2 emissions, energy consumption and economic growth in BRIC countries. Energy Policy 38 (12), 7850−7860.

Paul, S., Bhattacharya, R.N., 2004. Causality between energy consumption and economic growth in India: a note on conflicting results. Energy Economics 26 (6), 977−983.

Potrafke, N., 2015. The evidence on globalisation. World Econ. 38 (3), 509−552.

Rogers, P.J., 2008. Using programme theory to evaluate complicated and complex aspects of interventions. Evaluation 14 (1), 29−48.

Saudi, M.H.M., Sinaga, O., Roespinoedji, D., Ghani, E.K., 2019. The impact of technological innovation on energy intensity: evidence from Indonesia. Int. J. Energy Econ. Pol. 9 (3), 11.

Schumpeter, J.A., 1939. Business Cycles, vol. 1. McGraw-Hill, New York, pp. 161−174.

Shahbaz, M, Balsalobre-Lorente, D., Sinha, A., 2019a. Foreign direct investment−CO2 emissions nexus in Middle East and North African couies: importance of biomass energy consumption. J. Clean. Prod. 217, 603−614.

Shahbaz, M., Gozgor, G., Adom, P.K., Hammoudeh, S., 2019b. The technical decomposition of carbon emissions and the concerns about FDI and trade openness effects in the United States. Int. Econ. 159, 56−73.

Shahbaz, M., Khan, S., Iqbal, M., 2013. The dynamic links between energy consumption, economic growth, financial development and trade in China: fresh evidence from multivariate framework analysis. Energy Econ. 40, 8−21.

Shiu, A., Lam, P., 2004. Electricity consumption and economic growth in China. Energy Pol. 32, 47−54.

Sinha, A., Sengupta, T., Alvarado, R., 2020. Interplay between technological innovation and environmental quality: formulating the SDG policies for next 11 economies. J. Clean. Prod. 242, 118549.

Sinha, A., Shah, M.I., Sengupta, T., Jiao, Z., 2020. Analyzing technology-emissions association in Top-10 polluted MENA countries: how to ascertain sustainable development by quantile modeling approach. J. Environ. Manag. 267, 110602.

Sinha, A., Sengupta, T., Saha, T., 2020. Technology policy and environmental quality at crossroads: designing SDG policies for select Asia Pacific countries. Technol. Forecast. Soc. Change 161, 120317.

Sohag, K., Begum, R.A., Abdullah, S.M.S., Jaafar, M., 2015. Dynamics of energy use, technological innovation, economic growth and trade openness in Malaysia. Energy 90, 1497−1507.

Soytas, U., Sari, R., 2006. Can China contribute more to the fight against global warming? J. Pol. Model. 28 (8), 837−846.

Soytas, U., Sari, R., 2009. Energy consumption, economic growth, and carbon emissions: challenges faced by an EU candidate member. Ecol. Econ. 68 (6), 1667−1675.

Soytas, U., Sari, R., Ewing, B.T., 2007. Energy consumption, income, and carbon emissions in the United States. Ecol. Econ. 62 (34), 482−489.

Squalli, J., 2007. Electricity consumption and economic growth: bounds and causality analyses for OPEC members. Energy Econ. 29, 1192–1205.

Statista, 2020. https://www.statista.com/statistics/268171/index-of-economic-globalization. Accessed 7 July 2021.

Stern, D.I., 2004. The rise and fall of the environmental Kuznets curve. World Dev. 32 (8), 1419–1439.

Tang, C.F., Tan, E.C., 2013. Exploring the nexus of electricity consumption, economic growth, energy prices and technology innovation in Malaysia. Appl. Energy 104, 297–305.

Toda, Y.H., Yamamoto, T., 1995. Statistical inference in vector autoregressions with possibly integrated processes. J. Econ. 66, 225–250.

Yamada, H., Toda, Y.H., 1998. Inference in possibly integrated vector autoregressive models: some finite evidence. J. Econ. 86, 55–95.

Yoo, S., 2005. Electricity consumption and economic growth: evidence from Korea. Energy Pol. 33, 1627–1632.

Yuan, J.H., Kang, J.G., Zhao, C.H., Hu, Z.G., 2008. Energy consumption and economic growth: evidence from China at both aggregated and disaggregated levels. Energy Econ. 30, 3077–3094.

Zahra, S.A., Covin, J.G., 1994. The financial implications of fit between competitive strategy and innovation types and sources. J. High Technol. Manag. Res. 5 (2), 183–211.

Zhang, X.P., Cheng, X.M., 2009. Energy consumption, carbon emissions, and economic growth in China. Ecol. Econ. 68 (10), 2706–2712.

Zhang, D., Wang, J., Lin, Y., Si, Y., Huang, C., Yang, J., et al., 2017. Present situation and future prospect of renewable energy in China. Renew. Sustain. Energy Rev. 76, 865–871.

Zhang, M., Mu, H., Ning, Y., Song, Y., 2009. Decomposition of energy-related CO_2 emission over 1991–2006 in China. Ecological Economics 68 (7), 2122–2128.

Chapter 2

A long-run nexus of renewable energy consumption and economic growth in Nepal

Abiral Khatri[1], Nirash Paija[2]
[1]*Erasmus University, Rotterdam, Netherlands;* [2]*Tribhuwan University, Kirtipur, Kathmandu, Nepal*

1. Introduction

1.1 Background of the project

The Federal Republic of Nepal is a significant Himalayan country in South Asia having a gross domestic product (GDP) of $21.195 billion (WB, 2017) growing at 7.5% with per capita income of $853 as of 2017. The population is over 29 million (July 2017 estimate) and ranks 45th in the world (World-Factbook, n.d.). It joined the United Nations in 1955 and lies in the strategic location between the two largest economies, China and India, which are both growing annually around 7% growth rate (WB, 2017).

In 2015, Nepal was devastated by the aftermath of earthquake that damaged major infrastructures and cultural sites and set back economic development. One-quarter of Nepal's population are living under poverty, and the country is heavily dependent upon remittance, which accounts for 30% of the GDP, ranking third in the world in terms of its contribution to the economy (ILO, 2016). Almost 70% of the Nepalese people are directly engaged in agriculture, which provides a livelihood for almost two-thirds of the population, but it accounts for only one-third of the GDP (FAO, 2017). Industrial activity mainly involves the processing of agricultural products, including pulses, jute, sugarcane, tobacco, and grain.

Nepal has considerable opportunity to exploit its potentiality of hydropower sector, with an estimated theoretical potential of 83,000 MW from hydropower sector and economic feasibility of estimated 42,000 MW of capacity, while the growth of hydropower capacity is only at 8 MW per year (Er. Chhabi Raj Pokhrel, 2018). Despite being a water-rich country with advantageous topographical features and increasing domestic and regional demand, it is sluggish in development. The total energy consumption is

412 gigajoule (per capita 4167 kWh per annum), while the per capita net electricity consumption is 98 kWh per annum. It has over 6000 big and small rivers with an elevation difference of 70−8848 m within 193 km width. However, the average per capita water availability is −8170 m^3/year (FAO, 2005).

It is a well-known fact that the consumption is a huge component of the GDP. With energy being crucially important in the everyday livelihood of people, it is very relevant that the hydroelectricity consumption will have a positive impact on the economic growth of a country. Currently, nearly 17% of the world's total power generation is based on hydro resources, and its share to renewable power generation is 70%. Although hydropower is produced in 150 countries, Nepal's economically feasible hydropower generation capacity is one of the highest (IEA, 2016). However, this huge hydropower potential is still untapped. By harnessing the hydro resources, Nepal can meet its domestic demand, create a surplus for export, and generate employment for its citizens. Similarly, China has pledged a huge amount in the Nepal Investment Summit held in 2018 mostly in the hydropower sector, which has increased the scope of Nepal's development, but political uncertainty and a difficult business climate have hampered foreign direct investment.

Hydroelectricity consists almost all of the total electricity produced in the country. The slow progress of hydropower development is attributable to inadequate planning and investment in generation, transmission, and distribution capacity; concerns about the ability of the Nepal Electricity Authority (NEA) to honor take-or-pay contract obligations; and delays in project development, caused partly by legal and regulatory inadequacies. As a result, Nepal now suffers from a severe shortage of power. Load shedding is frequent, and the country ranks 137th out of 147 countries in quality of electricity supply. Furthermore, since most of the existing hydropower plants are of the run-of-the-river type, electricity generation fluctuates and is highly seasonal (ADB, 2015a,b).

While the overall quantity of energy that Nepal consumes is low, the amount of energy consumed relative to the economic output is very high. In fact, Nepal's energy intensity—the amount of energy consumed per unit of GDP—is 1.8 times higher than India and China, 4.5 times higher than Bangladesh, and 4.5 times the world average (NEEP, 2015). Nepal's high-energy intensity suggests that it has significant potential to increase both its use of energy for productive purposes and the energy efficiency of its production (ADB, 2015a,b). In 2013, transport, industry, and commercial and public services accounted for only 7%, 6%, and 2% of Nepal's energy consumption, respectively (IEA, 2016). With good governance and stability in the political scenario, Nepal can become one of the major renewable energy players in the world.

In the world, particularly in the developing countries, renewable energy resources appear to be one of the most efficient and effective solutions for sustainable energy. The chapter deals with policies to meet increasing energy and electricity demand with hydropower as a sustainable energy investment to strenghthen the economic growth in Nepal.

1.2 Alternative energy and hydropower potential

Alternative electricity technologies are effective for electrifying mountainous rural areas of Nepal and hold much greater potential. These technologies have helped to exploit numerous isolated power stations supplied by hydro, and a few others by powered diesel and solar (NEA, 2016). These have been instrumental in providing electricity to homes and businesses in areas that are difficult to reach with transmission lines (ADB, 2015a,b).

About 16% of the total population has access to electricity through renewable energy sources. Investments are being made in this sector with a target of making 10% of the total energy consumption available through alternative/renewable energy and 30% of energy consumed by the people with access to electricity produced through these sources within next 20 years and about 28% of the total population has been using clean renewable energy (GoN, 2013). In addition, such target aims at development and expansion of alternative energy that would help contribute to maintaining environmental balance, employment generation, and inclusive development.

In the current fiscal year, load shedding hour has been reduced through notable progress in the management and leakage control in electricity. Electricity leakage has dropped by 2.8% point in total. Such leakage has declined by 4.2% in the current fiscal year as compared to that of mid-January last year. Kathmandu and Pokhara, the two largest cities of Nepal, are free from load shedding, and load shedding hours have been reduced greatly in other areas as well, while 4 h of load shedding still persist in industrial areas during peak hours (Ministry of Finance, 2017). Recently, the domestic customers from across the country are free from load shedding. Industries in Biratnagar corridor have to face 4 h of load shedding, while those of Birgunj corridor are having 6 h of power outage. Industries from Bhairahawa corridor have been facing more than 6 h of power outage due to limited transmission capacity.

1.3 Hydropower and electricity consumption of Nepal

The Gross value added (GVA) of electricity, gas, and water group under secondary sector in Nepal is estimated to grow by 12.97% in the current fiscal year. Such growth rate of this group had remained negative by 7.40% in FY 2016. The contribution of this group to GDP that stood at 1.02 last year is estimated to reach 1.16% this year with 0.14% point growth.

The access of total population to electricity stands at 74%. GVA and contribution of the group are estimated to remain high in the current fiscal year as a result of significant level of electricity generation in this year compared to that of the previous fiscal year. An additional 105.3 MW of electricity has been generated in the first 8 months of current fiscal year in contrast to merely 18.5% electricity production in the same period of previous fiscal year (Ministry of Finance, 2017).

The total electricity production that had stood at 855.89 MW by the end of FY 2016 grew by 105.3 MW (12.3%) in the first 8 months of the current fiscal year and reached 961.2 MW. The total electricity generation in Nepal has reached 965.7 MW (production at full capacity) with the inclusion of 4.5 MW electricity generated through rural electrification projects that are not associated with National Integrated Power System, while the electricity consumption including all sectors of economy totaled 3043.35 GWh in the first 8 months of current fiscal year 2017. Such consumption stood at 3718.97 GWh in previous fiscal year 2016 (Ministry of Finance, 2017).

While analyzing the consumption trend from FY 2011/12 until mid-March of FY 2017, private household tops the electricity consumption table with an average consumption of 45.3% followed by industrial estate with 35.6% and other sectors with 11.2% and trade with 7.9% (Ministry of Finance, 2017).

The electricity consumption of household, industry, trade, and other sectors stood at 48.2%, 32.4% 7.7%, and 11.7%, respectively, in FY 2016, while such consumption figures are 45.4%, 36.0%, 7.4%, and 11.2%, respectively, in the first 8 months of the current fiscal year. Industrial sector had shared the most in the total electricity consumption in FY 2005, while household sector has been leaving all sectors behind in terms of such consumption in the years thereafter (Ministry of Finance, 2017).

Nepal's "national energy crisis" is partly caused by inconsistent hydropower whereby climate change will likely exacerbate. Nepal's river flows vary by season. The South—west monsoon delivers roughly 80% of Nepal's rainfall between June and September. In these months, the hydro plants installed are nearly sufficient to meet demand. In the dry season, however, Nepal's reduced supply falls far short of peak load. Most of existing hydro plants in Nepal are run-of-the-river type and lack reservoirs to enable storage (ADB, 2015a,b). As a result, NEA has been forced to impose widespread load shedding during the dry winter months, particularly in the evening hours when demand is highest. Reservoir-type hydro plants could help Nepal overcome this challenge (ADB, 2015a,b). However, these come with other social and environmental considerations.

Nepal declared a national energy crisis in 2008 after a flood of the Koshi River destroyed a key transmission line importing electricity from India, and drought in another part of the country reduced supply (World Bank, 2011). Similar extreme events, and the inconsistency of Nepal's hydro resources, may be exacerbated by climate change. While the impacts remain uncertain, likely effects include changes to patterns of precipitation and glacial retreat.

Projections show that Nepal's runoff could decline by as much as 14% due to climate change, reducing the generation capacity of existing plants and the economic feasibility of new ones (Pathak, 2010). Though there has been expansion in all service sectors, it could not be made reliable. The import of petroleum products declined drastically due to the border obstruction from India in 2016 right after the year when Nepal had a massive earthquake. India is by far the largest and most significant trade partner for Nepal as it accounts for 90% of the trade. In the fiscal year of 2017, import registered significant growth due to reopening of border and supply ease.

By first 8 months of the current fiscal year, transmission line has been extended to 3204 circuit kilometers against 2848.9 circuit kilometer extended in the same period previous year. The number of electricity consumers that stood at 2,969,576 last year grew by 5.1% and reached 3,121,902 by the first 8 months of the current fiscal year (NEA, 2017).

1.4 FDI and hydropower potentiality in Nepal

The current foreign direct investment (FDI) in Nepal is almost 600 billion USD. The Government of People's Republic of China has been providing assistance to the Government of Nepal under the bilateral agreement of economic and technical cooperative signed between two countries. China has always been a steady and reliable partner in Nepal's development endeavor, more specifically in the areas of infrastructure and human resources development, education, health, and food assistance, among others. China is top most investor in Nepal and Chinese investment in Nepal has been growing. Greater Chinese involvement particularly in the fields of water resources and infrastructure building would be extremely helpful potentially leading to a win−win outcome. By July 2013, 575 projects under Chinese investment have been approved in Nepal with investment of NPR 10,632 million, which helped to create 31,594 jobs (DCR, 2011/12).

In the recently completed Nepal Investment Summit 2017, out the total investment pledge, China topped the list with six Chinese companies committing more than $8.3 billion in sectors like hydropower, hospital, metro rail, airport, highway, mining and minerals, smart grid, and financial (Table 2.1).

1.5 Gap between energy demand and supply

Hydropower development is the best option for utilizing water, human, and financial resources. Successful implantation of these projects would be the gateway to FDI and means to generate employment among Nepalese youths who are turning into migrant workers abroad. The electricity demand has reached 1444.06 MW in the current fiscal year. The gap between demand and

TABLE 2.1 Nepal Investment summit 2017 pledge by countries.

Country	Pledge
China	8.3 billion
Bangladesh	2.4 billion
Japan	1 billion
United Kingdom	1 billion
Sri Lanka	500 million
India	317 million
Nepal	11.5 million
Total pledge	$13.52 billion

Based on Report from (Investment Board of Nepal, 2017). Opportunity for Growth. Investment Board of Nepal, Kathmandu.

supply has stood at 482.9 MW (NEA, 2017). The demand for electricity has continued to grow with the increased production of electricity. The gap between demand and supply between FY 2011/12 and FY 2015/16 has widened further. Such gap remained at 461.9 MW by the first 8 months of previous fiscal year, while this further grew by 21 MW reaching 482.9 MW in the review period (NEA, 2017).

About 1171.38 GWh of electricity has been imported from India in current fiscal year to reduce load shedding hour. Such import stood at 1782.86 GWh last year (NEA, 2017). The daily power shortage still remains close to 2.4 million units. The share of petroleum products and traditional sources of energy could be reduced if hydroelectricity production is increased. Through this, however, ever increasing trade deficit would decline, while the overall environment will receive positive impacts with reduction in growing deforestation for addressing energy crisis on the other. Domestically produced energy helps for the foundation of industrialization and brings positive changes in the lives of the people.

In the first 8 months of the FY 2015/16, the total energy consumption that stood at 10,970 tons of oil equivalent dropped by 24.7% and rested at 8257 tons of oil equivalent during the same period of the current fiscal year (Table 2.2).

According a recent report by World Bank, Nepal will probably not become a lower-middle—income country before 2030. The report tells us that a systematic assault is needed to break the vicious cycle and create the right balance between job creation at home and exports of labor (Cosic, 2017). As seen from the table above, the cost forecast for generating electricity and building

TABLE 2.2 Transmission infrastructure forecast.

SN	Year	Transmission infrastructure	Cost (billion USD)
1	2025	7000 MW	3.0
2	2030	13,000 MW	6.0
3	2040	42,000 MW	12.0

Reproduced from Secretariat, W.A., 2017b. Electricity Demand Forecast Report (2015−2040). Government of Nepal, Kathmandu.

transmission infrastructure in Nepal is very reasonable as a developing country. Therefore, foreign investors will have the benefit of cheap labor to minimize the cost and get maximum return from the investment in the hydropower sector.

FDI has become the major means to enhance development activities in developing and emerging markets such as China and India in past decades. The relevance of the role of FDI is growing with the onset of the 21st century. Now, it has become apparent that FDI is a critical and crucial source of capital and technology for the rational utilization of global untapped resources. In this context, it is obvious that harnessing Nepal's hydropower potential depends basically on the inflow of FDI.

1.6 Prospective models in building hydropower projects

With the importance of foreign investment to uplift the country's economy, several models have been formulated by the government with low interest rate loan over a long operation period for the investors: G2G (government to government), competitive bidding with models such as EPC (engineering, procurement, and construct), BT (build and transfer), EPC-F (engineering, procurement, construct and finance), and PPP (public−private partnership) under BOOT (build, own, operate, and transfer) model (National Transmission Grid Company Limited, 2018).

1.6.1 BT contract model

Privately financed infrastructure projects are generally developed on a BOOT modality and generally involve a concession period of 25−35 years. Given the long-term nature of such projects, the Project Development Agreement (PDA) is generally entered into between the government or its relevant authority (the "contracting authority") and private sector developer (the "concessionaire" or the "company") as a very useful tool of allocating responsibilities, risks, and rewards between the parties.

1.6.2 EPC model

The practice of EPC contracts is rare in Nepal. So far, projects in Nepal have usually been implemented by using traditional contract forms with design and procurement risks with the client. There are a few instances where new projects are adopting EPC contracts. In June 2015, the 37 MW Upper Trishuli 3B Hydroelectric Project issued invitation for prequalification to undertake EPC construction of the project. Similarly, the 102 MW Middle Bhotekoshi Hydroelectric Project is using an EPC contract for the construction of the civil works. For example, Lahmeyer International has been selected as the engineering firm because of its experience in EPC contracts worldwide and its decade-long experience in Nepal.

1.6.3 PPP model

Robust and reliable infrastructure is a key driver of economic growth and improved standards of living. Public infrastructures such as roads, railways, bridges, tunnels, water supply, sewers, and electrical grids are essential elements in all societies, providing connectivity and creating networks that facilitate business and remove barriers in access to jobs, markets, information, and basic services. Yet every region of the world faces a chronic infrastructure gap. In order to address this gap, governments around the world have turned to PPPs to design, finance, build, and operate infrastructure projects, while also addressing budgetary constraints. Internationally accepted definition of PPPs denotes "contractual arrangement which is made between a public entity and authority with the involvement of a private entity, in order to provide public asset or service, where the private party has to bear high risk with management responsibility" (PPPIRC, 2016).

Publication of project assessments and tender documents online leads to a greater predictability of the pipeline project quality. However, many economies still do not comply with this practice. Only 22% of the economies surveyed publish PPP proposal assessments online, while 60% publish PPP tender documents. Moreover, only one-third of the economies have developed standardized PPP model contracts (IBRD, 2018).

As emphasized by the United Nations' Sustainable Development Goals (SDGs), one of the targets of the SDGs is to "encourage and promote effective public, public-private and civil society partnerships, building on the experience and resourcing strategies of partnerships." Thus, reforms of the regulatory framework specific to PPPs have been implemented around the globe as economies' leaders have recognized that the "enabling of fair and transparent PPP legislation is vital to the development of a market economy" (UN, 2017).

1.7 Government policies and legal provisions for hydropower

Several policies and legal instruments passed by the GoN have encouraged more investors and allowed FDI in the hydropower sector of Nepal. The first

Hydro Power Development Policy in 1992 had opened doors for private sector involvement in power generation, Electricity Act 1992 Rules 1993 passed act and rules to implement Hydropower Development Policy 1992, and Hydropower Development Policy 2001 to accelerate generation and transmission of electricity for domestic use and export. Similarly, Foreign Investment and Technology Transfer Act 1992 has helped to facilitate foreign investment and technology in hydroelectricity of Nepal. After the millennia, Water Resources Strategy 2002 along with various periodic plans, annual budget speech, tax provisions, etc., has been passed while the Company Act was incorporated in 2006 to register, regulate, and facilitate companies.

Previously, there was a major constraint with unclear policy and legislation followed by the political instability in the country with no sectorial master plan with project prioritization and no manpower development plan for this sector. Recently, with adequate policy commitments for the enhanced participation of private sector government, engagements in large-scale storage and multipurpose projects and unbundling of NEA with Electricity Generation Company Ltd., National Transmission Grid Company Ltd., NEA Engineering Company Ltd., and Power Trade Company Ltd. have enhanced the capability to uplift the energy generation of Nepal. With Nepal Electricity Regulatory Commission Act 2017, special electricity tariff and power purchase agreement have also been formed.

Since the first hydropower Pharping establishment in 1911, which had a capacity of 500 KW, in 107 years of time, there has only been 960 MW electricity generated at present. Being a water-rich country with advantageous topographical features and increasing domestic and regional demand with correct policy implementation by the government, hydropower can lead Nepal to prosperity.

1.8 Problem statement energy demand and hydropower of Nepal

For any country to prosper and develop, economic growth is of vital importance. Considering GDP as a major indicator to measure how well a country is performing, energy consumption plays a crucial role. For a developing country like Nepal, the generation of electricity will help to thrive the industries and generate employment opportunities. With the rising issues of climate change and threat from global warming it has been necessary to invest in sustainable energy sectors. With tremendous potentiality in the hydropower sector, Nepal can attract huge amount of FDI and leap frog in development by generating electricity.

There are various problems facing the development of electricity in Nepal. The electricity sector faces high system losses, delays in completion of new plants, low installed plants efficiency, erratic power supply, frequent load shedding and blackouts, and low efficiency in power plant maintenance. Load shedding, at times, becomes a daily event in Nepal. The electric utility of the country has been characterized by the low management efficiency and low generating capacity that, in turn, has led to frequent load shedding. Power outages result in a loss of industrial output and reduce GDP growth. A major obstacle in efficiently delivering power is caused by the low generating capacity and an inefficient distribution system.

Both economic growth and hydropower electricity consumption are complementary to each other; hence, it is the crucial role of all the concerned stakeholders to raise issue on developing efficient hydropower plants in the country.

1.9 Objectives/research questions of the research

The primary objective of the research is to see how the hydroelectricity consumption impacts the economic growth of Nepal. However, the research has several more objectives:

a. To find out the determinants of hydropower potentiality of Nepal.
b. To examine the impact of hydroelectricity consumption on GDP and vice versa.
c. To examine relationship between real GDP (RGDP) and Hydroelectricity consumption (HEC) using time series econometric analysis.

The purpose of this research is to address the key issues as to how Nepal can become a hub for sustainable hydropower electricity generation and help the economy grow. Issues on government policies particularly in FDI and hydropower electricity have also been raised in order to help generate employment and economic growth in the country.

In Fig. 2.1, the theoretical framework of methodology of this research, the relationship of our research question, and objectives are shown using the econometric model that we have used.

1.10 Organization of the project report

The entire study of the research is outlined into five sections. Each section has its own significance and deals with important aspects of the study. Section 1 has the introduction about Nepal's economy, hydropower as the primary energy source and potentiality, alternative energy source, hydropower sector and situation in Nepal, role of FDI in hydropower, prospective models in building projects, and government laws and provisions to enhance hydropower sector in Nepal. Similarly, it includes problem statement and objective of the research. In Section 2, the literature review explores the development of hydropower electricity consumption and the economic growth around the world. It also shows several models used by researchers in similar topics which has guided us to identifying an appropriate model for this study. Similarly, Chapter III depicts the research methodology undertaken to conduct the

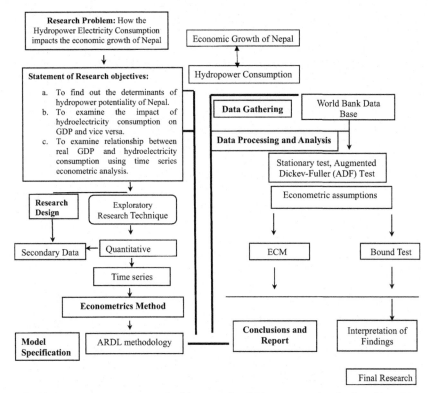

FIGURE 2.1   Theoretical framework of the research. *ARDL*, autoregressive distributed lag; *ECM*, error correction model *Model Made by Author.*

research including econometric modeling and explains the steps taken in brief. Furthermore, it provides an overview of the two variables we used in this research, hydropower consumption and RGDP, looking as to how each of them complements the other. Section 4 reveals the results using a descriptive analysis of the research shown by the collected data with their interpretation using autoregressive distributed lag (ARDL) and time series analysis. The final section houses the discussion of the entire research, its findings gained through analysis, and finally implications for the organization studied.

2. Literature review

Several literatures have been overviewed while conducting the analysis of hydropower energy consumption impact on economic growth. Kraft (1978), Berndt (1978), Akarca (1980), Hwang (1984), Choi (1985), and Yu (1987) were among the first researchers who examined the relation between GDP and energy consumption. In pursuit of their work, many other papers have analyzed the relationship between energy consumption and economic growth, obtaining different results on direction of causality. Different results on direction of causality allowed for four hypotheses: neutrality, conservation, growth hypothesis, and the feedback hypothesis.

In the case of Nepal, research conducted by Dhungel (2009) suggested that there is an unidirectional causality from per capita RGDP to per capita electricity consumption using a Granger causality test. The results show that GDP drives energy consumption, and per capita electricity consumption does not cause per capita RGDP. This was an important policy implication finding.

Similar to Nepal, Turkey also has some advantages on geography and opportunities to exploit its renewable energy (Turkey, 2017). A paper by Yuksel (2010) deals with policies to meet increasing energy and electricity demand for sustainable energy development in Turkey. According to the results presented by Melike Bildirici on the relationship between economic growth and hydropower energy consumption of the short-run causality, there is evidence to support the growth hypothesis in Organisation for Economic Co-operation and Development (OECD) countries with high incomes. Variables of hydro energy consumption and RGDP were used to conclude that the unidirectional causality goes from economic growth to energy consumption and suggest that the policy of conserving hydropower energy consumption may be implemented with little or no adverse effects on economic growth in less energy-dependent economies (Bildirici, 2016).

In Nigeria, the relationship between energy consumption and the Nigerian economy has been reviewed from the period of 1970–2005 (Gbadebo, 2009). By following a cointegration technique, the results infer that there exists a positive relationship between current period of energy consumption and economic growth. With exception of coal being positive, there was a negative relationship for lagged valued of energy consumption and economic growth. The study implicates that increased energy consumption is a strong determinant of economic growth, having an implicit effect in lagged periods and both an implicit and explicit effect over the period of Nigeria.

The paper by Stern (2010) shows that when energy is scarce, it imposes a strong constraint on the growth of the economy, but when energy is abundant, its effect on economic growth is much reduced. It also explained industrial revolution of the constraints on economic growth due to the development of methods of using coal and the discovery of new fossil fuel resources. Structural change toward more service-intensive economies tends to have less impact than is commonly thought because service industries in fact need energy-intensive infrastructures. By using series analysis, he showed that energy and GDP cointegrate and energy use Granger causes GDP when capital and other production inputs are included in the vector autoregression model.

Jin (1992) conducted the first cointegration study of the energy and GDP relationship. The results of this and subsequent studies differ according to the regions, time frames, and measures of inputs and outputs used. It has been argued that if a multivariate approach helps in uncovering the Granger causality relations between energy and GDP, a multivariate approach should also be used to investigate the cointegration relations among the variables. When multivariate cointegration methods are used, a picture emerges of energy playing a central role in determining output in a diverse set of developed and developing nations.

Warr (2010) replicated the same model with the United States using the measures of energy and useful work in place of Stern's Divisia index of energy use. They find both short- and long-run causality from either energy or useful work to GDP but not vice versa Oh (2004). and Ghali (2004) apply Stern's (Stern, 1993, 2000) methodology to Korea and Canada, respectively, coming to exactly the same conclusions, extending the validity of Stern's results beyond the United States. Lee (2008) used panel data cointegration methods to examine the relationship between energy, GDP, and capital in 16 Asian and 22 OECD countries over a three- and 4-decade period, respectively. They find a long-run causal relationship from energy to GDP in the group of Asian countries, while a bidirectional relationship in the OECD sample.

A meta-analysis of hydropower externalities carried out by Matteo-Mattmann (2016) used different meta-regression model specifications to test the robustness of significant determinants of nonmarket values, including different types of hydropower impacts. It found that the main positive externality of hydropower generation is the avoidance of greenhouse gas emission, which positively influences welfare estimates when combined with the share of hydropower in national energy production.

Energy is the primary factor of production. The output is the result of the combined effort of labor, capital, and energy. A number of studies such as Sharma (1994), Nourzad (2000), Paul (2004), Beaudreau (2005), Lee (2005), Thompson (2006), K.R (2014), and Sari (2007) have shown strong evidence of internalization of the important role of technology in the production function. Thus, in this respect, their argument seems justifiable to internalize the role of energy in the production function by giving a consideration to the external costs. This is a very relevant finding because unlike hydropower energy and other renewable energy, the nonrenewable energies such as fossils fuels and coal are often considered to be negative externalities as they have external cost to the environment. Externalities undermine efficiency because one party does not pay the costs or get all the (net) benefits of its actions. The solution to this is therefore to internalize the externality; hence, the government should prioritize on the hydropower and clean energy sectors providing welfare gain to the society.

In the report on hydropower development and economic growth of Nepal, Herat Guntilake (2013) evaluated the macroeconomic implication of a substantial hydroelectric build-out in the country, including generation and transmission resources that could serve external markets. With a state-of-the art dynamic forecasting model, it examines the consequences of Nepal realizing merely 20% of its theoretical and 40% of technically feasible hydro potential. The report's result suggested that the electric power sector at the moment is indeed a major economic impediment and could instead become a potent catalyst for growth, nearly doubling RGDP by 2030. Similarly, using various database from the government authorities like Ministry of Energy (MOE) and NEA, Ranjan Parajuli (2013) presented three scenarios where estimated electricity consumption in Nepal was estimated. In business-as-usual scenario, the estimated electricity consumption in 2030 would be 7.97 TWh (Electricity Demand Forecast Report, 2017), which is in fact 3.47 times higher than that of 2009, while during the medium growth scenario, it is estimated to increase by a factor of 5.71 compared to 2009. Likewise, in high-growth scenario, electricity consumption would increase by 10-fold until 2030 compared to 2009, demanding installed capacity of power plant at 6600 MW, which is solely from hydropower (NEA, 2017).

As a part of a policy recommendation on hydropower of Nepal, Bergner (2012) made an analysis where Nepal's electricity demand will continue to rise by around 7.5% annually until 2020. The analysis considers several alternatives in addressing the 300 MW imbalance between supply and demand for electricity given the feasibility and cost constraints. Among the alternatives are letting preset trends continue, developing micro-hydropower projects (<100Kw) in the short term and expanding capacity using midrange dams (1-1—MW); and pursuing large-scale hydroelectric projects.

While discussing on the issues about the advantages and disadvantages of hydropower, the Three Gorges Dam, the world's largest power station of 22,500 MW operated in 1993 and completed in 2009, is an example to make an assessment, which was designed to serve three main purposes: flood control, hydroelectric power production, and navigation improvement. However, there have been some problems because along with the river flowing, a large number of lands are experiencing erosion, which led to tons of pebbles into the Yangtze River, resulting in pollution. Similarly, there has been huge migration problem where it has inundated two cities, 11 counties, 140 town, 326 townships, and 1351 villages and about 23,899 ha; more than 1.1 million people will have to be resettled (Cruises, 2016). Therefore, in construction of such big power plants, all the externalities must be kept into consideration.

3. Methodology

3.1 Concept and theoretical framework

This research design is based upon econometric modeling considering time series secondary data, and the research design tries to direct both the structure of the research problem and the plan of study used to find empirical evidence on relations of the problem.

With the given evidence, following hypotheses have been made:

H_0: There is no significant relationship between hydropower and economic growth of Nepal.

H_1: There is a significant relationship between hydropower and economic growth of Nepal.

As it can be seen from Fig. 2.1, we have stated the problem statement of this research along with the research objectives we intend to accomplish. Using various techniques of data collecting and processing into our econometric model, we use time series analysis Augmented Dickey Fuller(ADF) test to analyzing and interpreting our findings. The above hypothesis will be further interpreted below in the econometric analysis, whether or not there is a cointegration among the variables in the long run and the short run.

3.2 Research design

To attain the objectives that have been proposed, total hydropower consumption and RGDP over the period from 1972 to 2014 have been analyzed using the data from World bank and changed into natural log data.

3.3 Data collection and sources

In this research, we use the logarithm of GDP as a proxy to represent income (y), and the logarithm of household final consumption as a proxy to represent the consumption expenditure (C). GDP is the total value of all final goods and services produced in a country in a given year, equal to total consumer, investment, and government spending, plus the value of exports, minus the value of imports.

UN data, International energy agency, world bank database, Economic stability report from various years of central bank of Nepal, world bank, Ministry of finance economic survey report, and annual reports have been thoroughly reviewed and examined by the authors.

3.4 Augmented Dickey–Fuller unit root test

In time series, new arriving observations are stochastically depended on the previously observed data. Therefore, time series are often are nonstationary or have means, variances, and covariance that vary over time due to trends, cycles, random walks, or combinations of the three. For econometric modeling, this is the greatest bliss as well as curse. This dependency makes the inferences spurious. Hence, in time series nonstationary must be transformed into stationary. The test that checks if the time series is stationary or not is known as unit root test (Econterms, 2017).

Suppose, y_t is an AR eq. (2.1).

$$y_t = \rho y_{t-1} + x_t' \beta + \varepsilon_t \tag{2.1}$$

Alternatively, the unit root test measure involves the following process:

$$\Delta(y)_t = \delta y_{t-1} + x_t' \beta + \varepsilon_t \tag{2.2}$$

where $\delta = \rho - 1$ and Δ is the difference operator.

Where x_t' indicates transpose of exogenous regressors that consists of a constant, or a constant and trend (slope) ρ and β which parameters that needs to be estimated are, and also assumed white noise (Stephania, 2016). For $|\rho| > 1$, y_t will be a nonstationarity series as the variance will also increase with the time and approach toward infinity, and if $|\rho| < 1$, it is a (trend) stationary series (Greene, 1997).

In unit root test the null hypothesis $H_0 : \rho = 1$ is tested against alternative hypothesis $H_1 : \rho \neq 1$. To test the unit root, Augmented dickey−fuller (ADF) test will be considered.

Time series data are usually nonstationary. ADF tests the stationarity or unit root of the series. The data then were made stationary by differing with necessary orders as per requirement.

3.5 ARDL bounds testing approach

To investigate, the traditional approach to determining long- and short-run relationship among variables has been to use the standard Johanson cointegration and vector error correction model framework; however, this approach suffers from serious flaws as discussed by Pesaran (1999a,b,c). We adopt the ARDL framework popularized by Pesaran (1997) and Pesaran (1999a,b,c) and establish the direction of causation between variables. The ARDL method provides consistent and robust results both for the long- and short-run relationship between HEC and RGDP variables we used. This approach does not involve pretesting variables, which means that the test for the existence of relationships between variables should be applicable no matter whether the underlying regressors are purely I (0), purely I (1), or a mixture of both (Pesaran, 1999a,b,c).

$$\text{LnRGDP}_t = \beta_0 + \beta_1 \text{LnHEC} + \mu_t \tag{2.3A}$$

$$\text{LnHEC}_t = \alpha + \alpha_1 \text{LnRGDP} + \mu_t \tag{2.3B}$$

where β_0 is drift component and μ_t is white noise error. In order to obtain robust results, we utilize the ARDL approach to establish the existence of long- and short-run relationships. ARDL is extremely useful because it allows us to describe the existence of an equilibrium/relationship in terms of long- and short-run dynamics without losing long-run information. The ARDL approach consists of estimating the following equations:

$$\Delta\ln(\text{RGDP})_t = \alpha_o + \sum_{i=1}^{n}\beta_i\Delta\ln(\text{RGDP})_{t-i} + \sum_{i=0}^{n}\delta_i\Delta\ln(\text{HEC})_{t-1}$$
$$+ \lambda_1\ln(\text{RGDP})_{t-1} + \lambda_2\ln(\text{HEC})_{t-1} + \varepsilon_t \tag{2.4A}$$

$$\Delta\ln(\text{HEC})_t = \alpha_o + \sum_{i=1}^{n}\beta_i\Delta\ln(\text{HEC})_{t-i} + \sum_{i=0}^{n}\delta_i\Delta\ln(\text{RGDP})_{t-1}$$
$$+ \lambda_1\ln(\text{HEC})_{t-1} + \lambda_2\ln(\text{RGDP})_{t-1} + \varepsilon_t \tag{2.4B}$$

where Δ is the first difference operator, α_0 is the drift component, and ε_t is the usual white noise residual. The coefficients β_i and δ_i represent the short-run dynamics of the model, whereas the parameters λ_1 and λ_2 represent the long-run relationship. The null hypothesis of the model is

H_0: $\lambda_1 = \lambda_2$ (there is no long-run relationship)

H_1: $\lambda_1 \neq \lambda_2$ (there is the presence of cointegration)

In order to investigate the existence of the long-run relationship among the variables in the system, a bounds test for the null hypothesis of no cointegration is done. The calculated F-statistic needs to be compared with the critical value as tabulated by Pesaran (1997) and Pesaran (2001). The null hypothesis with a no long-run relationship can be rejected no matter whether the underlying order of integration of the variables is 0 or 1 (Pesaran, 1999a,b,c), when the test statistics exceeds the upper critical value (Pesaran, 1997). In the same way, if the test statistic falls below a lower critical value, the null hypothesis is not rejected. However, when the test statistic falls between these two bounds, the result will be inconclusive (Pesaran, 1997). When order of integration of the variables is known given that all the variables are $I(1)$, the decision is made based on the upper bound. Similarly, when all the variables are $I(0)$, then the decision made will be based on the lower bound (Stern, 2010). Alternatively, as a cross-check, a bound t-test is conducted under the null hypothesis of H_0: $\lambda_1 = 0$ against H_1: $\lambda_1 < 0$ (Pesaran, 2015). When the t-statistics is greater than $I(1)$ upper bound, we are able to come to conclusion that there exist a long-run relationship between variables (Pesaran, 1999a,b,c). When f-statistics is less than $I(0)$ lower bound, we are able to that the variables are all stationary (Pesaran, 2001).

The ARDL method estimates $(z_i+1)^k$ number of regressions in order to obtain the optimal lag length for each variable, where z_i is the maximum number of lags to be used and k is the number of variables in the equation (Uko, 2016).

In the second step, if there is evidence of a long-run relationship (cointegration) among the variables, the following long-run model (Eqs. (2.4a) and (2.4b)) is estimated:

$$\ln(\text{RGDP})_t = \alpha 1 + \sum_{i=1}^{n} \beta_i \ln(\text{RGDP})_{t-i} + \sum_{i=0}^{n} \delta_i \ln(\text{HEC})_{t-i} + \varepsilon_t \quad (2.5A)$$

$$\ln(\text{HEC})_t = \alpha 1 + \sum_{i=1}^{n} \beta_i \ln(\text{HEC})_{t-i} + \sum_{i=0}^{n} \delta_i \ln(\text{RGDP})_{t-i} + \varepsilon_t \quad (2.5B)$$

If we find evidence of a long-run relationship, we then estimate the error correction model (ECM), which will indicate the adjustment speed back to long-run equilibrium after there is a short-run disturbance (Moon, 2005). The standard ECM involves estimating the following equation:

$$\Delta \ln(\text{RGDP})_t = \gamma 1 + \delta 1 (\text{ECM})_{t-1} + \sum_{i=1}^{n} \alpha_i \Delta \ln(\text{RGDP})_{t-i}$$
$$+ \sum_{i=0}^{n} \beta_i \Delta \ln(\text{HEC})_{t-i} + \varepsilon_t \quad (2.6A)$$

$$\Delta \ln(\text{HEC})_t = \gamma 1 + \delta 1 (\text{ECM})_{t-1} + \sum_{i=1}^{n} \alpha_i \Delta \ln(\text{HEC})_{t-i}$$
$$+ \sum_{i=0}^{n} \beta_i \Delta \ln(\text{RGDP})_{t-i} + \varepsilon_t \quad (2.6B)$$

To determine the goodness of fit of the ARDL model we used, we conduct diagnostic and stability tests (Pesaran, 1999a,b,c). The diagnostic test will be examining the serial correlation, functional form, normality that should exist, and heteroscedasticity if any associated along with the model (Thach, 2017). The structural stability test is conducted by using the cumulative residuals (CUSUM) (Turner, 2010), and also the cumulative sum of squares of recursive residuals (CUSUMSQ) (Turner, 2010).

4. Chapter IV: Results

4.1 Descriptive analysis

The mean is equal to the sum of all the values in the data set divided by the number of values in the data set. So, if we have n values in a data set and they have values $x_1, x_2 \ldots x_n$, the sample mean, usually denoted by $\bar{x}$, is

$$\bar{x} = \frac{(x_1 + x_2 + \ldots x_n)}{n} \quad (2.7)$$

This formula is usually written in a slightly different manner using the Greek capital letter, $\sum$, pronounced "sigma," which means "sum of"

$$\bar{x} = \frac{\Sigma x}{n} \tag{2.8}$$

Source (Laerd Statistics, 2013):

The median of a set of data is the middlemost number in the set. The median is also the number that is halfway into the set, and mode is the value that occurs most often. If no number is repeated, then there is no mode for the list (Purplemath, 2014).

Standard deviation (often abbreviated as "Std Dev" or "SD") is a measure of dispersion that provides an indication of how far the individual responses to a question vary or "deviate" from the mean (SurveyStar, 2013).

The Jarque—Bera test, a type of Lagrange multiplier test, is a test for normality. Normality is one of the assumptions for many statistical tests, like the t-test or F-test; the Jarque—Bera test is usually run before one of these tests to confirm normality. It is usually used for large data sets because other normality tests are not reliable when n is large (Stephanie, 2016).

The formula for the Jarque—Bera test statistic (usually shortened to just JB test statistic) is

$$JB = n\left[(\sqrt{b1})^2 / 6 + (b_2 - 3)^2 / 24\right] \tag{2.9}$$

where n is the sample size, $\sqrt{b_1}$ is the sample skewness coefficient, and b_2 is the kurtosis coefficient.

After the data transformation into natural form, we get the following result (Table 2.3).

Table 2.3 reveals that Logarithm of Real GDP (LRGDP) and Logarithm of hydroelectricity consumption (LHEC) fail to reject the null hypothesis of normality as the P-value of Jarque—Bera test is greater than 5% level of significance, which is very significant.

Table 2.3 reveals that average LRGDP is approximately 22.77% over the sample period, whereas LHEC is around 20.29% over the sample period; surprisingly, it is somehow equal. This can be debatable because the economic theory tells us that if a country faces trade deficit, current account will be negative. A diagrammatic representation of the RGDP having positive upward shift trend is exhibited in Fig. 2.2.

A correlation matrix is a table showing correlation coefficients between sets of variables. Each random variable (X_i) in the table is correlated with each of the other values in the table (X_j). This allows you to see which pairs have the highest correlation. The correlation matrix is simply a table of correlations. The most common correlation coefficient is Pearson's correlation coefficient, which compares two ratio variables. But there are many others, depending on the type of data you want to correlate (Stephanie, 2016).

TABLE 2.3 Summary of basics statistics.

Measures	Logarithm of Real GDP (LRGDP)	Logarithm of hydroelectribity consumption (LHEC)
Mean	22.77538	20.29410
Median	22.76474	20.41625
Maximum	23.67499	22.09259
Minimum	21.97095	17.86963
Standard deviation	0.530,415	1.216,932
Skewness	0.048546	−0.402,991
Kurtosis	1.677,071	2.051668
Jarque−Bera	3.152,558	2.775,185
Probability	0.206,743	0.249,676
Observations	43	43

Note: Sample period is from 1972 to 2014.
Calculated by Author Based on World Bank Data.

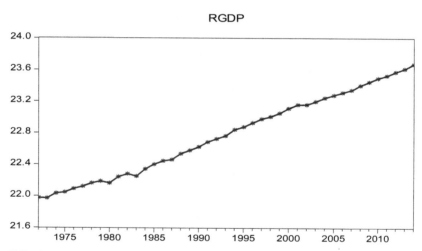

RGDP

FIGURE 2.2   Real gross domestic product (RGDP). *Calculated by Author Based on World Bank Data.*

Table 2.4 is a correlation matrix table. The table exhibits correlation value, t-statistics (which shows how many standard error is correlation value far away from zero), and P-value (which shows the probability value correlation being zero).

TABLE 2.4 Correlation matrix.

Correlation matrix	Logarithm of Real GDP (LRGDP)	Logoraithm of Hydroelectricity Consuption (LRHEC)
LRGDP	1.00	
LHEC	0.984,447	1.00
t-statistic	35.88007	–
Probability	0.00***	–

Note: *, **, and *** represent significance in 1%, 5%, and 10% level of significance respectively. Calculated by Author Based on World Bank Data.

Table 2.4 reveals that LHEC has a positive correlation to LRGDP, which is moderately far from the zero. However, LHEC is strongly correlated to LRGDP with positive sign (0.98), also P-value favor, and it is significantly beyond from the zero. The graph is shown in Fig. 2.2 and explained further.

Fig. 2.2 shows the index of RGDP of the Nepalese economy. A rising trend reflects that the country is striving toward development. The World Bank (World Bank, 2017) assessed that, as of 2013, Nepal's RGDP has broadly increased.

Fig. 2.3 shows the trend of hydropower consumption of Nepal, how it has increased gradually with the increase in the country's population, and energy demand over the period of time. By looking at Fig. 2.4, we can conclude that RGDP and HEC have been growing over the period of 30—40 years' time span.

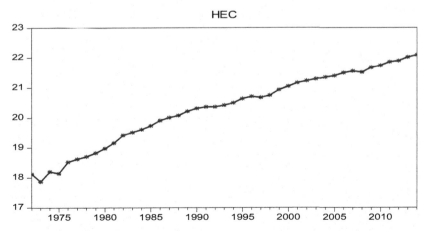

FIGURE 2.3   Hydropower electricity consumption (HEC). *Calculated by Author Based on World Bank Data.*

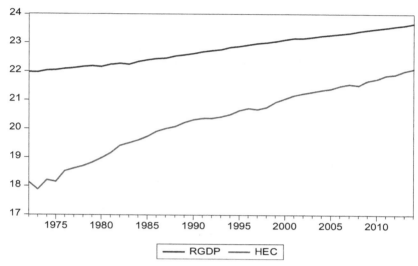

FIGURE 2.4 Real gross domestic product (RGDP) and hydropower electricity consumption (HEC).

4.2 ADF unit root tests

Here in this section, ADF unit roots are performed on the variable to identify their order of integration. The test allowed maximum lags of 9, and the optimum lags were automatically selected by minimizing Schwarz information criterion. The optimum lags whiten the errors. The test was imposed permitting an intercept and trend and intercept as exogenous variable(s) in Eq. (2.1).

Table 2.5 follows ADF unit roots test. Presence of unit root of failure to reject the null hypothesis (*P*-value >5% level of significance) indicates that the data are nonstationary. Nonstationary simply represents that the mean, variance, and covariance of sample data keep changing over the time period.

As shown in Table 2.5, LRGDP and LRHEC at level contain unit roots (nonstationary, i.e., *P*-value > 5%), but while differencing for the first time, both LRGDP and LHEC became free from unit root (stationary, i.e., *P*-value < 5%). While selecting in the intercept and in trend and intercept, the variables seem to be at 1% significance level.

With the inference I (1), we can say that the ADF model showed us a very good relationship between the variables we chose.

4.3 LRGDP as dependent variable

As the directionality is identified and it was represented as LRGDP = f_{RGDP} (LRGDP/HEC), functionality is estimated and different model selection criteria are used to justify the lag orders of each variable in the system. An appropriate lag selection criterion will be able to identify the true dynamics of

TABLE 2.5 Augmented dickeyfuller test.

Variables	Level data		First difference		Sample 1976:2013
	Intercept [K]	Intercept and trend [K]	Intercept [K]	Intercept and trend [k]	Inferences
LRGDP	1.36(0.9985)[2]	−3.157(0.64) [0]	−6.47(0.00)***[1]	−6.76(0.00)***[0]	I (1)
LHEC	−1.24(0.6443) [0]	−1.53(0.7997) [0]	−10.9(0.00)***[0]	−12.78(0.00)***[0]	I (1)
Test critical values					
	1% level			−4.226,815	
	5% level			−3.536,601	
	10% level			−3.20032	

Note: Reported are the test statistics. The number in the [] indicates the lag length. The Schwarz information criteria is used for automatic selection of the lag length (Econterms, 2017).*, **, and *** represent significance at 10%, 5%, and 1% level of significances, respectively.
Calculated by Author based on World Bank Data.

the model. The maximum lag order is set to 3 (Pesaran, 1999a,b,c) as the data are annual and there are only 40 observations. With this maximum lag order, the adjusted sample period for analysis becomes 1975 to 2014.

As we can see from Table 2.6, the intuition behind this table is that, if there is 1% increase in the hydroelectricity consumption, then the GDP will increase by 4% in the short run, and as observed, the coefficient of LHEC as 0.043424.

This setting also helps save the degree of freedom, as our available sample period for analysis is quite small. Using EViews 9, all the selection criteria have given the same results. EViews runs the (p+1)k numbers of regressions and selects the best model on the basis of different model selection criteria, where we have p as the maximum number of lags that is to be used, while k as

TABLE 2.6 Full-information ARDL estimate results.

Based on akaike information criterion, selected ARDL: (3,0)

Dependent variable used is LRGDP

Number of observations used is 40 to estimate from year 1975 to 2014

Regressors	Coefficients	Standard errors	T-statistics	Probability
$LRGDP_{t-1}$	0.498,585	0.151,615	3.288,504	0.0023
$LRGDP_{t-2}$	0.024873	0.168,094	0.147,974	0.8832
$LRGDP_{t-3}$	0.394,488	0.143,642	2.746,319	0.0095
LHEC	0.043424	0.015541	2.794,214	0.0084
C	1.055844	0.459,592	2.297,350	0.0277

Diagnostic checks				
R-squared	0.998,526	Mean dependent variable	22.83405	
Adjusted R-squared	0.998,357	SD dependent variable	0.502,301	
S.E. of regression	0.020359	Akaike information criterion	−4.834,075	
Sum squared residual	0.014508	Schwarz criterion	−4.622,965	
Log likelihood	101.6815	Hannan−Quinn criterion	−4.757,744	
F-statistic	5925.983	Durbin−Watson statistic	1.907,349	
Probability (F-statistic)	0.000000			

Note A: Lagrange multiplier test of residual serial correlation (Pesaran, 1997); B: Ramsey's RESET test with the use of the square of the fitted values (Pesaran, 1997); C: based on a test of skewness and kurtosis of residual; D: based on the regression of squared residual on squared fitted values (Pesaran, 2015); whereas *, **, and *** represent significance at 10%, 5%, and 1% level of significance. Calculated by Author based on World Bank Data.

the number of variables put in that equation (Association, 2017). The ARDL lag length (3, 0) is selected (Pesaran, 1997) on the basis of all criteria like adjusted R2, Schwarz Bayesian criterion, AIC and Hannan−Quinn criterion (Uko, 2016) for the real determinants of trade balance. According to Pesaran (1997), AIC performs relatively well in small samples. However, AIC is chosen to perform ARDL. Therefore, AIC has been used, as a criterion for the optimal lag selection, in all cointegration estimations (Jerome and Busemeyer, 2014).

Table 2.6 reveals that the overall goodness of fit of the estimated ARDL regression model is very high with the result of adjusted $R^2 = 0.99$. The diagnostic tests applied to the ECM indicate that there is no evidence of serial correlation and heteroskedasticity. Furthermore, the Ramsey Regression Equation Specification Error (RESET) test implies the correctly specified ARDL model, and the Jarque−Bera normality test indicates that the residuals are normally distributed.

4.3.1 Bounds test sample period (1972−2014)

We had set the null hypothesis that there does not exit any long-run relationship. This is the step where we estimate Eq. (2.5) through the Ordinary least squares (OLS) regression procedure to examine for the presence of long-run relationship among variables in the system, popularized by Pesaran (1997). In order to select the optimal lag length, we use the AIC and the lag length minimizing AIC.

$$\ln(\text{RGDP})_t = \alpha 1 + \sum_{i=1}^{n} \beta_i \ln(\text{RGDP})_{t-i} + \sum_{i=0}^{n} \delta_i \ln(\text{HEC})_{t-i} + \varepsilon_t$$

This will show us whether there is a relationship or not between the dependent and independent variables we have chosen.

The results in Table 2.7 suggest that the calculated F-statistics is 4.888, which is larger than the upper bound critical value of 4.78 at the 10% level of significance using the restricted intercept and no trend as reported by Pesaran (2007). It also indicates to us the null hypothesis of no cointegration. Thus, the long-run cointegrating relationship among the respective variables is recognized.

In the long run, independent variable will influence the dependent variable. In this case, our dependent variable is RGDP; hence, the F-statistics 4.88 is greater that the upper bound, so it is very significant at 10%.

4.3.2 Long- and short-run analysis (1972−2014)

From Table 2.8, we can conclude that if the hydropower consumption is increased by 1%, then there will be 0.52% increase in the RGDP.

Because a cointegration relationship is identified, this study proceeds to estimate Eq. (2.5a) following the lag specification (3, 0) (AIC) in Table 2.8

TABLE 2.7 ARDL bounds test for cointegration analysis.

Test statistic	Value	k
F-statistic	4.88	1
	Critical value bounds	
Significance	Lower bound I0	Upper bound I1
10%	4.04	4.78
5%	4.94	5.73
2.5%	5.77	6.68
1%	6.84	7.84

Note: The relevant critical value bounds are with intercept and no trend with three regressors. Calculated by Author based on World Bank Data.

TABLE 2.8 Estimated Long-Run Coefficient using the ARDL Approach.

Variable	Coefficient	Standard Error	t-Statistic	Probability
HEC	0.529,218	0.048914	10.819,440	0.0000
C	12.867,762	0.804,534	15.994,063	0.0000

Calculated by Author Based on World Bank Data.

(Stats, 2017). Interestingly, the long-run estimation of the ARDL framework shows that the coefficient of HEC is 0.529, which is also significant at the 1% level. This finding imparts that a 1-unit increase in HEC is associated with an increase in RGDP of 0.529 units, all other things constant. This finding is consistent from both theoretical, which is the Marshallian demand theory, and the empirical points of view (Policonomics, 2016).

In the paper by Kazi Sohag (2015), they also used a Marshallian demand framework to investigate the effects of technology innovation on energy use in Malaysia. Using an ARDL bounds testing approach for the sample period 1985−2012, this study confirmed both short- and long-run theoretical predictions. Research conducted by Bildirici (2016) forecasted hydropower consumption relationship with the economic growth in the long- and short-run dynamic effects by using the ARDL approach as well. According to the results of the short-run causality, there is evidence to support the growth hypothesis in the OECD countries with high income, and in the long-run causality result, there is evidence to support bidirectional causality for all countries except Brazil and France.

Another study conducted by Ersinb (2015) used the ARDL method to analyze the relationship between Biomass Energy Consumption (BMC), Oil Prices (OP), and economic growth in some European countries and evaluated the long-run relationship between the variables using the ARDL approach. The paper found out that there is a unique long-term or equilibrium relationship between BMC, OP, and economic growth in those countries. Similarly, the relation between woody biomass energy consumption and economic growth was discussed by examining the ARDL model for selected sub-Saharan African countries for the period of 1980−2013, which showed unidirectional causality from woody biomass energy consumption (Fulya Özaksoy, 2016).

This study (Jerome and Busemeyer, 2014) estimates the short-run impact of the respective variables on energy consumption along with conducting an error correction mechanism (Moon, 2005) using Eq. (2.6a). Similar to the long-run phenomenon, a unit increase in HEC leads to an approximately 0.043-unit increase in energy use (Table 2.9).

The significance of an error correction term (ECT) in the table, ECT_{t-1}, shows causality in at least one direction. The lagged error term (ECM_{t-1}) in our results is negative and highly significant. The coefficient of -0.082053 indicates a high rate of convergence to equilibrium, which implied that deviation from the long-term equilibrium is corrected by 8% each year over time while the lag length of the short-run model has been selected on the basis of using AIC (Stats, 2017). In other words, the negative and highly significant coefficient of the error correction mechanism implies that short-run disequilibrium adjusts by 8% toward the long-run equilibrium.

TABLE 2.9 Short-run estimated coefficient with ARDL.

Selection: ARDL (3,0) on the basis of 1 (AIC)				
Dependent variable: LRGDP				
40 observations from 1972 to 2014 (for estimation)				
Regressor	Coefficient	Standard error	t-Statistic	Probability
$\Delta RGDP_{t-1}$	−0.419,361	0.146,218	−2.868,064	0.0070
$\Delta RGDP_{t-2}$	−0.394,488	0.143,642	−2.746,319	0.0095
ΔHEC	0.043424	0.015541	2.794,214	0.0084
ECT_{t-1}	−0.082053	0.033596	−2.442,385	0.0198
Cointeq = RGDP − (0.5292*HEC + 12.8678)				
Calculated by Author based on World Bank Data.				

4.3.3 Diagnostic test result

From the diagnostic result, we can know that our model gave us a very significant result as the *P*-value is more than 0.05, which is 0.26.

Table 2.10 shows the results of estimating Eq. (2.1a) by using the ordinary least squares method. AIC is used to select the optimum number of lags in the ARDL model. The Breusch—Godfrey test (LM test) indicates that null hypothesis of presence from serial correlation in the residuals is rejected (Thach, 2017) in both estimates at 5%.

Breusch—Godfrey test result shows there is no serial correlation for all equations in 5% significance level for any of the equations. Lagrange multiplier tests indicate that there is no autoregressive conditional heteroscedasticity (ARCH) (Pesaran, 1999a,b,c). Breusch—Pagan—Godfrey test signifies there is not heteroskedasticity. The JB test result reveals that all residuals presume normality.

Fig. 2.5 reveals the CUSUM test of ARDL period (1972—2014) is stable as the values lie within the critical lines at the 5% significance level, which means long-run coefficient stability.

4.4 LHEC as dependent variable

In this part, we take HEC as our dependent variable as from our initial representation, $\mathrm{LnHEC} = \beta_0 + \beta_1 \mathrm{LnRGDP}_t + \mu_t$.

As we can see from this table, if there is a 1% increase in the GDP, then hydroelectricity consumption will increase by 0.11% in the short run.

Table 2.11 reveals that the overall goodness of fit of the estimated ARDL regression model is very high with the result of adjusted $R^2 = 0.99$. The diagnostic tests applied to the ECM indicate that there is no evidence of serial correlation and heteroskedasticity. Furthermore, the RESET test implies the correctly specified ARDL model, and the Jarque—Bera normality test indicates that the residuals are normally distributed.

TABLE 2.10 Result from diagnostic test.

Diagnostic tests	F-statistic	*P*-value
Breusch—Godfrey serial correlation LM test	0.217,579	0.6439
Heteroskedasticity test: Breusch—Pagan—Godfrey test	2.016029	0.1136
Heteroskedasticity test: ARCH	0.483,652	0.4911
Normality test		0.26

Calculated by Author Based on World Bank Data.

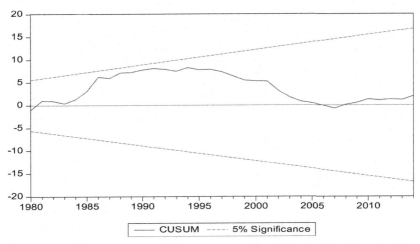

FIGURE 2.5 Stability check LRGDP as dependent variable. *Calculated by Author Based on World Bank Data.*

4.4.1 Bounds test for HEC sample period (1972–2014)

We had set the null hypothesis that there does not exit any long-run relationship. We will now follow the same procedure that we used for the bound test when RGDP was dependent. In order to select the optimal lag length, we use the AIC and the lag length minimizing AIC.

$$\Delta\ln(\text{HEC})_t = \alpha_o + \sum_{i=1}^{n} \beta_i \Delta\ln(\text{HEC})_{t-i} + \sum_{i=0}^{n} \delta_i \Delta\ln(\text{RGDP})_{t-1}$$

$$+ \lambda_1 \ln(\text{HEC})_{t-1} + \lambda_2 \ln(\text{RGDP})_{t-1} + \varepsilon_t$$

This will show us whether there is a relationship we have between the dependent hydropower consumption and independent variable RGDP.

The results in Table 2.12 show us that the F-statistics we calculated is 10.83, which is larger than the upper bound critical value of 4.78 at the significance level of 10% using the restricted intercept and no trend as reported by Pesaran (2001). It also indicates to us the null hypothesis of no cointegration. Thus, the long-run cointegrating relationship among the respective variables is recognized. In this case, our dependent variable is HEC; hence, the F-statistics 10.825 is greater than the upper bound, which is 7.84 at 1% level of significance.

In the long run, independent variable will influence the dependent variable.

TABLE 2.11 Full-information ARDL estimate results.

Based on aAkaike information criterion (AIC) selected ARDL: (3,0)

Dependent variable: LHEC

Number of observations used for estimation: 39 (from year 1976 to 2014)

Regressors	Coefficients	Standard errors	t-Statistics	Probability
$LHEC_{t-1}$	0.774,981	0.144,805	5.351,899	0.0000
$LHEC_{t-2}$	−0.038198	0.172,396	−0.221,569	0.8260
$LHEC_{t-3}$	−0.105,749	0.136,986	−0.771,969	0.4456
$LHEC_{t-4}$	0.241,011	0.111,984	2.152,185	0.0388
RGDP	0.186,973	0.108,174	1.728,448	0.0933
C	−1.512,589	1.527,604	−0.990,170	0.3293

Diagnostic checks			
R-squared	0.997,367	Mean dependent variance	20.52052
Adjusted R-squared	0.996,968	SD-dependent variance	1.034144
S.E. of regression	0.056941	Akaike information criterion	−2.752,948
Sum squared residual	0.106,997	Schwarz criterion	−2.497,016
Log likelihood	59.68249	Hannan–Quinn criterion	−2.661,122
F-statistic	2500.193	Durbin–Watson statistic	1.587,685
Probability (F-statistic)	0.000000		

Note: A: Lagrange multiplier test of residual serial correlation; B: Ramsey's RESET test using the square of the fitted values; C: based on a test of skewness and kurtosis of residual; D: based on the regression of squared residual on squared fitted values (Pesaran, 2015); *, **, and *** represent significance at 10%, 5%, and 1% level of significance.
Calculated by Author Based on World Bank Data.

4.4.2 Long- and short-run analysis (1972–2014)

From Table 2.13, we can conclude that if RGDP is increased by 1%, then there will be 1.46% increase in the HEC.

From Table 2.14 below, HEC is significant at lag 3. The significance of an ECT in the table, ECM_{t-t-1}, shows causality in at least one direction. The lagged error term (ECM_{t-1}) in our results is negative and highly significant.

The coefficient of −0.127,954 indicates a high rate of convergence to equilibrium, which implied that deviation from the long-term equilibrium is corrected by 13% every year over a period. On the basis of the AIC, we selected the lag length of the short-run model (Thach, 2017). It also implies

TABLE 2.12 ARDL bounds test for cointegration analysis.

Test statistic	Value	k
F-statistic	10.82500	1
Critical value bounds		
Significance	Lower bound I0	Upper bound I1
10%	4.04	4.78
5%	4.94	5.73
2.5%	5.77	6.68
1%	6.84	7.84

Note: The relevant critical value bounds are with intercept and no trend with three regressors.

TABLE 2.13 ARDL approach for estimated long-run coefficient.

Variable	Coefficient	Standard Error	t-Statistic	Probability
RGDP	1.461,254	0.326,661	4.473,300	0.0001
C	−11.821,349	7.925,796	−1.491,503	0.1453

Calculated by Author Based on World Bank Data.

TABLE 2.14 Estimated short-run coefficient using the ARDL approach.

Selection based on akaike information criterion: ARDL (4,0)

Dependent variable: HEC

Number of observations used: 39, from year 1972 to 2014

Regressors	Coefficients	Standard errors	t-Statistics	Probability
ΔHEC_{t-1}	−0.097065	0.139,293	−0.696,841	0.4908
$\Delta RGDP_{t-2}$	−0.135,262	0.140,977	−0.959,461	0.3443
ΔHEC_{t-3}	−0.241,011	0.111,984	−2.152,185	0.0388
$\Delta RGDP$	0.186,973	0.108,174	1.728,448	0.0933
ECM_{t-1}	−0.127,954	0.048687	−2.628,096	0.0129

Cointeq = HEC − (1.4613 × RGDP − 11.8213)

Calculated by Author Based on World Bank Data.

TABLE 2.15 Results from diagnostic test.

Diagnostic tests	F-statistics	P-value
Breusch–Godfrey serial correlation LM test	2.088012	0.1410
Heteroskedasticity test: Breusch–Pagan–Godfrey	1.015001	0.4246
Heteroskedasticity test: ARCH	0.914,715	0.3452
Normality test		0.519,258

Calculated by Author Based on World Bank Data.

that the negative and highly significant coefficient of the error correction mechanism during the short-run disequilibrium adjusts by 12% toward the long-run equilibrium (Pesaran, 1997).

4.4.3 Diagnostic test result

From the diagnostic result, we can know that our model has normality as it gave us a very significant result, with the P-value being more than 0.05, which is 0.519 (Table 2.15).

From the diagnostics test, we intend to check the normality test so that it is higher than 5% level of significance. In this case, the P-value is almost 52%; hence, our result shows a very significant relationship.

Fig. 2.6 shows that the CUSUM test of ARDL (1972:2014) is stable as the values lie within the upper and lower bound. Moreover, Fig. 2.6 shows that the CUSUMSQ test of ARDL period is stable as the values lie within the lower and upper bound at the 5% significance level.

5. Chapter V: Discussions, implications, and conclusion

Hydropower is a proven, well-understood form of energy based on more than a century of application. Its schemes have the lowest operating costs and longest plant lives. Upgrades and refurbishment generating plant can readily extend scheme life. Hydropower plants also provide the most efficient energy conversion process (Yüksel, 2009). Due to steep gradient and mountainous topography, Nepal is blessed with the abundant hydro resources. The country's three major river systems and their smaller tributaries offer Nepal to produce economically and technically feasible nearly 50,000 MW power. Nepal can potentially generate over 90,000 MW hydropower (Bergner, 2013). The 99-point document was unveiled in February 2016, with the target of eradicating load shedding in 2 years, and generating 10,000 MW in the next 10 years.

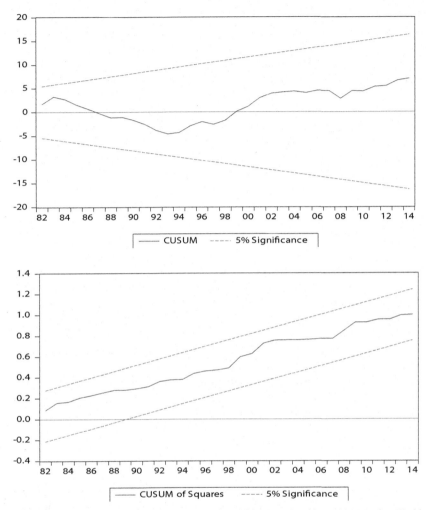

FIGURE 2.6 Stability check LHEC as dependent variable. *Calculated by Author Based on World Bank Data.*

Energy crisis was at its peak when the year 2016 began. Amid shortage of cooking gas, people switched to electricity for cooking. Now, life has improved with the end of load shedding, which was been in place for over a decade. With load shedding coming to end, industries, though barred to operate during evening peak hours, are mulling over expanding their capacity. Similarly, hotels have reported rise in profits with diesel consumption to power big generators coming to zero with the end of load shedding. MOE has estimated that saving from diesel will be worth $ 0.16 billion a year. Taking a leave from the fuel crisis that the country saw during 2016 Indian economic blockade, the government has pledged to generate more energy for self-reliance. In the decade-long plan announced earlier this year, the government announced to end load shedding in 2 years and generate 10,000 MW within a decade.

Production and consumption from nonrenewable energy like fossil fuels and coal result to a negative externality to the society; therefore, in order to reduce the external costs from these, the government should invest in renewable energy like hydropower, which will provide welfare gain to the society and internalize the marginal costs. Similarly, with hydropower like runoff river types in Nepal, great amount of fresh water could be saved and used for other purposes like drinking water as well as to save the aquatic plants and fishes.

Although Nepal's hydropower development started with 0.5 MW plant in Pharping near Kathmandu, 103 years ago (one of the earliest in Asia), as mentioned earlier, Nepal's total hydropower is around 847 MW today. The power shortage is so acute that the load shedding is over 4—5 h each day. The shortfall of power (to meet the grid connected load requirements) is over 500 MW as peak power demand reaches over 1300 MW. During lean session, the shortage of power becomes more severe. Nepal's national transmission grid capacity is only 132 kV. The NEA has planned to upgrade the capacity of the line to 220 kV by the end of 2016 and finally its full capacity of 400 kV by the end of 2017. The World Bank has made a commitment of US $138 million for this interconnector project (Association, 2016).

The year 2017 will set a record in power generation from the hydropower sector. Two plants of NEA, having combined capacity of 44 MW, are starting generation in 2017, while projects developed by independent power producers will add a minimum of 200 MW to national grid this year, according to a projection made in the National Energy Crisis Prevention and Electricity Development Decade (2016—26) document forwarded by the MOE.

We used the ARDL method to analyze the relationship between hydro-electricity consumption and RGDP. To examine the complementary relationship, we used the two-step procedure from ADF unit root test to identify their order of integration. We choose both HEC and RGDP as dependent and independent variable. Then for each of our cases, firstly, the long- and short-run relationship between the variables are evaluated by using the ARDL approach, and then a diagnostic test is done to see the normality of these variables. With our result, we accept the alternate hypothesis stated in our methodology that there is a significant relationship between HEC and the RGDP in both the short run and the long run.

The empirical results of this study provides policymakers a better understanding between hydroelectricity consumption and economic growth nexus to formulate better energy policy in Nepal. Therefore, the ARDL approach might lead to problematic errors in policies. These findings demonstrate that energy policies aimed at improving the energy infrastructure and increasing the energy supply are the appropriate options.

Upper Tamakoshi Hydropower Project of 456 MW will likely be added to the national grid by coming fiscal year, ending the country's reliance on power import at least during the rainy season. The Asian Development Bank has forecasted 5.5% of GDP growth in Nepal in the fiscal year 2018/19 (ADB, Macroeconomic Update of Asia, 2018). With consumption being a big component of GDP, it is also a no exception of hydroelectricity consumption contributing to the economic growth of Nepal.

Given the vast potentiality of Nepalese hydro energy and domestic constraints for financing capability, it would be an opportunity for Chinese investors as a part of the Belt and Road Initiative to engage in Nepal's hydro energy sector in terms of financing, construction, and transmission. Nepal needs big lump of investment in developing transmission infrastructure, which is not sufficient with the regular budget. The support from the Chinese counterpart and government would be a huge contribution for Nepal's economy. Similarly, bilateral ties between the companies are also essential to exchange the knowledge, collaborative works, and capacity building. For large-scale investment in hydropower projects, Nepal needs to attract foreign sovereign and private investments as well as markets for power sale.

Climate Minister of Sweden, Isabella Lövin, announced recently that the target of the country is to be an entirely fossil fuel−free welfare state by the year 2045 (Farand, 2017). Similarly, 200 countries have agreed to limit temperature rise above the 1990s industrial levels by 2C and work toward a 1.5C target. All these show a great indication that renewable energy is the future. Nepal as a developing economy must learn from the new trend in energy economics and focus on renewable energy like hydropower. There needs to be more alternatives explored and policies implemented. Similarly, modern hydropower plants can convert 95% of renewable to electricity while for fossil fuel plants are only 60% (Energy, 2018). Hydropower also has the highest energy payback ratio. During the lifetime of a scheme, it can produce more than 200 times the energy required to build it (Agency, 2000). So, renewable energy is both efficient as well as sustainable. In a time when renewable energy is the trending issue around the world, Nepal could indeed take a leap in hydropower and really contribute a lot to the sustainable development of the region.

Several countries led by India and China have been negotiating with Nepal for possible investment in large-scale hydropower projects. The economic development in the region (North India, Bangladesh, and South Central China) requires power, especially green power. Nepal's hydropower suits their needs well. As hydropower projects require huge capital investment, Nepal should pursue joint venture hydro projects with India and Bangladesh. This will allow to peacefully develop Nepal's hydropower as well secured power purchase agreement with India and Bangladesh for sustained power trade.

Nepal can be one of the major green power exporters in the region. The revenue from power export will help to achieve economic prosperity and generate funds for education, health care, housing, agriculture, and infrastructures. Realization of Nepal's hydro potential will not only meet up the domestic suppressed demand but also create a surplus that could be exported to neighboring countries in South Asia. Thus, tapping the vast hydropower resources of Nepal will be critical to meet the rapidly growing demand of the country and the region. The available hydropower resources could provide a large surplus if strategically developed with a view to foster regional energy trade.

As a well-known Chinese proverb by Lao Tzu goes (Tzu, 2016), "the journey of a 1000 miles begins with a single step," Nepal has tremendous potentiality in the hydropower energy sector, and the stakeholders of its investment can get huge benefit for themselves and for the economy as a whole.

References

ADB, 2015a. Energy Outlook for Asia and the Pacific. Asian Development Bank, Manila.

ADB, 2015b. Financing Asia's Future Growth. Asian Development Outlook, Manila.

ADB, 2018. Macroeconomic Update of Asia. Asia Development Bank.

Agency, I.E., May 2000. Hydropower and the Environment: Present Context and. International Energy Agency.

Akarca, A.L., 1980. On the relationship between energy and GNP: a re-examination. Energy Dev. 5, 326–331.

Association, I.H., November 26, 2016. Hydropower Status Report (. Retrieved from Hydropower.org. https://www.hydropower.org.

Association, I.R., 2017. Renewable and Alternative Energy: Concepts, Methodologies, Tools, and Applications. IGI Global, Hershey, PA.

AQUASTAT, FAO, 2005 "Water availability information for Nepal". Retrieved from: https://www.greenfacts.org/en/water-resources/figtableboxes/aquastat124.htm.

Beaudreau, B., 2005. Engineering and Economic Growth. Structural Change and Economic Dynamics, pp. 211–220.

Bergner, M., 2012. Developing Nepal's Hydroelectric Resources: Policy Alternatives. University of Virginia's Frank Batten School of Leadership and Public Policy, Virginia.

Bergner, M., 2013. Development of Nepal's Hydroelectric Resources. University of Virginia, Virginia.

Berndt, E., 1978. The Demand for Electricity:Comment and Further Results. MIT Energy. Laboratory, WP. No. MIT-EL78-021WP.

Bildirici, M., 2016. The Relationship between Hydropower Energy Consumption and. Istanbul Conference of Economics and Finance. Elsevier B.V, Instanbul, p. 7.

Choi, Y.a., 1985. The causal relationship between energy and GNP : an international comparison. J. Energy 10 (2), 249.

Cosic, D.D., 2017. Climbing Higher : Toward a Middle-Income Nepal. World Bank.

Cruises, Y.R., 2016. Advantages & Disadvantages of the Three Gorges Dam. Retrieved from yangtze river cruises. https://www.yangtze-river-cruises.com/three-gorges/pros-cons.html.

DCR, 2011/12. Development Cooperation Report. Ministry of Finance.

Dhungel, K.R., 2009. Does Economic Growth in Nepal Cause Electricity Consumption. Kathmandu: Hydro Nepal Issue, vol. 5.

Econterms, September 3, 2017. The Augmented Dickey-Fuller Test. Retrieved from thoughtco.com. https://www.thoughtco.com/the-augmented-dickey-fuller-test-1145985.

Electricity Demand Forecast Report, 2017. Water and Energy Commission Secretariat, Nepal. Retrieved from. https://moewri.gov.np/storage/listies/May2020/electricity-demand-forecast-report-2014-2040.pdf.

Energy, A., 2018. Hydroelectric Power. Retrieved from Alternative Energy. http://www.altenergy.org/renewables/hydroelectric.html.

Er. Chhabi Raj Pokhrel, 2018. Invesetment Opportunity in Hydropower Sector in Nepal. Embassy of Nepal. Hydroelectricity Investment and Development Company Ltd, Beijing, p. 42.

Ersinb, M.B., 2015. An investigation of the relationship between the biomass energy. Proc. Soc. Behav. Sci. 203−212.

FAO, 2017. Nepal at a Glance. Food and Agriculture Organisation of the United Nations. Retrieved from. http://www.fao.org/nepal/fao-in-nepal/nepal-at-a-glance/en/.

Farand, C., February 3, 2017. Independent. Retrieved from independent.co.uk. https://www.independent.co.uk/news/science/sweden-pledges-greenhouse-gas-emissions-zero-2045-paris-agreement-a7561111.html.

Fulya Özaksoy, M.B., 2016. Woody biomass energy consumption and economic growth in Subsaharan Africa. Proc. Econ. Finance 287−293.

Gbadebo, O.O., 2009. Does energy consumption contribute to economic performance? Empirical evidence from nigera. East West J. Econ. Bus. 37−59.

Ghali, K.H.-S., 2004. Energy use and output growth in Canada. Energy Econ. 225−238.

GoN, 2013 "Subsidy Policy for Renewable Energy 2013" published by Ministry of Science, Technology and Environment, Government of Nepal, Retrieved from: https://policy.asiapacificenergy.org/sites/default/files/Subsidy%20Policy%20for%20Renewable%20Energy%202069%20BS%20282013%29%20%28EN%29.pdf.

Greene, W.H., 1997. Augmented dickey-fuller test. In: Greene, W.H. (Ed.), Econometric Analysis. Macmillan Publishing Company, pp. 190−230.

Herat Guntilake, D.R.-H., 2013. Hydropower Development and Economic Growth in Nepal. Asian Development Bank.

Hwang, Y.a., 1984. The relationship between energy and GNP: further results. Energy Econ. 6, 186−190.

IBRD, 2018. Procuring Infrastructure Public-Private Partnerships Report. International Bank for Reconstruction and Development,World Bank, Washington DC.

IEA, 2016. Nepal. International Energy Association. Retrieved from. http://www.iea.org/countries/non-membercountries/nepal/.

ILO, 2016. Promoting informed policy dialogue on migration, remittance and development in Nepal, published by International Labour Organization, Kathmandu. Retrieved from http://ilo.org/wcmsp5/groups/public/—asia/—ro-bangkok/—ilo-kathmandu/documents/publication/wcms_541231.pdf

Investment Board of Nepal, 2017. Opportunity for Growth. Investment Board of Nepal, Kathmandu.

Jerome, R., Busemeyer, A.D., 2014. Estimation and testing of computational psychological models. In: Jerome, A.D., Busemeyer, R. (Eds.), Neuroeconomics, second ed. Elsevier.

Jin, Y., 1992. Cointegration tests of energy consumption, income, and employment. Resour. Energy 259−266.

Kazi Sohag, R.A., 2015. Dynamics of energy use, technological innovation, economic growth. Energy 7.

K.R, D., 2014. On the relationship between electricity consumption and selected macroeconomic variables: empirical evidence from Nepal. Mod. Econ. 360−366.

Kraft, J.K., 1978. On the relationship between energy and GNP. J. Energy Dev. 401−403.

Laerd Statistics, 2013. Measures of Central Tendency. Retrieved from Laerd Statistics. https://statistics.laerd.com/statistical-guides/measures-central-tendency-mean-mode-median.php.

Lee, C., 2005. Energy consumption and GDP in developing countries: a co-integrated panel analysis. Energy Econ. 415−427.

Lee, C.P.-F., 2008. Energy-income causality in OECD countries: the key role of capital sotcks. Energy Econ. 2359−2373.

MatteoMattmann, I.L., 2016. Hydropower externalities: a meta-analysis. Energy Econ. 66−77.

Ministry of Finance, 2017. Economic Survey. Ministry of Finance, Government of Nepal, Kathmandu.

Moon, T.K., 2005. Error Correction Coding: Mathematical Methods and Algorithms. John Wiley and Sons.

National Transmission Grid Company Limited, 2018. Contract Model. Investment Opportunities in Hydropower Projects in Nepal. Government of Nepal, Beijing.

NEA (2016). A Year in Review-Fiscal Year 2015/16, published by NEA, Kathmandu, Nepal. Retrieved from: https://www.nea.org.np/admin/assets/uploads/supportive_docs/35953868.pdf.

NEA, 2017. Reports. Nepal Electricity Authority. Retrieved from. http://www.nea.org.np/.

NEEP, 2015. Nepal Energy Efficiency Programme. Retrieved from energyefficiency.gov. http://energyefficiency.gov.np/article-component3.

Nourzad, F., 2000. The productivity effect of government capital in developing and industrialized countries. Appl. Econ. 1181−1187.

Oh, W.a., 2004. Causal relationship between energy consumption and GDP. Energy Econ. 51−59.

Pathak, M., 2010. Climate change: uncertainty for hydropower development in Nepal. Hydro Nepal 6, 31−36.

Paul, S.B., 2004. Causality between energy consumption and economic growth. Energy Econ. 977−983.

Pesaran, H., 1997. The role of economic theory in modelling the long- run. Econ. J. 107, 178−191.

Pesaran, H.a., 1999a. Autoregressive distributed lag modelling approach to cointegration analysis, chapter 11, in storm. In: Pesaran, H.a. (Ed.), Econometrics and Economic Theory in the 20th Century: The Ragnar Frisch Centennial Symposium (P. Chapter 11). Cambridge University Press, Cambridge.

Pesaran, H.a., 1999b. Chapter 11. In: Storm, S. (Ed.), Econometrics and Economic Theory in the 20th Century: The Ragnar Frisch Centennial Symposium, Autoregressive Distributed Lag Modelling Approach to Cointegration Analysis. Cambridge University Press, Cambridge. H. a. Pesaran.

Pesaran, H.M., 1999c. Bounds testing approached to the analysis of level relationships. J. Appl. Econom. 326, 289−326.

Pesaran, H.M., 2001. Bounds testing approaches to the analysis of level relationships. J. Appl. Econom. 289−326.

Pesaran, M.H., 2015. M. Hashem pesaran. In: Pesaran, M.H. (Ed.), Time Series and Panel Data Econometrics. Oxford University Press, p. 521.

Policonomics, 2016. Marshallian and Hicksian Demands. Retrieved from Policonomics. http://policonomics.com/marshallian-hicksian-demand-curves/.

PPPIRC, 2016, 02 16. A World Bank Resource for PPPs in Infrastructure. Public-Private-Partnership in Infrastructure Resource Center. Retrieved from. http://ppp.worldbank.org/public-private-partnership/spanhomespanimg-althome-srcpppsitesworldbankorgpppfileshome-icon-redpng/home.

Purplemath, 2014. Mean. Median, Mode and Range. Retrieved from Purplemath. http://www.purplemath.com/modules/meanmode.htm.

Ranjan Parajuli, P.A., 2013. Energy consumption projection of Nepal: an econometric approach. Renew. Energy 432−444.

Sari, R.S., 2007. The growth of income and energy consumption in six developing countries. Energy Pol. 889−898.

Secretariat, W.a., 2017b. Electricity Demand Forecast Report (2015−2040). Government of Nepal, Kathmandu.

Sharma, S.D., 1994. Causal analysis between exports and economic growth in developing countries. Appl. Econ. 1145−1157.

Stats, 2017. Time Series, How to Extract Long Run and Short Run Coefficients from ARDL (UECM) Estimates. Retrieved from stats.stackexchange.com. https://stats.stackexchange.com/questions/94353/how-to-extract-long-run-and-short-run-coefficients-from-ardl-uecm-estimates.

Stephania, June 7, 2016. ADF. Retrieved from statisticshowto.com. http://www.statisticshowto.com/adf-augmented-dickey-fuller-test/.

Stephanie, May 6, 2016a. Correlation Matrix. Retrieved from Statistics How To. http://www.statisticshowto.com/correlation-matrix/.

Stephanie, May 7, 2016b. Jarque-Bera Test. Retrieved from Statistics How To. http://www.statisticshowto.com/jarque-bera-test/.

Stern, D.I., 1993. Energy use and economic growth in the USA, A multivariate approach. Energy Econ. 137−150.

Stern, D.I., 2000. A multivariate cointegration analysis of the role of energy in the U.S. macro-economy. Energy Econ. 267−283.

Stern, D.I., 2010. The Role of Energy in Economic Growth, vol. 52. Centre for Climate Economics & Policy.

SurveyStar, 2013. How to Interpret Standard Deviation. SurveyStar.

Thach, L.H., 2017. Econometrics for Financial Applications. Springer International Publishing, Cham.

Thompson, H., 2006. The applied theory of energy substitution in production. Energy Econ. 410−425.

Turkey, B., 2017. The Important Location of Turkey. Retrieved from bigloveturkey. http://www.bigloveturkey.com/pages/36-geographical-importance-of-turkey.asp.

Turner, P., 2010. Power properties of the CUSUM and CUSUMSQ tests for parameter instability. J. Appl. Econ. Lett. 17 (Issue 11), 1049−1053.

Tzu, L., 2016. BBC Learning English. Retrieved from bbc.co.uk. http://www.bbc.co.uk/worldservice/learningenglish/movingwords/shortlist/laotzu.shtml.

Uko, E.N., 2016. Autoregressive distributed lag (ARDL) cointegration technique. J. Stat. Econom. Methods 5 (4), 63−91, 1792-6602 (print), 1792-6939 (online).

UN, 2017. United Nations, Sustainable Development Goals. Retrieved from un.org. http://www.un.org/sustainabledevelopment/.

Warr, B.R., 2010. Energy use and economic development: a comparative analysis. Ecol. Econ. 1904−1917.

WB, 2017. Data. World Bank. Retrieved from. http://data.worldbank.org/country.

World Bank, June 21, 2011. World Bank SUpports Cross-Border Energy Cooperation between India and Nepal. Retrieved from worldbank.org. http://www.worldbank.org/en/news/press-release/2011/06/21/world-bank-supports-cross-border-energycooperation-.

World Bank, 2017. Database World Bank. World Bank. Retrieved from. http://data.worldbank.org/country/nepal.

WorldFactbook, C., n.d. CIA World Factbook. Retrieved from Central Intelligent Agency. https://www.cia.gov/library/publications/the-world-factbook/geos/np.html.

Yu, E.a., 1987. On the relationship between electricity and income for industrialized countries. J. Electr. Employ. 13, 113–122.

Yüksel, I., 2009. Dams and hydropower for sustainable development. Energy Sources 100–110.

Yuksel, I., 2010. As a renewable energy hydropower for sustainable development in Turkey. Renew. Sustain. Energy Rev. 3213–3219.

Chapter 3

The energy consumption-growth nexus in Jamaica: does globalization matter?

Adian McFarlane[1], Anupam Das[2], Kaycea Campbell[3]

[1]*School of Management Economics and Mathematics, King's University College at Western University Canada, London, ON, Canada;* [2]*Department of Economics, Justice, and Policy Studies, Mount Royal University, Calgary, AB, Canada;* [3]*Department of Economics, Los Angeles Pierce College, Woodland Hills, CA, United States*

1. Introduction

In this chapter, we discuss the issues related to the energy-growth nexus in the context of globalization, conduct a related empirical exercise for Jamaica, and discuss the policy implications of our findings. Since the seminal paper of Kraft and Kraft (1978), the energy-growth nexus has been a popular topic of investigation among energy economists. Scholars have studied this nexus, the relationship between energy consumption and economic growth, with a view to inform their attendant policies.[1] At the same time, the increasing, and sometimes uneven, global integration of domestic and international economies adds another layer of complexity in trying to comprehend and characterize the empirical relationship between energy consumption and economic growth. Yet, a dearth in the literature exists on the possible role and impact of global integration on this nexus for some countries. Unfortunately, energy policy decisions that do not account for the role of global integration may not be most effective, so not having information about this can be a hindrance to good policy.

A greater degree of globalization, as reflected in a higher level of global integration of domestic and foreign economies, has the potential to increase domestic economic activities.[2] This increase can lead to a boost in the derived demand for energy in developing countries. Therefore, it is instructive to

---

1. See Ozturk et al., (2010); Paul and Bhattacharya (2004); Odhiambo (2009); Shahbaz et al. (2012), (2013a); Soytas and Sari (2009).
2. See for example Das and Paul (2011).

Energy-Growth Nexus in an Era of Globalization. **https://doi.org/10.1016/B978-0-12-824440-1.00008-4**
67

investigate how energy consumption interplays with changing levels of global integration and the impact this has on gross domestic product (GDP) growth. Further, it is also of value to assess the extent to which empirical findings can inform policies for energy consumption. Given the structural differences across developing countries due to variations in income levels, the history of colonization, and institutional developments, country-specific studies are warranted to examine how global integration affects the relationship between energy consumption and economic growth.

These country-specific studies are of particular importance for emerging countries such as Jamaica, which embraced globalization only three decades ago. Jamaica's participation in globalization along with its economic growth and energy consumption trajectories all are closely connected. For Jamaica, globalization has meant expanding its own carbon footprint for several reasons. Jamaica has one of the largest bauxite reserves in the world. Bauxite mining is a carbon-intensive process that has caused Jamaica to expand its overall energy consumption. Both manufacturing and tourism industries are also energy intensive. Finally, as a developing country, Jamaica has an expanding middle class whose energy needs are also greater than they were in the pre-globalization era when Jamaica was primarily an agricultural economy. Insofar as globalization has expanded Jamaica's ability to tap into tourism, bauxite, and manufacturing demand, and insofar as each of these three kinds of demands is associated with energy emissions, it is reasonable to believe that there may be a potential connection between globalization and energy consumption in Jamaica. Although there exist no time-series studies on Jamaica, several scholars have noted that globalization can impact the demand for energy in both developed and developing countries (e.g., Baek et al., 2009; Shahbaz et al., 2016, 2018a). Noteworthy, in terms of policy implications, this impact may differ in the short run and long run. For example, Shahbaz et al. (2018b) found that for Ireland and the Netherlands over the period 1970–2015, globalization and energy consumption had a positive relationship in the long run, but that there was no significant impact of globalization on energy consumption in the short run. Their results highlight that there could be differences in the impacts of energy on economic growth due to globalization, as the intensity of global integration varies across economies due to differences in the political and institutional arrangements of countries.

There is a gap in the literature as it relates to empirical studies on the linkages between energy consumption, economic growth, and globalization in Jamaica. We contribute to filling this gap in the literature by addressing the following research question: What are the long run and short run causal relations between GDP and energy-global openness intensity in Jamaica? Energy-global openness intensity is energy consumption per capita interacted with trade openness (volume of trade divided by GDP). To answer this research question, we apply an augmented Solow growth model that accounts for changes in economic conditions as reflected in the evolution of total factor productivity, the labor force, and capital stock. We also control for business cycle GDP growth asymmetries by accounting for the tendency for GDP to grow at a faster pace when rebounding from a recession in comparison to when it is growing in an expansion, the so-called current depth regression (CDR) effect. We use data for the period 1971−2017 and apply the autoregressive distributed lag (ARDL) model and Granger causality testing.

There are two key findings. First, we find that energy-global openness intensity matters in the evolution of the GDP of Jamaica, after controlling for changes in the macroeconomic environment and business cycle economic growth asymmetries. There is a positive cointegrating long run relationship running from energy-global openness intensity to GDP. In this relationship, a 1% increase in energy-global openness intensity is associated with a 0.13% increase in GDP. Second, there is bidirectional short run causality, in the Granger sense, between energy-global openness intensity and GDP. The rest of this chapter is as follows. Section 2 presents a brief on the macroeconomic profile, literature, and our contribution. Section 3 outlines the data sources that will be used and the methods to be employed. Section 4 presents the econometric findings. Section 5 concludes with a discussion of the results and their policy implications.

2. Macroeconomic profile, literature, and contribution

2.1 Macroeconomic profile

Jamaica achieved its independence from the British Empire in 1962. For much of the 19th century, the Jamaican economy was dominated by sugar, which, in the early 20th century, was complemented by bananas (Hancock, 2017). After independence, the Jamaican economy was defined primarily by its reliance on tourism, which boomed after the popularization of long-haul commercial air travel. From independence onwards, Jamaica promoted itself as a Caribbean destination of choice—particularly in the context of near-shore tourism from the United States—defined by beach culture and reggae music (Johnson and Bartlett, 2013). Currently, roughly 70% of Jamaican GDP is based in the services sector, which, in turn, is highly reliant on international tourism (Apergis and Payne, 2012). Jamaican tourism, like the tourism industry of many other countries, was slow to reflect evolving preferences toward the so-called green tourism (Chen and Tung, 2014; Han et al., 2010). Beginning in the 1990s, green tourism began to call attention to (a) tourism experiences that had a large carbon footprint, as in the case of large hotels; and (b) manifestations of tourism that did not make contributions to the sustainability of local economies and cultures but, rather, enriched multinational tourism companies (Fortenberry, 2021).

In 2005, the publication of Jamaica's *Master Plan for Sustainable Tourism Development* served as perhaps the first indication that the country's policy and economic planning cadres were carefully considering relationships between tourism, the environment, and Jamaica's economy (McFarlane-Morris, 2019). Three years later, the Jamaican Ministry of Tourism held an event centered on the impact of climate change. Since the mid-2000s, therefore, Jamaica has attempted to strike a balance between the need to attract more tourists and the need to protect the environment—for example, by conserving energy, water, and air quality. For many years, the energy consumption model of Jamaica was divided between (a) the low levels of energy consumption of the island's populace and (b) the tourism economy's carbon-intensive energy policies. In particular, hotels and resorts were substantial consumers of energy resources in Jamaica (Fortenberry, 2021).

A sharp rise in per capita energy use in Jamaica in the 1990s makes it an intriguing study country to investigate the relationship between energy consumption and economic growth. Per capita energy use was falling throughout the 1970s and mid-1980s. However, since the late 1980s and early 1990s, per capita energy use has shown an upward trend until the onset of the *Great Recession*. Jamaican per capita GDP also exhibited a similar trend (see Fig. 3.1). In the 1990s, Jamaica underwent a process of trade and financial liberalization, a period that could be characterized as the start of an intensification of global integration for Jamaica (Bloom et al., 2001). Therefore, at least from the data, there seems to exist correlations between energy consumption, GDP, and the onset of a more globally integrated Jamaican economy.

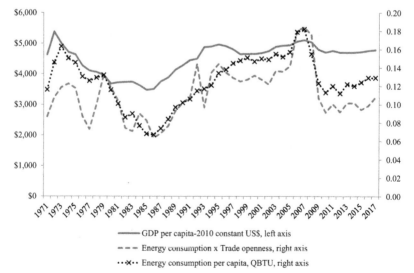

——— GDP per capita-2010 constant US$, left axis

– – – Energy consumption x Trade openness, right axis

··×·· Energy consumption per capita, QBTU, right axis

Notes: (1) Author's calculation based on World Bank (2020) and Federal Reserve Bank of St. Louis (2020)

FIGURE 3.1 **Evolution of energy use and gross domestic product in Jamaica.** (1) Author's calculation based on World Bank (2020) and Federal Reserve Bank of St. Louis (2020).

2.2 What we know about the energy-growth nexus and globalization?

In the literature, the analysis of the energy-growth nexus is grouped into four streams according to the findings of long run and short run causal relationships between economic growth and aggregated and disaggregated energy consumption (Das et al., 2013; Shahbaz and Lean, 2012; Wolde-Rufael, 2004; Yuan et al., 2008). These streams are the growth hypothesis, conservation hypothesis, feedback hypothesis, and the neutrality hypothesis. In the growth hypothesis stream, the findings support a unidirectional causality from energy consumption to economic growth (e.g., Shahbaz et al., 2013b). Here energy consumption increases economic growth by augmenting labor and capital in production. For the conservation hypothesis, the unidirectional causality runs from economic growth to energy consumption (e.g., Apergis and Payne, 2010). Thus, conserving energy consumption is not an impediment to economic growth. With respect to the feedback hypothesis, there is bidirectional causality between energy consumption and economic growth (e.g., Belke et al., 2011). In this case, energy consumption and economic growth influence each other. In the neutrality hypothesis, there is no causal relationship between energy consumption and economic growth (Payne, 2009).

Most developing countries went through the process of globalization in the late 1980s and early 1990s. Shahbaz et al. (2016) studied the relationship between globalization and energy consumption for India for the period 1971−2012. Using the cointegration method, they found that globalization led to a decline in energy consumption in India. Saud and Chen (2018) found similar results for China from 1980 to 2016. On the contrary, Marques et al. (2017) found that globalization played an important role in promoting both energy consumption and economic growth. They applied the ARDL technique to a dataset that consisted of a panel of 43 countries over the period 1971−2013. Other scholars who found similar results include Tansuchat and Khamkaew (2011). Clearly, there is no consensus among scholars on how globalization may have played a role in the energy-growth nexus.

2.3 Situating our contribution to the literature

Few empirical studies on energy-related issues in Jamaica exist. The first study that examined the energy-GDP nexus for Jamaica was by Ramcharran (1990). Using the dataset from 1970 to 1986, he argued that electricity consumption had an impact on economic growth in the island nation. Later on, Francis et al. (2007) supported the earlier findings of Ramcharran (1990) and found evidence of bidirectional causal relations between economic growth and energy consumption over the sample period 1971−2002. Bildirici (2013) used the ARDL approach of cointegration and error correction techniques and showed that there was unidirectional causality from biomass energy consumption to GDP in Jamaica from 1980 to 2009. More recently, Das and McFarlane (2019) examined the dynamic relationships between GDP, electric power losses, and electricity consumption in Jamaica over the 1971–2014 period. They found long run cointegrating relationships between the variables. Further, in the short run, they found evidence of unidirectional causal relation from electricity consumption to economic growth. Therefore, there is some evidence of the impact of energy consumption on economic growth in Jamaica.

What is missing from the literature is whether Jamaica's integration in the rest of the world has played any role in its energy-growth nexus. Theoretically, it can be argued that with the growth of international trade due to globalization, export industries are likely to use more energy. Thus, the association between energy use and trade may have a positive impact on the overall economic activities in Jamaica. Simultaneously, a rise in income due to globalization may induce capital accumulation in the exported goods and service industry and increase overall demand for energy. Given the contributions of tourism and bauxite industries to Jamaica's export earnings, it is also likely that economic output will cause movements in energy consumption due to globalization. The evolution of energy demand, economic growth, and the interplay between energy and trade makes Jamaica an intriguing case to examine the role of globalization in understanding the relationships between energy consumption and economic growth. We precisely address this issue, for the first time for Jamaica.

3. Data and methods

3.1 Data

The data are annual and cover the period 1971–2017 for Jamaica. They were obtained from the World Development Indicators published by the World Bank (2020), the database of the Federal Reserve Bank of St. Louis (2020), and energy statistics from the U.S. Energy Information Administration (2020). Capital stock (CAP) is measured net of depreciation on constant 2011 US dollar per capita basis. Total factor productivity (TFP) is measured at constant national prices with 2011 set equal to 100. Total energy consumption includes the consumption of dry natural gas, petroleum, coal, net nuclear, hydroelectric, nonhydroelectric renewable electricity, net electricity imports (electricity imports–electricity exports) and net coke imports (coke imports–coke exports). Energy consumption is measured in quadrillion British thermal units per capita. GDP is measured in 2010 US constant dollar terms. Our metric for the intensity of globalization on energy we denote as ENG. It is measured as energy per capita times the ratio of the sum imports and exports divided by GDP. We call this measure energy-global openness intensity. The sum of exports and imports can be thought of as a measure of how much an economy is open or integrated in the global economy both monetarily and in the trade in goods and services.

In our analysis, we account for the contractionary and expansionary phases of economic activity and potential nonlinear growth impacts of GDP as it proceeds thorough these phases. To this, we include the CDR effect as an exogenous parameter in the ARDL estimations. The CDR effect is positive when GDP contracts and zero when it expands and is calculated as shown in Eq. (3.1).

$$\text{CDR}_t = \frac{\max(\text{GDP}_s)_{s=0}^t - \text{GDP}_t}{\text{GDP}_t} \tag{3.1}$$

where:

GDP_t is the logarithm of the level of real GDP at time t.

$\max(\text{GDP}_t)_{s=0}^t$ is the historical maximum of GDP_t level from time 0 to t.

The nonlinear growth impacts refer to the tendency for GDP to grow at a faster pace when rebounding from a contractionary phase in comparison to its growth rate when the economy is already in an expansionary phase (Altissimo and Violante, 2001; Beaudry and Koop, 1993).

3.2 Methods

For the statistical analysis, all variables are transformed by applying the natural logarithm and are denoted with an LN prefix applied to the afore-mentioned variables, so that they will now be referred to as LNCAP, LNTFP, LNENG, and LNCDP. We apply the logarithm transformation to stabilize the variances of the series and allow for ease of interpretation of the coefficients on the independent variables as elasticities. Our methods are based on an augmented Solow type production function where LNGDP is related to the other variables within an ARDL bounds testing framework alongside Granger causality testing. To start, we assess the orders of integration of our variables before applying the ARDL bounds testing and Granger causality testing. We do this using the Augmented Dickey Fuller (ADF) tests for unit roots with specified intercept, intercept and trend, and without intercept and trend. The ARDL bounds testing for long run level relation among variables LNCAP, LNTFP, LNENG, and LNGDP requires that none of them be integrated or order two or more. This testing framework is illustrated without the loss of generality by considering the case of three variables x, y, and z where y is treated as the dependent variable. In this case, we estimate Eq. (3.2) as shown below.

$$\Delta y_t = \beta_0 + \sum_1^m \beta_{1i} \Delta y_{t-i} + \sum_0^n \beta_{2i} \Delta x_{t-j}$$
$$+ \sum_0^p \beta_{3i} \Delta z_{t-k} + \gamma_1 y_{t-1} + \gamma_2 x_{t-1} + \gamma_3 z_{t-1} + e_t \tag{3.2}$$

The lag lengths m, n, and p are chosen so that the residuals are namely serially uncorrelated, normally distributed, and homoscedastic. An initial starting point for lag selection is based on minimizing the Akaike information criterion. The existence of a long run relationship running from x to y and z involves an F-test and t-test (Pesaran and Shin, 1999; Pesaran et al., 2001).

The F-test assesses the joint significance of the restriction in the null hypothesis of $\gamma_1 = \gamma_2 = \gamma_3 = 0$ at a given level of statistical significance that there is evidence in favor of the existence of a long run relationship. The F-test procedure involves contrasting the F-statistic obtained from imposing the restriction with upper, I(1), and lower, I(0), bound critical values. The decision rule is as follows. When the F-test statistic is greater, more extreme, than the upper bound critical value, there is evidence in favor of long run level of relationship at the chosen level of statistical significance. Therefore, we would in this case reject the null hypothesis. When the F-test statistic is between the lower and upper bound critical values, the test fails as this result is inconclusive on the existence of a long run level relationship. On the other hand, when the F-test statistic is below the lower critical value, we do not reject the null of no long run levels relationship. The t-test is a cross-check of the F-test. For the t-test, the null hypothesis is $\gamma_1 = 0$ and the alternative hypothesis is $\gamma_1 < 0$. The null hypothesis is rejected if the computed t-test statistic is more extreme than the t-test statistic's upper bound critical values. For us to conclude that there is an overall long run level relationship running from the independent variables to the dependent variable, both the F-test and t-test must provide evidence at the chosen level of statistical significance that allows us to reject their respective null hypotheses.

For a long run level relationship to be a stable cointegrating one, the coefficient on the error correction term, the speed of adjustment parameter, must be negative and less than two in absolute terms (Johansen, 1995). In addition, we must have overall model statistical adequacy before making any inference from the results. In this regard, we first check for well-behaved errors. This is done by testing the residuals for normality with the Jarque-Bera test, serial correlation with the Breusch-Godfrey Lagrange Multiplier test, and heteroskedasticity with the Breusch-Pagan-Godfrey test. Second, we test for structural stability. This is done by charting the Cumulative Sum of the Recursive Residuals (CUSUM) and the Cumulative Sum of the Recursive Residuals Squares (CUSUM squares). In addition, we also report the Ramsey Regression Specification test (RESET).

For the Granger causality part of our empirical analysis, we use the testing framework of Toda and Yamamoto (1995) and Dolado and Lütkepohl (1996). Without loss of generality, the framework can be illustrated by considering x, y, and z as variables to which causal relationships are to be established. The starting point of this approach is the estimation of a basic vector autoregression (VAR) with $p + d$ lags as shown below in Eqs. (3.3)−(3.5).

$$\ln(x_t) = k_1 + \sum_{i=1}^{p+d} \alpha_{1i}\ln(x_{t-i}) + \sum_{i=1}^{p+d} \beta_{1i}\ln(y_{t-i}) + \sum_{i=1}^{p+d} \gamma_{1i}\ln(z_{t-i}) + e_1 \quad (3.3)$$

$$\ln(y_t) = k_2 + \sum_{i=1}^{p+d} \alpha_{2i}\ln(x_{t-i}) + \sum_{i=1}^{p+d} \beta_{2i}\ln(y_{t-i}) + \sum_{i=1}^{p+d} \gamma_{2i}\ln(z_{t-i}) + e_2 \quad (3.4)$$

$$\ln(z_t) = k_3 + \sum_{i=1}^{p+d} \alpha_{3i}\ln(x_{t-i}) + \sum_{i=1}^{p+d} \beta_{3i}\ln(y_{t-i}) + \sum_{i=1}^{p+d} \gamma_{3i}\ln(z_{t-i}) + e_3 \quad (3.5)$$

In this VAR, p is the optimal order of the VAR, d is the highest order of integration among the variables, and the error terms are e_1, e_2, and e_3. Lag p is determined on the basis of information criteria and so that the VAR system has residuals that are serially uncorrelated. For the Granger causality test among the variables, the first p lags are treated as endogenous parameters. The last d lags are treated as exogenous parameters. Only the endogenous lags of the variables are tested for causality using the Wald test. One important advantage of this testing framework comes from using an augmented VAR lag approach based on possibly integrated processes. In this regard, the Granger causality tests will be valid regardless of the order of integration of the variables or the number of cointegrating vectors that exist among them.

4. Results

4.1 Unit root and autoregressive distributed lag bounds tests

The ADF unit root test results are reported in Table 3.1. The null hypothesis for this test is that the series being tested has a stochastic trend, which implies its mean and variance are changing over time. Three different versions of the test are reported in Table 3.1, with intercept, with intercept and trend, and without intercept and trend. From this table we find the consistent result that the LNGDP, LNCAP, LNPOP, LNTFP, and LNENG are nonstationary in levels in the intercept, with intercept and trend, and without intercept and trend versions of the ADF tests. However, we find that their first differences are all stationary and, therefore, all the series are integrated of order one.

4.2 Autoregressive distributed lag bounds testing

Having established that none of the variables LNGDP, LNCAP, LNPOP, LNTFP, and LNENG are integrated of order two or greater, we can proceed to ARDL bounds testing to determine the short run and long run causal relations among them. These results are reported in Table 3.2. We report the ARDL bounds test with each of LNGDP, LNCAP, LNPOP, LNTFP, and LNENG taking on the role of the dependent variable as shown in first (level equation) column of Table 3.2. To start, we examine the results from the case of taking

TABLE 3.1 Augmented Dickey Fuller unit root test.

Variable	No intercept or trend		With intercept		With intercept and trend	
LNGDP	0.99		−0.10		−3.09	
ΔLNGDP	−6.92	***	−6.96	***	−7.04	***
LNCAP	1.52		0.40		−0.87	
ΔLNCAP	−2.73	***	−3.04	**	−4.10	**
LNPOP	0.29		−3.23	**	−2.68	
ΔLNPOP	−2.35	**	−3.56	**	−3.87	**
LNTFP	−1.22		−2.51		2.16	
ΔLNTFP	−5.64	***	−5.84	***	−6.18	***
LNENG	−0.45		−2.26		−2.63	
ΔLNENG	−6.48	***	−6.41	***	−6.32	***

Note: (1) ADF t-statistic provided. (2) *** and ** indicate significance at the 1% and 5% level, respectively. (3) The null hypothesis of the ADF test is that the series has a unit root.

TABLE 3.2 Autoregressive distributed lag bounds test.

Level equation	ARDL	F- statistic		t- Statistic	
LNGDP = F(LNCAP, LTFP, LNPOP, LNENG)	(1,3,5,5,4)	8.40	***	−6.20	***
LNCAP = F(LNGDP, LTFP, LNPOP, LNENG)	(5,4,5,4,5)	1.86		−1.13	
LNTFP = F(LNCAP, LNGDP, LNPOP, LNENG)	(1,2,1,2,2)	0.36		−0.14	
LNPOP = F(LNTFP, LNCAP, LNGDP, LNENG)	(2,1,0,2,3)	5.97	***	−1.07	
LNENG = F(LNTFP, LNCAP, LNGDP, LNENG)	(2,3,3,2,2)	3.74		−3.92	

Statistical significance	I(0)	I(1)	I(0)	I(1)
10%	2.66	3.84	−2.57	−3.21
5%	3.20	4.54	−2.86	−3.53
1%	4.43	6.25	−3.43	−4.10

Note: (1) *** indicates statistical significance at the 1% level for the rejection of the null hypothesis. (2) The null hypothesis is that there is a long run level relationship as specified in the given equation.

LNGDP as the dependent variable. In this case, we find that there is strong statistical evidence in favor of concluding that there is a level long run relationship running from the other variables to LNGDP. This is because the F-test statistic has a value of 8.40 and that is higher than the 1% level of statistical significance upper bound F-test critical value of 6.25. In addition, we also find that the t-statistic has a value of -6.20, and this is more extreme that the t-test critical value at the 1% level of statistical significance of -4.10. When we consider the cases of LNCAP, LNTFP, and LNENG as the dependent variables, we find that the conventional levels of statistical significance support rejecting the null hypothesis of a long run level relationship. We observe from Table 3.2 that in these cases, all the respective F-statistics and t-statistics are less extreme than the F-test and t-test critical lower bound critical values at the 10% level of statistical significance or less. In the case of LNPOP as dependent variable, we find that the F-statistic is greater than the upper bound F-test critical value (5.97 vs. 4.54) at the 5% level of statistical significance, which is suggestive of a long run level relationship with the other variables. However, when we examine the t-test statistic, we find that this value of -1.07 is less extreme than the t-test upper bound critical at any of the conventional levels of statistical significance. Consequently, in this case also we fail to reject the null hypothesis of no level relationship running from the other variables to LNPOP.

4.3 Autoregressive distributed lag cointegration and long run form

First, we consider the long run dynamics. From the results of the bounds test, there is only one long run level relationship, and this runs from LNCAP, LNPOP, LNTFP, and LNENG to LNGDP. Consequently, we report the cointegrating and long run forms in Table 3.3 to determine the nature of this long run relationship. This table shows that capital, total factor productivity, and population all have positive long run elasticities with respect to their impacts on GDP. However, only the coefficients on LNCAP and LNENG are statistically significant at conventional levels (p-value <0.01). A 1% increase in TFP leads to 0.82% increase in GDP. For POP, a 1% increase leads to a marginally higher 0.84% increase in GDP. LNENG is the main variable of interest. The coefficient on this variable is also positive (0.13) and statistically significant at conventional levels (p-value <0.05). The value of the coefficient on LNENG means that a 1% increase in the energy-trade openness intensity leads to, on average, a 0.13% increase in GDP in the long run.

We next turn to the analysis of the short run dynamics. From Table 3.3, we observe that the coefficient on the CDR effect term is negative and statistically significant at the less than 1% level. This indicates that accounting for the nonlinear growth dynamics is important in the evolution of the relationships among LNCAP, LNPOP, LNTFP, and LNENG to LNGDP. The coefficient on the short run error correction term, the speed of adjustment parameter, has a value of -1.26 and is statistically significant at the less than 1% level. The fact that this parameter is less than two in absolute value means that the level

TABLE 3.3 Autoregressive distributed lag cointegrating and long-run forms.

Equation: LNGDP = F(LNCAP, LTFP, LNPOP, LNENG)

Variable	Coefficient		Standard error
LNCAP	0.18		0.12
LNTFP	0.82	***	0.13
LNPOP	0.84	***	0.16
LNENG	0.13	**	0.01
Short run error correction term	−1.26	***	0.17
CDR	−0.90	***	0.16
ARDL	(1,3,5,5,4)		
Adjusted R-squared	0.88		
Akaike info criterion	−5.87		
Number of oobservations	42		
Sample	1971−2017		

*** and ** indicate statistical significance level at the 1% and 5% level respectively.

relationship established by the ARDL bounds testing framework also has a stable cointegrating relationship. The −1.26 value of the coefficient on the error correction term indicates that when the relationship among the variables is in long run equilibrium and some short run shock disturbs this equilibrium, there will be a rapid and dampened oscillatory adjustment process back to the long run equilibrium path.

4.4 Robustness checks

The robustness of inferences drawn from the ARDL depends on whether the residuals are well-behaved with parameters of the cointegrating relationship that are stable over the period examined. Table 3.4 reports the results from the

TABLE 3.4 Diagnostics checks.

	Test statistic	P-value
Normality[a]	1.63	0.44
Serial correlation[b]	0.59	0.57
Heteroskedasticity[c]	1.37	0.25
Ramsey RESET, stability[d]	1.80	0.20

(1) The null hypotheses are respectively: (a) normally distributed residuals (Jarque-Berra), (b) no serial correlation of the residuals (Breusch-Godfrey LM), (c) no heteroskedasticity of the residuals (Breusch-Pagan-Godfrey) and (d) correct functional form.

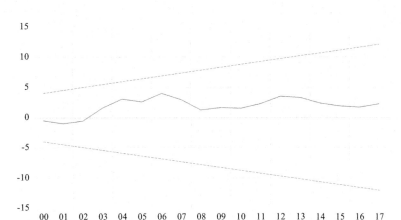

           CUSUM ----- 5% Significance

Notes: The CUSUM provides evidence that supports an absence of coefficient instability in the long run level relationship.

FIGURE 3.2 Cumulative sum of the recursive residuals. The CUSUM provides evidence that supports an absence of coefficient instability in the long run level relationship.

test for serial correlation, normality, heteroskedasticity, and the Ramsey RESET test for coefficient stability and model misspecification. From this table, we note that the residuals are serially uncorrelated, normally distributed, and not heteroskedastic. Similarly, the Ramsey RESET test indicates no parameter instability from the perspective of functional form correctness.

Stability of the cointegration is also checked by considering CUSUM of the recursive residuals and the CUSUM of the recursive residual squares. A CUSUM or the CUSUM of squares outside of the 5% significance confidence band is suggestive of an ARDL model with parameter and variance instability. The CUSUM and the CUSUM squares are reported in Figs. 3.2 and 3.3, respectively. From these figures, we can conclude that there is no parameter of variance instability.

4.5 Granger causality testing

We consider causality testing from an appropriately well-specified VAR model in the levels of the variables, namely LNCAP, LNPOP, LNTFP, and LNENG to LNGDP. The VAR is chosen such that the residuals are serially uncorrelated and multivariate normally distributed. The final VAR settled upon was one with the first four lags being endogenous and the fifth lag treated as exogenous. In addition, the CDR term is also included as an exogenous parameter. Table 3.5 reports the results from the serial correlation tests. From this table we observe that both the likelihood ratio statistic and Rao F-statistic indicate that we cannot reject the null hypothesis of no serial correlation at lag orders one, two, three, four, and five. Table 3.6 reports the results from the multivariate

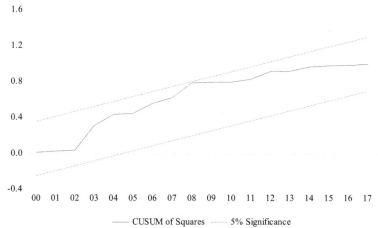

1.6

1.2

0.8

0.4

0.0

-0.4

00 01 02 03 04 05 06 07 08 09 10 11 12 13 14 15 16 17

—— CUSUM of Squares ----- 5% Significance

Notes: The CUSUM of squares provides evidence that supports an absence of coefficient or variance instability in the long run level relationship.

FIGURE 3.3 Cumulative sum of the recursive residuals squares. The CUSUM of squares provides evidence that supports an absence of coefficient or variance instability in the long run level relationship.

TABLE 3.5 Vector autoregression model serial correlation Lagrange multiplier test.

Lag	LRE statistic	Rao F-Statistic
1	32.53	1.44
2	28.96	1.21
3	39.32	1.95
4	29.96	1.27
5	30.25	1.29

(1) VAR has first four lags endogenous and the fifth lag treated as exogenous. (2) Null hypothesis: no serial correlation at lag order indicated. (3) The Edgeworth expansion corrected likelihood ratio statistic (LRE) and Rao F-statistics both indicate no serial correlation at the 5% level of statistical significance or less.

normality test. The results residuals from each equation of the VAR system are normally distributed.

With an optimal VAR established, we proceed to report the Granger causality tests. These results are reported in Table 3.7. From this table, the results from LNGDP and LNENG taking the role of the dependent and independent variables are the perspectives of interest. In this regard, we note that LNGDP is Granger caused by LNENG at the 10% level of statistical significance ($\chi^2 = 8.07$). Table 3.7 also reveals that LNENG is Granger caused by LNGDP

TABLE 3.6 Vector autoregression model multivariate residual normality test.

Equation	Jarque-Berra	Probability
LNGDP	0.66	0.72
LNCAP	1.34	0.51
LNTFP	1.53	0.46
LNPOP	1.97	0.37
LNENG	0.82	0.66
Joint	6.32	0.79

(1) VAR has first four lags endogenous and the fifth lag treated as exogenous. (2) Null Hypothesis is that the residuals are multivariate normal is not rejected. (3) The null is not rejected for the residuals associated which equation and all equations.

TABLE 3.7 Toda and Yamamoto vector autoregression Granger causality test.

Dependent variable	Excluded	Chi-squared (χ^2)		Probability
LNGDP	LNCAP	1.31		0.89
	LNTFP	5.37		0.25
	LNPOP	11.07	**	0.03
	LNENG	8.07	*	0.09
	All	79.08		0.23
LNCAP	LNGDP	3.71		0.45
	LNTFP	4.69		0.32
	LNPOP	3.36		0.50
	LNENG	4.17		0.38
	All	14.95		0.53
LNTFP	LNGDP	4.79		0.31
	LNCAP	20.35	***	0.00
	LNPOP	10.90	**	0.03
	LNENG	2.92		0.57
	All	30.42		0.02

Continued

TABLE 3.7 Toda and Yamamoto vector autoregression Granger causality test.—cont'd

Dependent variable	Excluded	Chi-squared (χ^2)		Probability
LNPOP	LNGDP	0.80		0.94
	LNCAP	2.03		0.73
	LNTFP	0.96		0.92
	LNENG	5.00		0.29
	All	18.70		0.28
LNENG	LNGDP	10.40	**	0.03
	LNCAP	6.12		0.19
	LNTFP	12.14	**	0.02
	LNPOP	4.49		0.34
	All	42.35	***	0.00

(1) VAR has first four lags endogenous and the fifth lag treated as exogenous. (2) ***, **, and * indicate statistical significance at the 1%, 5%, and 10% level, respectively, that the indicated variable should not be excluded.

at the 5% level of statistical significance ($\chi^2 = 10.40$). On basis of these results, we conclude that there is bidirectional causality between output and our measure of energy-trade openness.

5. Discussion and concluding remarks

In this chapter, we discuss the energy-growth nexus within the context of globalization. We then examine this nexus for Jamaica for the period 1971–2017. We interact trade openness with energy consumption per capita to produce a measure of energy-global openness intensity. We then establish causal relations in the short run and long run using ARDL bounds and Granger causality testing. Importantly, we control for potentially confounding factors resulting from changes in the macroeconomic conditions (total factor productivity, capital stock, labor force) and business cycle economic growth asymmetries over the sample period. Our analysis is a contribution to the literature on the analysis of the energy-growth nexus for Jamaica taking into account a global perspective.

Our empirical analysis of the relationship between energy-global openness intensity and economic growth in Jamaica indicate two key findings. First, energy-global openness intensity has a positive impact on economic growth in the long run. The short run results, as suggested by the Granger causality, within the Toda and Yamamoto and Dolado and Lütkepohl framework, show bidirectional causality between the variables. Therefore, in the short run, we find evidence of the feedback hypothesis. Overall, the short run results echo the findings of Ramcharran (1990), while the long run results are similar to that of Das and McFarlane (2019). For Jamaica, the value added of our findings is that we have considered, for the first time to our knowledge, the dynamics of econmic growth and energy consumption accounting for the degree of global integration, the latter being reflected in our energy-global openness intensity measure.

Although our findings are based on several statistical assumptions, they have important policy implications for Jamaica. Since energy-global openness intensity plays a role in determining economic growth in Jamaica, the government should design policies to meet the growing energy demand with a view to make energy more affordable, given that energy consumption has a positive impact on GDP. This is important as Jamaica's energy utility rates are higher than that of the average rates in the Caribbean region (Energy Transition Initiative, 2015). Making energy affordable could be done by providing subsidies to export-oriented industries. The government should also consider giving tax breaks to those industries. Further, as suggested by Das and McFarlane (2019), the government could construct power plants, build capacity, and develop infrastructure to ensure increasing supply of energy, and thereby increase the likelihood of sustained economic growth. It has been noted that Jamaica is heavily dependent on imported fossil fuels. For example, more than 94% of Jamaica's total electricity is produced from imported petroleum-based fuels (Energy Transition Initiative, 2015). There are problems with such dependency. One such problem is that the energy industry will become more vulnerable to exogenous shocks, such as oil price and exchange rate fluctuations. Another is problem is that emissions from burning fossil fuels are responsible for environmental degradation due to their carbon footprints. Given these problems, it is important for Jamaica to consider developing clean energy projects such as expanding wind, solar, and hydro-electric energy sources.

References

Altissimo, F., Violante, G.L., 2001. The nonlinear dynamics of output and unemployment in the U.S. J. Appl. Econom. 16 (4), 461−486.

Apergis, N., Payne, J.E., 2010. Energy consumption and growth in South America: evidence from a panel error correction model. Energy Econ. 32 (6), 1421−1426.

Apergis, N., Payne, J.E., 2012. Tourism and growth in the Caribbean—evidence from a panel error correction model. Tourism Econ. 18 (2), 449−456.

Baek, J., Cho, Y., Koo, W.W., 2009. The environmental consequences of globalization: a country-specific time-series analysis. Ecol. Econ. 68 (8−9), 2255−2264.

Beaudry, P., Koop, G., 1993. Do recessions permanently change output? J. Monetary Econ. 31 (2), 149−163.

Belke, A., Dobnik, F., Dreger, C., 2011. Energy consumption and economic growth: new insights into the cointegration relationship. Energy Econ. 33 (5), 782–789.

Bildirici, M.E., 2013. Economic growth and biomass energy. Biomass Bioenergy 50, 19–24.

Bloom, D.E., Mahal, A., King, D., Henry-Lee, A., Castillo, P., 2001. Globalization, Liberalization and Sustainable Human Development: Progress and Challenges in Jamaica. United Nations Conference on Trade and Development/United Nations Development Programme. Available at: https://unctad.org/en/docs/poedmm176.en.pdf.

Chen, M.-F., Tung, P.-J., 2014. Developing an extended Theory of Planned Behavior model to predict consumers' intention to visit green hotels. Int. J. Hospit. Manag. 36, 221–230.

Das, A., Paul, B.P., 2011. Openness and growth in emerging Asian economies: evidence from GMM estimations of a dynamic panel. Econ. Bull. 31 (3), 2219–2228.

Das, A., McFarlane, A.A., Chowdhury, M., 2013. The dynamics of natural gas consumption and GDP in Bangladesh. Renew. Sustain. Energy Rev. 22, 269–274.

Das, A., McFarlane, A., 2019. Non-linear dynamics of electric power losses, electricity consumption, and GDP in Jamaica. Energy Econ. 84, 104530.

Dolado, J.J., Lütkepohl, H., 1996. Making Wald tests work for cointegrated VAR systems. Econom. Rev. 15 (4), 369–386.

Energy Transition Initiative, 2015. Energy Snapshot: Jamaica. Available at: https://www.nrel.gov/docs/fy15osti/63945.pdf.

Federal Reserve Bank of St. Louis, 2020. FRED Economic Data. Available at: https://stlouisfed.org/.

Fortenberry, B.R., 2021. Heritage justice, conservation, and tourism in the greater Caribbean. J. Sustain. Tourism 29 (2–3), 253–276.

Francis, B.M., Moseley, L., Iyare, S.O., 2007. Energy consumption and projected growth in selected Caribbean countries. Energy Econ. 29 (6), 1224–1232.

Han, H., Hsu, L.-T.J., Sheu, C., 2010. Application of the theory of planned behavior to green hotel choice: testing the effect of environmental friendly activities. Tourism Manag. 31 (3), 325–334.

Hancock, J.F., 2017. Plantation Crops, Plunder and Power: Evolution and Exploitation. Taylor & Francis, New York.

Johansen, S., 1995. Identifying restrictions of linear equations with applications to simultaneous equations and cointegration. J. Econom. 69 (1), 111–132.

Johnson, K.R., Bartlett, K.R., 2013. The role of tourism in national human resource development: a Jamaican perspective. Hum. Resour. Dev. Int. 16 (2), 205–219.

Kraft, J., Kraft, A., 1978. On the relationship between energy and GNP. J. Energy Dev. 3 (2), 401–403.

Marques, L.M., Fuinhas, J.A., Marques, A.C., 2017. Augmented energy-growth nexus: economic, political and social globalization impacts. Energy Procedia 136, 97–101.

McFarlane-Morris, S., 2019. 'Come this close, but no closer!' Enclave tourism development and social change in Falmouth, Jamaica. J. Tourism Cult. Change 1–15 (Published online).

Odhiambo, N.M., 2009. Energy consumption and economic growth nexus in Tanzania: an ARDL bounds testing approach. Energy Pol. 37 (2), 617–622.

Ozturk, I., Aslan, A., Kalyoncu, H., 2010. Energy consumption and economic growth relationship: evidence from panel data for low and middle income countries. Energy Pol. 38 (8), 4422–4428.

Paul, S., Bhattacharya, R.N., 2004. Causality between energy consumption and economic growth in India: a note on conflicting results. Energy Econ. 26 (6), 977–983.

Payne, J.E., 2009. On the dynamics of energy consumption and output in the US. Appl. Energy 86 (4), 575–577.

Pesaran, M.H., Shin, Y., 1999. An autoregressive distributed lag modelling approach to cointegration analysis. In: Strom, S. (Ed.), Chapter 11 in Econometrics and Economic Theory in the 20th Century: The Ragnar Frisch Centennial Symposium. Cambridge University Press, Cambridge.

Pesaran, M.H., Shin, Y., Smith, R.J., 2001. Bounds testing approaches to the analysis of level relationships. J. Appl. Econom. 16 (3), 289–326.

Ramcharran, H., 1990. Electricity consumption and economic growth in Jamaica. Energy Econ. 12 (1), 65–70.

Saud, S., Chen, S., 2018. An empirical analysis of financial development and energy demand: establishing the role of globalization. Environ. Sci. Pollut. Control Ser. 25 (24), 24326–24337.

Shahbaz, M., Lean, H.H., 2012. The dynamics of electricity consumption and economic growth: a revisit study of their causality in Pakistan. Energy 39 (1), 146–153.

Shahbaz, M., Zeshan, M., Afza, T., 2012. Is energy consumption effective to spur economic growth in Pakistan? New evidence from bounds test to level relationships and Granger causality tests. Econ. Modell. 29 (6), 2310–2319.

Shahbaz, M., Hye, Q.M.A., Tiwari, A.K., Leitão, N.C., 2013a. Economic growth, energy consumption, financial development, international trade and CO_2 emissions in Indonesia. Renew. Sustain. Energy Rev. 25, 109–121.

Shahbaz, M., Khan, S., Tahir, M.I., 2013b. The dynamic links between energy consumption, economic growth, financial development and trade in China: fresh evidence from multivariate framework analysis. Energy Econ. 40, 8–21.

Shahbaz, M., Lahaini, A., Homudeh, S., Abosedra, S., 2018a. The role of globalization in energy consumption: a quantile cointegrating regression approach. Energy Econ. 71, 161–170.

Shahbaz, M., Mallick, H., Mahalik, M.K., Sadorsky, P., 2016. The role of globalization on the recent evolution of energy demand in India: implications for sustainable development. Energy Econ. 55, 52–68.

Shahbaz, M., Shahzad, S.J.H., Mahalik, M.K., Sadorsky, P., 2018b. How strong is the causal relationship between globalization and energy consumption in developed economies? A country-specific time-series and panel analysis. Appl. Econ. 50 (13), 1479–1494.

Soytas, U., Sari, R., 2009. Energy consumption, economic growth, and carbon emissions: challenges faced by an EU candidate member. Ecol. Econ. 68 (6), 1667–1675.

Tansuchat, R., Khamkaew, T., 2011. The impact of energy consumption on economic performance in the era of globalization. In: The Scale of Globalization, Think Globally, Act Locally, Change Individually in the 21st Century, pp. 346–358.

Toda, H.Y., Yamamoto, T., 1995. Statistical inferences in vector autoregressions with possibly integrated processes. J. Econom. 66 (1), 225–250.

U.S. Energy Information Administration, 2020. Annual Energy Review. Available at: https://www.eia.gov/totalenergy/data/annual/index.php.

Wolde-Rufael, Y., 2004. Disaggregated industrial energy consumption and GDP: the case of Shanghai, 1952–1999. Energy Econ. 26 (1), 69–75.

World Bank, 2020. World Development Indicators. Available at: https://data.worldbank.org/.

Yuan, J.H., Kang, J.G., Zhao, C.H., Hu, Z.G., 2008. Energy consumption and economic growth: evidence from China at both aggregated and disaggregated levels. Energy Econ. 30 (6), 3077–3094.

Chapter 4

Return and volatility spillovers between fossil oil and seafood commodity markets

Akram Shavkatovich Hasanov[1], Walid Mensi[2,3],
Yessengali Oskenbayev[4,5]

[1]*Department of Econometrics and Business Statistics, Monash University, Subang Jaya, Selangor, Malaysia;* [2]*Department of Finance and Accounting, University of Tunis El Manar, Tunis, Tunisia;* [3]*Department of Economics and Finance, College of Economics and Political Science, Sultan Qaboos University, Muscat, Oman;* [4]*Business School, Suleyman Demirel University, Kaskelen City, Kazakhstan;* [5]*Department of Finance and Accounting, University of International Business, Almaty, Kazakhstan*

1. Introduction

A considerable body of literature has examined the linkage between the fossil oil and major agricultural commodities (e.g., cereals and oilseeds) that have both food and industrial uses (i.e., as feedstock in the production of biofuels) (see, inter alia, Abdelradi and Serra, 2015; Du et al., 2011; Hameed and Arshad, 2009; Hasanov et al., 2016, Nazlioglu, 2011; Nazlioglu and Soytas, 2011; Nazlioglu et al., 2012; and Mensi et al., 2014). Most notably, this is because the upward movements in oil prices are believed to result in the growth of biofuels (e.g., biodiesel) production that drives up demand for the biofuel feedstock commodities and thereby impact on the food commodity prices as well as land-use change (Chen et al., 2010). Moreover, as noted by Serra and Zilberman (2013), biofuel feedstock commodities (e.g., corn, vegetable oils) are used to produce other essential food products (e.g., meat, fish, and fishery, and others) for human consumption in addition to being used in renewable energy production. For example, potential alternative agricultural feed ingredients for farmed fish and shrimp include barley, canola, corn, peas, soybeans, other various plant proteins, and wheat (National Marine Fisheries Service (NMFS), 2016). These alternative feeds have accelerated the attempts toward reducing the dependence on fishmeal and fish oil use in aquaculture feeds while retaining the essential human health benefits of seafood consumption (NMFS, 2016). Thus, unlike many earlier studies in this context,

Energy-Growth Nexus in an Era of Globalization. https://doi.org/10.1016/B978-0-12-824440-1.00009-6
87

we have examined a somewhat different aspect by focusing on the price level change and volatility transmission between the energy oil market and seafood commodity markets. There has been limited research on how fossil oil price changes and volatility are transmitted to the food market chain (Serra and Zilberman, 2013). Hence, an interesting question is worth being explored: how significant is the role of the mineral oil market in terms of information (e.g., volatility) transmission in commodity markets (i.e., seafood) where biofuel feedstocks are also used as essential nutritional feeds? An analysis of volatility spillover between commodity markets is vital for discovering price volatility behavior of commodities and making global across-market hedging decisions (Ji and Fan, 2012). Besides, financial market participants have been considering commodity markets as alternative investment areas for diversification and risk-minimization purposes (see, for example, Nazlioglu et al., 2012). Therefore, the volatility transmission analysis and research had been paid full attention for portfolio diversification strategies and extensively employed among financial markets (Ji and Fan, 2012).

Fish and fishery products continue to be among the most traded food commodities all over the world. World fish production has expanded at an average annual rate of 3.2%, rising faster than world population growth [Food and Agriculture Organization (FAO), 2014]. The per capita fish production consumed directly by global human population increased from about 9.9 kg in the 1960s to approximately 19.2 kg in 2012—driven by higher demand from increasing population, more efficient distribution systems, and economic growth (FAO, 2014). In such an impressively developing sector, accurate information on the future price and volatility movements of seafood products is essential for agricultural policy and financial decision-makers. They are particularly interested in the following questions: Are marine commodity markets independent of the fossil oil market? Is there any evidence of volatility transmission from mineral oil to seafood commodities? Does the nonrenewable energy market contain useful information to determine the future movements of seafood prices? To the authors' knowledge, no study has explored the above questions for the seafood markets. This paper addresses this gap by investigating the information transmission from fossil oil markets to widely traded and consumed seafood commodities (i.e., salmon and shrimp).

It is worth noting that the world salmon and shrimp markets have expanded significantly over the last few decades as a result of expanding cultured activities. Warm-water shrimps are produced mainly in Asia, Central America, and South America. The shrimp export market has considerably grown since the early 1980s, and this expansion is primarily due to increased production, especially in Asia and South America. According to the FAO (FAO, 2014), shrimp and salmon continue to be the essential fish products traded in value terms, representing approximately 15% and 14% of the total value of internationally traded fishery products, respectively. The increased production of shrimp with the expansion in the world export market has resulted in a decline

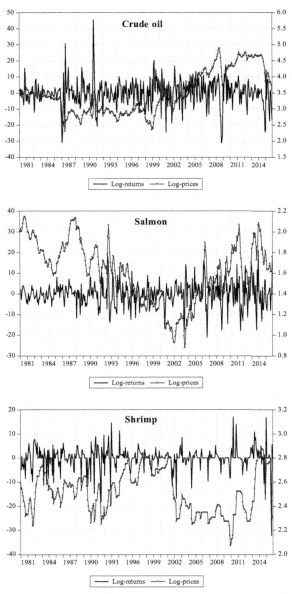

FIGURE 4.1   Prices for the period from January 1980 to December 2015 [the right vertical label (gray in print version)] and log returns from February 1980 to December 2015 [the left vertical label (dark gray in print version)].

in the world export prices during 2001–09 (see Fig. 4.1). Most of the shrimp industry growth comes from a substantial productivity growth which has considerably reduced unit costs over time. Thus, the average farmed shrimp prices continue to decline in all major markets from late 2000 until recently.

All major markets have been affected by the dramatic growth in world production. The reduction in production costs is due to two main reasons. First, fish farmers can produce more with a given amount of inputs, and second, improved input factors have made the production process cheaper (for example, the development of better feed and feeding technology).

The total salmon trade increased almost fourfold between 1997 and 2011 throughout the world (United Nations commodity trade database). The volume and sources of world salmon supply have changed dramatically over the past two decades. In 1980, total world salmon supply was less than 550,000 metric tons (mt). By 2004, world supply has increased to more than 2.4 million mt (Oglend, 2010). The rapid growth of salmon farming has taken place in Norway, which is the leading producer of salmon accounting for around 40% of the world production (Oglend, 2010). The achievement of significant growth of salmon production is due to biological improvement and the development of cultivation technology (FAO, 2012). World salmon consumption is divided among five major markets: the European Union fresh and frozen market, the Japanese fresh and frozen market, the US fresh and frozen market, canned salmon markets, and other markets. There are significant differences between these markets in their sources of supply, species and products consumed, and short-run market conditions (Bjørndal et al., 2003).

Our paper adds to and differs from the related literature in the following regards. First, it analyzes the price level changes and volatility transmission between crude oil and the two most traded seafood commodity markets. Second, our analysis employs an asymmetric model which allows determining whether fossil oil price increases have the same causal influences on agricultural price volatility than price declines. Previous volatility studies in this context have relied on the assumption that negative and positive shocks have a symmetric effect on volatility (Serra and Zilberman, 2013). Third, we apply a relatively recent econometric method developed by Hafner and Herwartz (2004), which is superior to a test proposed by Cheung and Ng (1996). Fourth, we adopt a heavy-tailed conditional density (bivariate Student t) in the causality analysis. Finally, as the existing literature suggests, a dominant consensus regarding the linkage between crude oil and food commodities has not been reached yet. Therefore, this necessitates analyzing the issue within the context of various methodological approaches (see Nazlioglu, 2011).

The main aim of this paper is to examine the causality in mean and variance from the crude oil price to two widely traded seafood commodity prices (i.e., salmon and shrimp) using a dynamic causality analysis based on a multivariate GARCH model. Specifically, we employ the bivariate asymmetric BEKK model of Engle and Kroner (1995) to parameterize the conditional variance-covariance matrix. We consider the mean and variance causality analysis, which relies on the relatively new statistical procedure suggested by Hafner and Herwartz (2004). The multivariate GARCH family model allows spillovers and time variation in the conditional variance-covariance equations. Additionally, this methodology accounts for the possible asymmetry in the price volatility.

The remainder of this paper is organized as follows. The next section presents the methodology used in this study. Section 3 describes the data, and Section 4 reports the empirical results. Finally, Section 5 concludes the paper.

2. Econometric methodology

2.1 The GARCH-BEKK model

The empirical analysis relies on the estimations of a bivariate GARCH-BEKK model for crude oil and each seafood commodity price returns under concern, $r_{c,t}$ and $r_{f,t}$, respectively. Here, the subscript c denotes crude oil while f indicates an individual seafood commodity (i.e., salmon fish and shrimp price returns). Therefore, the modeling approach adopted in this study carries a pair-wise nature. It is worth noting that pair-wise analysis is commonly applied in the price transmission literature and serves as a potential and natural avenue for investigating the links between prices (see, among others, Balcombe and Rapsomanikis, 2008; Serra, 2011). More importantly, while the econometric model employed in this study can be extended to higher-order settings including other variables, as commonly noted in the literature, these extensions usually lead to an issue of "curse of dimensionality." This issue appears not only in nonparametric methods (see Serra, 2011) but also arises in a parametric multivariate statistical analysis which implies that the number of parameters of a model rises dramatically with the increase of time series dimension (Tsay, 2010). Hence, simpler models (e.g., pair-wise analysis) or some restrictions are often sought to overcome this issue. Thus, two pairs of price returns: salmon-crude oil and shrimp-crude oil are considered in this study. Consequently, we compactly express the conditional mean equations for each pair of price returns with the use of matrices and vectors as follows

$$r_t = c + \sum_{i=1}^{p} \Psi_i r_{t-i} + u_t \tag{4.1}$$

$$u_t \Big| \Omega_{t-1} \sim (0, H_t), \quad H_t = \begin{bmatrix} h_{ff,t} & h_{cf,t} \\ h_{fc,t} & h_{cc,t} \end{bmatrix},$$

where Ω_{t-1} represents the information set available at time $t-1$, H_t is the conditional covariance matrix, 0 is the null vector, and

$$r_t = \begin{bmatrix} r_{f,t} \\ r_{c,t} \end{bmatrix}; \quad u_t = \begin{bmatrix} u_{f,t} \\ u_{c,t} \end{bmatrix}; \quad \psi_i = \begin{bmatrix} \psi_{11}^{(i)} & \psi_{12}^{(i)} \\ \psi_{21}^{(i)} & \psi_{22}^{(i)} \end{bmatrix};$$

In this analysis, we exploit an asymmetric BEKK model of Grier et al. (2004), which is given as follows

$$H_t = C'C + A'u_{t-1}u'_{t-1}A + B'H_{t-1}B + D'\xi_{t-1}\xi'_{t-1}D \qquad (4.2)$$

where $C = \begin{bmatrix} c_{11} & 0 \\ c_{21} & c_{22} \end{bmatrix}$; $A = \begin{bmatrix} a_{11} & a_{12} \\ a_{21} & a_{22} \end{bmatrix}$; $B = \begin{bmatrix} b_{11} & b_{12} \\ b_{21} & b_{22} \end{bmatrix}$;

$D = \begin{bmatrix} d_{11} & d_{12} \\ d_{21} & d_{22} \end{bmatrix}$; $\xi_{t-1} = \begin{bmatrix} \xi_{f,t-1} \\ \xi_{c,t-1} \end{bmatrix}$;

This parameterization considers lagged conditional variances and co-variances, H_{t-1}, as well as past values of $u_{t-1}u_{t-1}$ and $\xi_{t-1}\xi_{t-1}$, in joint estimations of contemporaneous volatilities of fossil oil and the individual seafood commodities under concern. Here, the term, $\xi_{t-1}\xi_{t-1}$, accounts for potential asymmetric responses. The asymmetric impact refers to different influences of negative and positive returns of a similar magnitude on conditional volatility. To capture the possible asymmetries, we define both $\xi_{f,t}$ and $\xi_{c,t}$ as $\min\{u_{f,t}, 0\}$ and $\min\{u_{c,t}, 0\}$, respectively.

2.2 Quasi-maximum likelihood estimate

The bivariate asymmetric and unrestricted VAR-(p)-GARCH-BEKK is estimated by employing the quasi-maximum likelihood (QML) method with multivariate Student t distribution. The QML estimator is proven to be consistently provided that the conditional mean and variance equations are specified correctly. For statistical inference on the estimated models, we consider the robust standard errors proposed by Bollerslev and Wooldridge (1992). Given T observations of r_t, the following likelihood function is optimized assuming multivariate Gaussian distribution.

$$\max_\theta \log L_T(\theta) = \sum_{t=1}^{T} l_t(\theta) \qquad (4.3)$$

where θ is a vector of parameters

$$\sum_{t=1}^{T} l_t(\theta) = c - \frac{1}{2}\sum_{t=1}^{T}\ln|H_t| - \frac{1}{2}\sum_{t=1}^{T}u'_t H_{t-1}^{-1}u_t \qquad (4.4)$$

Furthermore, the multivariate Student's t density is a natural alternative to the multivariate normal density (see Fiorentini et al., 2003; Rossi and Spazzini, 2010). Unlike Gaussian distribution, the Student t density has an additional scalar parameter, tail parameter, denoted v hereafter. As noted by Rossi and Spazzini (2010), the value of v shows the order of the existence of the moments. For example, if $v = 2$, the second-order moments do not exist, but the first-order moments do. Thus, tail parameter v is commonly assumed to be higher than 2, so that $\boldsymbol{H}_t$ can always be interpreted as a conditional covariance matrix. The multivariate Student t density for stochastic error process can be defined as

$$f(u_t; \boldsymbol{H}_t, v) = \frac{\Gamma\left(\dfrac{v+N}{2}\right)}{\Gamma\left(\dfrac{v}{2}\right)[\pi(v-2)]^{N/2}}|\boldsymbol{H}_t|^{-1/2}\left[1 + \frac{u_t'\boldsymbol{H}_t^{-1}u_t}{v-2}\right]^{-\frac{N+v}{2}} \qquad (4.5)$$

Therefore, the log-likelihood contributions for observation t are of the form

$$l_t(\boldsymbol{\theta}) = \ln\left[\frac{\Gamma\left(\dfrac{v+N}{2}\right)}{\Gamma\left(\dfrac{v}{2}\right)[\pi(v-2)]^{N/2}}\left(1 + \frac{u_t'\boldsymbol{H}_t^{-1}u_t}{v-2}\right)^{-\frac{N+v}{2}}\right] - \frac{1}{2}\ln(|\boldsymbol{H}_t|) \qquad (4.6)$$

where $\boldsymbol{H}_t$ is the conditional variance-covariance matrix of errors, $\Gamma(\cdot)$ is the gamma function, N is the number of variables in the system, and $\boldsymbol{\theta}$ is a vector of parameters to be estimated by the maximum likelihood function given by

$$L(\boldsymbol{\theta}) = \sum_{t=1}^{T} l_t(\boldsymbol{\theta}) \qquad (4.7)$$

The parameters for both the conditional mean and covariance equations together with a distributional (tail) parameter, v, are estimated jointly. Note that the tail (shape) parameter v is included in model estimations based on multivariate Student t density. We employ the numerical algorithm so-called Broyden, Fletcher, Goldfarb, Shanno (BFGS) quasi-Newton method explained in Green (2012), and Press et al. (2007) to conduct an optimization. Due to a large number of parameters as well as a complicated nonlinear structure of log-likelihood functions, the multivariate GARCH model estimations are computationally more difficult compared to their univariate counterparts. Therefore, as mentioned in Green (2012), it is useful to exploit a simplex method in the initial steps for several iterations before shifting to standard numerical optimization algorithms. Hence, as the simplex method helps to obtain better starting values for log-likelihood functions, we also use this method to choose the initial parameters.

3. Data and stochastic properties

Our analysis uses monthly data for crude oil spot prices (average of U.K. Brent, Dubai, and West Texas Intermediate, expressed in US dollars per barrel), salmon (farm-bred Norwegian salmon, export price, USD per kilogram), shrimp (New York port, US cents per pound) for the period ranging from January 1980 to December 2015. This period has been characterized by high levels of volatility and an upward trend in prices and also covers all episodes of sharp fluctuations in crude oil markets. It also includes several events of wide instabilities and crises (e.g., the 2001 US terrorist attacks, the 2001 Dot-com bubble, the 2003 Gulf wars, the 2011 Libyan revolution, the food price surge of 2007—08, the 2008—09 global financial crisis, and the 2009—12 Eurozone debt crisis). The monthly data are obtained from the primary commodity price database of the International Monetary Fund. The statistical properties of the return behavior of the commodity markets are formally shown in Table 4.1. The monthly averages of these return series are smaller compared to the computed standard deviations for three commodities under study. As can be seen from the Table entries, crude oil has positive average returns (0.0058), while salmon and shrimp have the negative mean returns which are −0.0582 and −0.0635, respectively. The computed standard deviations indicate that the unconditional volatility of return series is substantial for all commodities. The standard deviation and the range of seafood commodity price returns under study are much smaller than those of crude oil

TABLE 4.1 Summary statistics for price returns.

	Crude oil	Salmon	Shrimp
Mean	0.0058	−0.0582	−0.0635
Median	0.4246	0.0805	0.0000
Maximum	45.668	21.093	16.950
Minimum	−31.184	−21.044	−32.158
Std. Dev.	8.2774	5.8299	4.5178
Skewness	−0.1680	−0.0292	−1.3615
Kurtosis	6.5449	4.5543	11.689
J—B	227.71 [0.000]	43.447 [0.000]	1489.2 [0.000]
A—D	3.8361 [0.000]	3.5016 [0.000]	25.387 [0.000]
BDS(2)	0.0763 [0.000]	0.0680 [0.000]	0.0296 [0.011]
Q(12)	76.808 [0.000]	122.52 [0.000]	31.271 [0.002]
$Q^2(12)$	102.72 [0.000]	67.227 [0.000]	37.256 [0.000]
Modified Q(12)	32.150 [0.000]	66.450 [0.000]	14.910 [0.010]
ARCH-LM (4)	19.981 [0.000]	9.7400 [0.000]	4.2620 [0.002]
ARCH-LM (8)	11.059 [0.000]	6.2910 [0.000]	2.7420 [0.005]

Notes: Basic statistics of sample returns and their stochastic features are computed over the period from February 1980 to December 2015. The figures in the square brackets are p-values. J—B and A—D denote Jarque—Bera and Anderson—Darling tests, respectively, for normality. Q(12) and Q2(12) denote Ljung—Box Q test statistics for serial correlation up to lag 12 for returns and squared returns, respectively. Modified Q(12) refers to the empirical statistics of modified Ljung—Box up to lag 12. The BDS test is for series independence with the embedding dimension of 2, whereas the distance between pairs of consecutive observations is set to be 1. ARCH-LM is the Lagrange multiplier test of autoregressive conditional heteroskedasticity applied to the log returns.

price returns, implying that salmon, as well as shrimp prices, are less volatile than crude oil prices. All the return series are negatively skewed and exhibit significant excess kurtosis, which shows that they have heavier tails and longer left tails than a Gaussian distribution. The significant excess kurtosis and nonzero skewness indicate that the return series is conditionally hetero-skedastic. Apart from that, both Jarque—Bera and Anderson—Darling test statistics consistently reject the null hypothesis of normality at 1% significance level. Moreover, the BDS[1] test statistics advocated by Brock et al. (1996) support the hypothesis that all the return series are not independently and identically distributed at the high significance level (i.e., i.i.d process), un-derlying the presence of conditional heteroskedasticity for all return series.

As reported in Table 4.1, the Ljung—Box serial correlation test statistics advocated by Ljung and Box (1978) suggest that the null hypotheses of no serial correlation up to the 12th lag order are rejected for all return and squared return series. We also carry out the ARCH heteroskedasticity test of Engle (1982) to further analyze the distributional features of the return series (see Table 4.1). This test provides a visible indication of ARCH effects in the series considered. A modified version of the Ljung—Box test proposed by West and Cho (1995), which is robust to conditional heteroskedasticity, also confirms the serial dependence in the log returns. Graphically, Fig. 4.1 displays the dynamics of price and log return series for the commodities under concern. The presence of volatility clustering and persistence is evident in the series. Overall, these results justify the use of a GARCH-family model is appropriate.

4. Results and discussion

We jointly estimate the bivariate asymmetric and unrestricted VAR-(p)-GARCH-BEKK models and explore the links between logarithmic differences of monthly prices for the salmon-crude and shrimp-crude pairs. The models have been estimated assuming multivariate Student t distribution. Although we have estimated many variants of the model discussed in Section 2, we only report a subset of these results. Also, we have evaluated the conditional mean with the VARMA specification [i.e., VARMA(2,2), VARMA(2,1), VARMA(1,2), and VARMA(1,1)] with both multivariate normal and Student t distributions. However, these estimated models have been found to be inferior compared to the VAR specifications of conditional mean in terms of diagnostic checks, selection criteria, and numerical convergence.

We follow Lütkepohl (2005) to select the optimal lags for unrestricted VARs for the pairs of commodities considered. More precisely, we apply four criteria, namely, final prediction error (FPE.), Akaike information criterion (AIC), Schwarz criterion (SC), and Hannan—Quinn criterion (HQC). For the

---

1. The acronym BDS comes from the names (Brock, Dechert, Scheinkman) of authors who published the paper.

salmon-crude and shrimp-crude pairs, FPE and AIC indicate a preference for a specification with five lags, while SC and HQC prefer the specification with lower lag orders. Since the different selection criteria suggest different optimal VAR lag lengths for the models considered, we further analyze residual diagnostics and log-likelihood values to select a relatively competent model among estimated VAR (p)-GARCH-BEKK models. Specifically, an asymmetric VAR (5)-GARCH-BEKK model with multivariate t distribution is preferred for the salmon-crude pair while a VAR (4)-GARCH-BEKK model with multivariate t distribution has been selected for the shrimp-crude couple.

We report the estimated parameters of conditional mean and variance-covariance equations in separate tables due to presentation convenience, although they are estimated jointly. Table 4.2 presents QML parameter estimates for the conditional mean of salmon-crude and shrimp-crude pairs. Table 4.3 shows the estimated parameters for conditional asymmetric BEKK outlined in Eq. (4.2).

TABLE 4.2 Parameter estimates for the conditional mean of salmon-crude and shrimp-crude pairs.

$$r_t = c + \sum_{i=1}^{p} \Psi_i r_{t-i} + u_t$$

Salmon-crude				Shrimp-crude			
Salmon Eq. ($r_{f,t}$)		Crude Eq. ($r_{c,t}$)		Shrimp Eq. ($r_{f,t}$)		Crude Eq. ($r_{c,t}$)	
c_f	−0.1238 [0.584]	c_c	−0.1877 [0.574]	c_f	0.1549 [0.138]	c_c	0.0834 [0.814]
$\psi_{11}^{(1)}$	0.3531 [0.000]	$\psi_{21}^{(1)}$	0.0485 [0.331]	$\psi_{11}^{(1)}$	0.2998 [0.000]	$\psi_{21}^{(1)}$	0.0446 [0.543]
$\psi_{12}^{(1)}$	0.0351 [0.139]	$\psi_{22}^{(1)}$	0.2509 [0.000]	$\psi_{12}^{(1)}$	−0.0101 [0.375]	$\psi_{22}^{(1)}$	0.2776 [0.000]
$\psi_{11}^{(2)}$	−0.0598 [0.236]	$\psi_{21}^{(2)}$	0.0054 [0.906]	$\psi_{11}^{(2)}$	0.0402 [0.133]	$\psi_{21}^{(2)}$	−0.0290 [0.715]
$\psi_{12}^{(2)}$	0.0454 [0.029]	$\psi_{22}^{(2)}$	−0.1257 [0.022]	$\psi_{12}^{(2)}$	0.0136 [0.208]	$\psi_{22}^{(2)}$	−0.1376 [0.011]
$\psi_{11}^{(3)}$	−0.0880 [0.084]	$\psi_{21}^{(3)}$	−0.0194 [0.705]	$\psi_{11}^{(3)}$	0.0239 [0.259]	$\psi_{21}^{(3)}$	−0.0899 [0.114]
$\psi_{12}^{(3)}$	−0.0211 [0.301]	$\psi_{22}^{(3)}$	0.0342 [0.423]	$\psi_{12}^{(3)}$	−0.0114 [0.272]	$\psi_{22}^{(3)}$	0.0498 [0.277]
$\psi_{11}^{(4)}$	0.0214 [0.664]	$\psi_{21}^{(4)}$	−0.0049 [0.923]	$\psi_{11}^{(4)}$	0.0082 [0.655]	$\psi_{21}^{(4)}$	−0.1032 [0.071]
$\psi_{12}^{(4)}$	0.0545 [0.007]	$\psi_{22}^{(4)}$	−0.0694 [0.098]	$\psi_{12}^{(4)}$	0.0362 [0.001]	$\psi_{22}^{(4)}$	−0.0523 [0.181]

Continued

TABLE 4.2 Parameter estimates for the conditional mean of salmon-crude and shrimp-crude pairs.—cont'd

$$r_t = c + \sum_{i=1}^{p} \Psi_i r_{t-i} + u_t$$

	Salmon-crude			Shrimp-crude		
$\psi_{11}^{(5)}$	−0.1098 [0.005]	$\psi_{21}^{(5)}$	0.0069 [0.894]			
$\psi_{12}^{(5)}$	−0.0523 [0.046]	$\psi_{22}^{(5)}$	−0.0020 [0.963]			

Notes: p-values are in the bracket.

TABLE 4.3 Parameter estimates for conditional asymmetric BEKK variance-covariances.

$$H_t = C'C + A'u_{t-1}u'_{t-1}A + B'H_{t-1}B + D'\xi_{t-1}\xi'_{t-1}D$$

	Salmon-crude	Shrimp-crude
c_{11}	0.2446 [0.294]	2.9023 [0.000]
c_{21}	−1.3987 [0.144]	0.3618 [0.390]
c_{22}	−0.0001 [0.999]	1.5168 [0.163]
a_{11}	0.1913 [0.000]	0.6218 [0.007]
a_{12}	0.0827 [0.233]	−0.0191 [0.881]
a_{21}	−0.0294 [0.234]	0.0550 [0.152]
a_{22}	0.4310 [0.000]	0.6996 [0.000]
b_{11}	0.9692 [0.000]	0.3277 [0.000]
b_{12}	−0.0001 [0.999]	0.0907 [0.429]
b_{21}	0.0384 [0.005]	−0.0078 [0.478]
b_{22}	0.8432 [0.000]	0.8521 [0.000]
d_{11}	0.0917 [0.229]	−1.0746 [0.000]
d_{12}	−0.3105 [0.044]	0.0405 [0.821]
d_{21}	−0.0738 [0.023]	−0.0812 [0.288]
d_{22}	0.5568 [0.000]	0.7638 [0.001]
Tail parameter	7.8850 [0.000]	2.6444 [0.000]

Notes: p-values are in the bracket. The parameters a_{12} and a_{21} represent the mean spillover effects between oil and each of the seafood commodity while diagonal elements a_{11} and a_{22} represent the impact of its past shocks for each of the seafood prices and the crude oil, respectively. The elements b_{12} and b_{21} capture the cross-market volatility spillover effects between the oil and each of the seafood market returns. Diagonal elements b_{11} and b_{22} measure the effect of its own lagged volatility on the contemporaneous volatility. Parameters d_{11} and d_{22} capture own variance asymmetry while d_{12} and d_{21} cross-market asymmetric responses.

While the diagonal elements in matrix A measure the own ARCH effects (i.e., the influences of its own past shocks), off-diagonal elements capture the cross-market impacts of shocks. The subscript i ($=1$, 2) refers to each commodity under concern. The elements a_{ij} in the matrix A show the impact of shocks arising from market i on market j. Because the coefficient estimates for a_{ij} are insignificant at conventional levels in estimated models for both pairs, the volatilities of seafood commodity markets under consideration do not seem to be influenced by shocks from the fossil oil market. Parameter estimates for a_{11} and a_{22} are statistically significant at the 1% level in estimated models for both pairs, indicating that the volatilities of salmon, shrimp, and crude oil markets are affected by the shocks from their market. Hence, the ARCH effects are highly evident in models for both pairs.

Empirical estimates of matrix B measure the impact of past conditional variances on current conditional volatilities. In matrix B, the diagonal elements, b_{11} and b_{22}, measure the own GARCH effects while off-diagonal elements, b_{12} and b_{21}, capture the effect of fossil oil lagged volatility on the current volatility of seafood commodity price returns and vice versa. The findings suggest that the volatility of past returns in crude oil does not seem to impact on the current volatility of shrimp prices. On the other hand, the recent conditional volatility of mineral oil strongly affects the current fluctuations of salmon prices. The high statistical significance of b_{11} and b_{22} in both models generally indicates that a strong GARCH(1,1) process drives the conditional variances.

In panel A: Ljung−Box Q(12) statistic is a univariate version of serial correlation test for the standardized residuals, z_f and z_c, up to lag 12. McLeod-Li test is for serial dependence in the univariate squared standardized residuals. The ARCH LM (5) test statistic inspects the remaining heteroskedasticity up to lag 5.

In panel B: Multivariate Ljung−Box Q and ARCH tests are reported for serial correlation and heteroskedasticity, respectively, which have been applied to the vector of series as a whole.

In panel C: The values denote the Wald test statistics of parameter restriction for the underlying BEKK model. The test statistic has an asymptotic χ^2 distribution with a degree of freedom that equals the number of restricted parameters. Null hypotheses are constructed as follows:

Diagonal VAR: $H_0 : \psi_{12}^{(i)} = \psi_{21}^{(i)} = 0$ where i is the VAR order
No GARCH: $H_0 : a_{ij} = b_{ij} = d_{ij} = 0$ for all $i,j = 1,2$
Diagonal GARCH: $H_0 : a_{12} = a_{21} = b_{12} = b_{21} = d_{12} = d_{21} = 0$
No asymmetry: $H_0 : d_{11} = d_{12} = d_{21} = d_{22} = 0$

To analyze the robustness of estimated models, we perform several serial dependence and heteroskedasticity tests on the standardized and squared

TABLE 4.4 Diagnostic and specification tests for the unrestricted GARCH-BEKK.

	Salmon-crude	Shrimp-crude
Panel A: Univariate tests		
Ljung–Box Q(12) z_f	18.342 [0.105]	10.393 [0.581]
Ljung–Box Q(12) z_c	11.513 [0.485]	12.346 [0.418]
McLeod-Li (12) z_f	10.839 [0.543]	11.779 [0.463]
McLeod-Li (12) z_c	6.8258 [0.868]	9.0893 [0.693]
ARCH LM (5) z_f	0.3310 [0.894]	1.2120 [0.303]
ARCH LM (5) z_c	0.4010 [0.840]	0.5900 [0.707]
Panel B: Multivariate tests		
Multivariate Q(12)	52.012 [0.320]	45.542 [0.574]
Multivariate ARCH (4)	44.710 [0.484]	40.150 [0.291]
Panel C: Model specification tests		
Diagonal VAR	21.532 [0.017]	20.385 [0.008]
No GARCH	171,686 [0.000]	1136.1 [0.000]
Diagonal GARCH	15.815 [0.014]	3.6313 [0.726]
No asymmetry	26.408 [0.000]	23.921 [0.000]

Notes: Figures in square bracket are p-values.

standardized errors.[2] The panel A of Table 4.4 presents univariate test statistics including McLeod-Li test of McLeod and Li (1983); Ljung–Box Q test of Ljung and Box (1978); and the ARCH-LM test of Engle (1982). Besides, the multivariate versions of univariate dependence tests (Hosking, 1980) are also reported in panel B, which are applied to the vector of series as a whole.[3] Some review of multivariate tests is detailed in Lütkepohl (2005) and Bauwens et al. (2006). The computed univariate, as well as multivariate serial correlation test statistics, suggest that the standardized and squared standardized errors in both models are not serially dependent up to lag 12. Moreover, none of the ARCH-LM tests rejects the null hypothesis of homoscedasticity at conventional significance levels. In sum, these tests strongly support the specifications for conditional mean and variance-covariance equations used in this study.

2. The standardized errors are defined as $z_{k,t} = u_{k,t} h_{k,t}^{-1/2}$ for $k = r_{f,t},\ r_{c,t}$.
3. In Table 4.4 (panel B), we have also reported the multivariate version of ARCH LM heteroskedasticity test using the statistical routine proposed by the developers of RATS statistical package.

Concerning the adequacy of the estimated model specifications, we closely follow Grier et al. (2004) and rely on several specification tests. The results are reported in Table 4.4 (in panel C). Here, some points are worth noting. First, the diagonal VAR version of the conditional mean is rejected at 5% and 1% significance levels in salmon-crude and shrimp-crude pairs, respectively. Second, as the summary statistics in Table 4.1 shows, there is a significant conditional heteroskedasticity in the considered return series. The joint significance of all the estimates of parameter matrices (A, B, and D) also confirms the adequacy of the variance-covariance processes in both models. Third, the hypothesis of the nondiagonal variance-covariance process requires that the off-diagonal elements of above-mentioned parameter matrices are jointly significant. The computed test statistic suggests that the diagonal GARCH is rejected in a salmon-crude model. Fourth, the hypothesis of an asymmetric covariance process requires the elements of parameter matrix D are statistically significant jointly, and the overall asymmetric parameter matrix is highly significant in both models.

To examine the Granger-causality in mean and variance between the changes in the price of crude oil and the selected seafood commsodity price changes, we employ the bivariate version of asymmetric VAR-(p)-GARCH-BEKK model. While the mean causality can be performed directly from the joint significance of variables in the conditional mean equations, variance causality proposed by Hafner and Herwartz (2004) is carried out by using the estimated parameters of the BEKK variance-covariance matrix. Consider two conditionally heteroskedastic and stationary return series ($r_{c,t}$ and $r_{f,t}$). One may define a return series $r_{c,t}$ to be causal for $r_{f,t}$ in mean, if

$$E\left(r_{f,t+1}\middle|r_{f,t}, r_{f,t-1}, \ldots\right) \neq E\left(r_{f,t+1}\middle|r_{f,t}, r_{f,t-1}, \ldots, r_{c,t}, r_{c,t-1}, \ldots\right)$$

Further, as noted by Lütkepohl (2005), this definition can be directly extended to higher-order conditional moments. We define $r_{c,t}$ Granger cause $r_{f,t}$ in k-th moment if

$$E\left(r_{f,t+1}^{k}\middle|r_{f,t}, r_{f,t-1}, \ldots\right) \neq E\left(r_{f,t+1}^{k}\middle|r_{f,t}, r_{f,t-1}, \ldots, r_{c,t}, r_{c,t-1}, \ldots\right)$$

Thus, this defines causality in variance when one considers the central second moments ($k = 2$). In addition, the interpretation of Granger causality in variance is similar to that of causality in mean. Hence, if $r_{c,t}$ Granger cause $r_{f,t}$ in variance, the conditional variances of $r_{f,t}$ can be predicted more accurately by taking into consideration the information set of $r_{c,t}$ than without accounting for this information (Lütkepohl, 2005). Also, as shown by Comte and Lieberman (2000), if the conditional covariance formulation can be specified by multivariate GARCH models, the variance causality procedure is analogous to that of mean causality in VAR and VARMA. Thus, the null hypothesis for no variance causality from the price returns of fossil oil to each of seafood commodity price returns under concern in BEKK covariance structure outlined by equation two can be constructed as $a_{21} = b_{21} = 0$.

TABLE 4.5 Results for mean and variance noncausality.

From	To	Test statistic	p-value
Panel A: H_0 : no causality in variance from crude oil to salmon/shrimp			
Crude oil	Salmon	12.222	0.002
Crude oil	Shrimp	2.2010	0.332
Panel B: H_0 : no causality in variance from salmon/shrimp to crude oil			
Salmon	Crude oil	2.1417	0.342
Shrimp	Crude oil	0.6314	0.729
Panel C: H_0 : no causality in mean from crude oil to salmon/shrimp			
Crude oil	Salmon	18.902	0.002
Crude oil	Shrimp	16.942	0.002
Panel D: H_0 : no causality in mean from salmon/shrimp to crude oil			
Salmon	Crude oil	1.2977	0.935
Shrimp	Crude oil	5.2732	0.260

Notes: Table entries report the Wald test statistics of the Granger noncausality. The test statistic has an asymptotic χ^2 distribution with a degree of freedom that equals the number of restricted parameters.

To test the hypotheses, we employ the Wald test statistic proposed by Hafner and Herwartz (2004). As Table 4.5 entries suggest, the mean transmission from fossil oil price changes to shrimp price changes appears to be more prevalent than the variance transmission in the shrimp-crude model. The volatility transmission from the changes in crude oil prices to shrimp price changes seems to be less evident for the considered sample in the current study. In contrast, there are both mean and variance transmissions from the crude oil market to the salmon market. Therefore, in general, these results indicate that the changes in crude oil market appear to have explanatory power in determining the future movements of selected seafood commodity price changes. The only exception is that the volatility of crude oil prices seems to be less useful in predicting movements in the volatility of shrimp prices and might be omitted as forecasting variables in volatility forecasting models. As seen from Table 4.5, the reverse does not hold (i.e., seafood prices and volatility do not appear to have any predictive power in the fossil oil market movements).

5. Conclusion

This study has investigated whether information transmission from the crude oil market to two widely traded commodity markets of salmon and shrimp markets occurs. To this end, the study has conducted a Granger-causality test in mean and variance proposed by Hafner and Herwartz (2004). Investigating the linkage between oil and food commodity markets has been important mainly due to following reasons. First, upward and downward movements in the fossil oil price affect the production costs of agricultural commodities through the use of energy-intensive inputs. Second, as stressed by Ji and Fan (2012), an analysis of information transmission between commodity markets is essential for uncovering price volatility behavior of commodities and making global across-market hedging decisions. Moreover, the presence of volatility transmissions from fossil oil prices to agricultural commodity markets is problematic for financial portfolios (Gardebroek and Hernandez, 2013). Third, over the last two decades, agricultural commodity markets have been viewed as alternative investment sectors by financial market participants for portfolio diversification purposes.

Several exciting conclusions have emerged from this study. First, the salmon market is susceptible to the changes in the fossil oil market in mean and volatility. Second, despite the evidence of causality-in-mean from crude oil market to the shrimp market, volatility does not seem to transmit from the crude oil market to the shrimp market. In general, analyses of information transmission in both models support the notion that seafood commodity market policies should be designed with consideration given to movements in prices as well as the shifts in volatility in nonrenewable energy commodity prices. All of this suggests that the crude oil market information improves the forecast accuracy of prices and volatilities for individual seafood commodities.

Acknowledgments

The authors would like to thank Robert Terpstra, Gary Rangel, and Ashutosh Sarker for their insightful suggestions on the earlier versions of the paper.

References

Abdelradi, F., Serra, T., 2015. Food-energy nexus in Europe: price volatility approach. Energy Econ. 48, 157—167.

Balcombe, K., Rapsomanikis, G., 2008. Bayesian estimation and selection of non-linear vector error correction models: the case of the sugar-ethanol-oil nexus in Brazil. Am. J. Agric. Econ. 90 (3), 658—668.

Bauwens, L., Laurent, S., Jeroen, V., Rombouts, K., 2006. Multivariate GARCH models: a survey. J. Appl. Econ. 21, 79—109.

Bjørndal, T., Knapp, G.A., Lem, A., 2003. Salmon: A Study of Global Supply and Demand. FAO/GLOBEFISH, Rome. http://www.globefish.org/upl/publications. (Accessed 17 June 2013).

Bollerslev, T., Wooldridge, J.M., 1992. Quasi-maximum likelihood estimation and inference in dynamic models with varying covariances. Econ. Rev. 11, 143—172.

Brock, T., Dechert, W.D., Scheinkman, J.A., 1996. A test for independence based on the correlation dimension. Econ. Rev. 15, 197—235.

Chen, S.-T., Kuo, H.-I., Chen, C.-C., 2010. Modeling the relationship between the oil price and global food prices. Appl. Energy 87, 2517—2525.

Cheung, Y.W., Ng, N.K., 1996. A causality-in-variance test and its application to financial market prices. J. Econom. 72, 33—48.

Comte, F., Lieberman, O., 2000. Second-order non-causality in multivariate GARCH processes. J. Time Anal. 21, 535—557.

Du, X., Yu, C.L., Hayes, D.J., 2011. Speculation and volatility spillover in the crude oil and agricultural commodity markets: a Bayesian analysis. Energy Econ. 33 (3), 497—503.

Engle, R.F., 1982. Autoregressive conditional heteroscedasticity with estimates of the variance of UK inflation. Econometrica 50, 987—1008.

Engle, R.F., Kroner, K.F., 1995. Multivariate simultaneous generalized ARCH. Econom. Theor. 11, 122—150.

Fiorentini, G., Sentana, E., Calzolari, G., 2003. Maximum likelihood estimation and inference in multivariate conditionally heteroskedastic dynamic regression models with Student t innovations. J. Bus. Econ. Stat. 21, 532—546.

Food and Agriculture Organization (FAO), 2012. GLOBEFISH Highlights. Rome. http://www.infofish.org/pdf/GSH/GSH%201-2012.pdf. (Accessed April 2013).

Food and Agriculture Organization (FAO), 2014. http://www.fao.org/fishery/sofia/en. (Accessed July 2016).

Gardebroek, C., Hernandez, M.A., 2013. Do energy prices stimulate food price volatility? Examining volatility transmission between US oil, ethanol and corn markets. Energy Econ. 40, 119—129.

Green, W., 2012. Econometric Analysis, seventh ed. Prentice-Hall, Boston.

Grier, K.B., Henry, Ó.T., Olekalns, N., Shields, K., 2004. The asymmetric effects of uncertainty on inflation and output growth. J. Appl. Econom. 19, 551—565.

Hafner, C.M., Herwartz, H., 2004. Testing for Causality in Variance Using Multivariate GARCH Models, Working Paper. Christian-Albrechts-Universität, Kiel.

Hameed, A.A., Arshad, F.M., 2009. The impact of petroleum prices on vegetable oil prices: evidence from co-integration tests. Oil Palm Ind. Econ. J. 9, 31—40.

Hasanov, A.S., Do, H.X., Shaiban, M.S., 2016. Fossil fuel price uncertainty and feedstock edible oil prices: evidence from MGARCH-M and VIRF analysis. Energy Econ. 57, 16—27.

Hosking, J., 1980. The multivariate portmanteau statistic. J. Am. Stat. Assoc. 75, 602—608.

Ji, Q., Fan, Y., 2012. How does oil price volatility affect non-energy commodity markets? Appl. Energy 89, 273—280.

Ljung, G.M., Box, G.E.P., 1978. On a measure of lack of fit in time series models. Biometrika 67, 297—303.

Lütkepohl, H., 2005. New Introduction to Multiple Time Series. Springer, Berlin.

McLeod, A., Li, W., 1983. Diagnostic checking ARMA time series models using squared residual autocorrelations. J. Time Anal. 4, 269—273.

Mensi, W., Hammoudeh, S., Nguyen, D.K., Yoon, S., 2014. Dynamic spillovers among major energy and cereal commodity prices. Energy Econ. 43, 225—243.

National Marine Fisheries Service (NMFS). http://www.nmfs.noaa.gov/aquaculture. (Accessed 29 July 2016).

Nazlioglu, S., 2011. World oil and agricultural commodity prices: evidence from non-linear causality. Energy Pol. 39, 2935—2943.

Nazlioglu, S., Soytas, U., 2011. World oil prices and agricultural commodity prices: evidence from an emerging market. Energy Econ. 33, 488—496.

Nazlioglu, S., Erdem, C., Soytas, U., 2012. Volatility spillover between oil and agricultural commodity markets. Energy Econ. 36, 658—665.

Oglend, A., 2010. An Analysis of Commodity Price Dynamics with Focus on the Price of Salmon: PhD Dissertation. University of Stavanger, Norway.

Press, W.H., Flannery, B.P., Teukolsky, S.A., Vettering, W.T., 2007. Numerical Recipes, third ed. Cambridge University Press, New York.

Rossi, E., Spazzini, F., 2010. Model and distribution uncertainty in multivariate GARCH estimation: a Monte Carlo analysis. Comput. Stat. Data Anal. 54, 2786−2800.

Serra, T., 2011. Volatility spillovers between food and energy markets: a semiparametric approach. Energy Econ. 33, 1155−1164.

Serra, T., Zilberman, D., 2013. Biofuel-related price transmission literature: a review. Energy Econ. 37, 141−151.

Tsay, R., 2010. Analysis of Financial Time Series, third ed. John Wiley & Sons, New Jersey.

West, K.D., Cho, D., 1995. The predictive ability of several models of exchange rate volatility. J. Econom. 69, 367−391.

Chapter 5

Econometric analysis of the economic growth-energy consumption nexus in emerging economies: the role of globalization

Alex O. Acheampong[1,2], Elliot Boateng[1,2], Mary Amponsah[1,2]

[1]*Newcastle Business School, University of Newcastle, NSW, Australia;* [2]*Centre for African Research, Engagement and Partnerships (CARE-P), University of Newcastle, NSW, Australia*

1. Introduction

This chapter examines the linear and nonlinear effects of economic, social, and political globalization on economic growth-energy consumption nexus in 23 emerging economies. In recent years, emerging economies have become increasingly more integrated globally (Arslan et al., 2018). Evidence shows that not only has increased globalization caused trade volume to increased more than double between 1970 and 2016 (Arslan et al., 2018), but also the tighter integration of emerging economies has raised both economic growth and energy consumption. For instance, it is estimated that emerging economies have accounted for approximately two-thirds of the World's gross domestic product (GDP) for the past 15 years (Woetzel et al., 2018). At the same time, recent report shows that economies such as Brazil, China, India, Indonesia, Mexico, and South Africa together consume one-third of the world's energy (International Energy Agency [IEA], 2018). Thus, emerging economies have continued to experience rapid globalization with increasing economic growth and energy consumption. Despite these evidence, the unresolve question is whether globalization explains the differences in economic growth-energy consumption nexus across regions. This chapter provides a detailed analysis of the role of globalization in driving economic growth and energy consumption.

Energy-Growth Nexus in an Era of Globalization. https://doi.org/10.1016/B978-0-12-824440-1.00011-4

105

A large body of studies have emerged examining the relationship between the different aspects of energy consumption and economic growth in emerging economies (Sadorsky, 2010). The findings from these studies have been conflicting and conclusions remain unresolved (Adewuyi and Awodumi, 2017; Omri, 2014; Ozturk, 2010). For instance, it is argued, on the one hand, that there is not much of a relationship between economic growth and energy consumption whereas other studies reveal a possible causal relationship between economic growth and energy consumption, leading to the development of four hypotheses, namely the neutrality, feedback (see Belke et al., 2011; Dogan, 2015a; Nasreen and Anwar, 2014; Ouedraogo, 2013), conservation (see Narayan and Narayan, 2010), and growth (see Acaravci et al., 2015; Dogan, 2014; Ozturk, 2010) hypotheses. Recent empirical studies have augmented the economic growth-energy consumption models with additional variables such as financial development, capital, urbanization, carbon emissions, industrialization, and energy price, among others to prevent variable omission bias (see Acheampong, 2018; Apergis and Payne, 2014, 2015; Huang et al., 2008; Karanfil, 2009; Mahadevan and Asafu-Adjaye, 2007; Marques et al., 2017; Sadorsky, 2010, 2011, 2013). Nevertheless, there are scanty studies on the essential role of political, economic, and social globalization on economic growth-energy consumption relationship. In this study, we argue that omitting globalization from economic growth-energy consumption model could constrain differing characteristics in the modelling process, which could limit the futurist understanding of how energy demand and economic growth in emerging economies respond to changes in global shocks.

Globalization promotes the transfer of technology, the use of energy and economic growth (Cole, 2006; Marques et al., 2017; Shahbaz et al., 2016). Additionally, globalization increases the demand for factors of production, which subsequently increase the use of energy and boost economic growth (Cole, 2006; Sadorsky, 2011, 2012). It also promotes specialization in production, leading to higher economies of scale and economic output (Heckscher, 1919; Ohlin, 1933). Some studies have attempted to use trade and FDI as proxies for globalization to investigate their impact on either energy consumption or economic growth (Cole, 2006; Lean and Smyth, 2010; Sadorsky, 2011, 2012). However, it is important to emphasize that proxing globalization by trade and FDI omits the impact of political and social globalization on energy consumption that could have several policy ramifications.

This chapter contributes to the debate on globalization, economic growth, and energy consumption in emerging economies by addressing three crucial issues. First, it employs a multivariate approach to explore the impact of economic, political, and social globalization on both energy consumption and economic growth in 23 emerging economies for the period 1970–2015. The emerging economies are studied because these economies have been experiencing rapid economic growth, high energy use, and bear the higher effect associated with globalization (Sadorsky, 2010; Shahbaz et al., 2016).

Therefore, investigating the role of globalization on energy consumption and economic growth in the emerging economies will contribute significantly to the literature and sustainable development policy. Second, unlike the existing studies, we present a comprehensive analysis of the nonlinear effect of economic, social, and political globalization on economic growth and energy consumption. Finally, we utilize the Pooled Mean Group (PMG) estimator of the dynamic heterogeneous panel to examine the linear and nonlinear effect of economic, political, and social globalization on the energy consumption and economic growth nexus. Alternatively, we employ the Pedroni (2001a,b) fully modified ordinary least square (OLS) and dynamic OLS to test the consistency of the empirical estimates.

Our results from the PMG estimator reveal four (4) empirical evidence. First, economic, social, and political globalization stimulate economic growth in the long-run. Second, we find robust evidence suggesting that economic, social, and political globalization directly reduce energy consumption. Third, we find that economic growth contributes significantly to energy consumption while energy consumption only contributes significantly to economic growth in the presence of economic and social globalization. Finally, while globalization factors significantly influence energy consumption and economic growth, the effect is nonlinear and varies substantially across energy consumption and economic growth models. The rest of the paper is organized as follows: Section 2 presents an extensive literature review related to the subject matter; Section 3 presents the estimation strategy; Sections 4 And 5 present the discussion and policy of the implication of the results, respectively.

2. Literature review

The section presents the literature review on the role of globalization on economic growth-energy consumption nexus. The literature is categorized into three main segments. The first discusses the relationship between economic growth and energy consumption, followed by literature on the relationship between economic growth and globalization. The final segment presents the literature on the relationship between energy consumption and globalization. Each subsection reports inconsistent findings in the literature. The empirical literature for each segment is summarized in a table.

2.1 Economic growth-energy consumption nexus

A plethora of literature has examined the relationship between economic growth and energy consumption, albeit the unresolved theoretical and empirical evidence. Mainly, while some studies conclude that there is not much of a relationship between economic growth and energy consumption (see Śmiech and Papież, 2014; Yu and Choi, 1985), other scholars have unearthed a possible causal relationship, which has been grouped into four central hypotheses (see Shahbaz et al., 2012; Shahbaz et al., 2018a,b,c). First, the growth hypothesis indicates that an increase in energy consumption leads to an increase in economic growth (see Acaravci et al., 2015; Dogan, 2014; Ozturk, 2010). Second, the conservative hypothesis opines that an increase in economic growth leads to energy consumption (see Narayan and Narayan, 2010). Some studies support a feedback hypothesis where a bidirectional causality exists between economic growth and energy consumption (see Belke et al., 2011; Dogan, 2015a; Nasreen and Anwar, 2014; Ouedraogo, 2013). The neutrality hypothesis diverges from these consensuses to show an independent relationship between energy consumption and economic growth (see Jafari et al., 2012; Menegaki and Tugco, 2016; Tang and Shahbaz, 2011). The findings for these hypotheses have varied across energy dimensions, countries, regions and income groupings, data set and econometric methodologies (see Ozturk, 2010; Payne, 2010; Shahbaz and Feridun, 2012; Shahbaz et al., 2014). Table 5.1 shows a summary of the empirical literature on the economic growth-energy consumption nexus.

TABLE 5.1 Economic growth–Energy consumption nexus.

Authors	Study period	Countries	Estimation strategy	Findings
Kraft and Kraft (1978)	1947–74	United States	Correlation and Granger causality test	GNP → EC for the postwar period No causality from energy to GNP.
Ferguson et al. (2000)	1970–2007	7 SSA countries	Bounds testing approach to cointegration	EC is cointegrated with EG (Congo, Cameroon, Cote d'Ivoire and South Africa). EG has a significant positive impact on EC in these countries. EG ↔ EC in Cote d'Ivoire EG → EC in Congo.
Aqeel and Butt (2001)	1955–56 and 1995–96	Pakistan	Cointegration and Hsiao's version of Granger causality	EG → PC EL → EC No causality between GC and EG. No feedback effect between EL and EC.

TABLE 5.1 Economic growth—Energy consumption nexus.—cont'd

Authors	Study period	Countries	Estimation strategy	Findings
Ghosh (2002)	1950—51 to 1996—97	India	Cointegration and VECM	No long-run equilibrium relationship between EL and EG. EG → EL No feedback effect
Shiu and Lam (2004)	1971—2000	China	Granger causality test and ECM	EL → EG No reverse causality
Altinay and Karagol (2005)	1950—2000	Turkey	Dolado-Lutkepohl and standard Granger causality test	EL → EG
Yoo (2006)	1971—2002	4 ASEAN countries	Granger causality test	EL ↔ EC (Singapore and Malaysia) EG → EL (Indonesia and Thailand); no feedback effect
Wolde-Rufael (2006)	1971—2001	17 African countries	The modified version of Granger causality test by Toda and Yamamoto (1995) and the newly developed cointegration technique by Pesaran et al. (2001)	Long-run: the relationship between EG and EC exists in nine of the selected countries with evidence of Granger causality in 12 of the selected countries. EG → EL (six countries) EL → EG (three countries) EL ↔ EC (three countries)
Alinsato (2009)	1973—74 to 2005—06	Benin and Togo	ARDL bounds testing approach of cointegration	Long-run and short-run: EG → EL (Benin). Short-run: EG → EL (Togo)

Continued

TABLE 5.1 Economic growth—Energy consumption nexus.—cont'd

Authors	Study period	Countries	Estimation strategy	Findings
Sqauli (2007)	1980—2003	OPEC countries	Bounds test of cointegration based on the unrestricted ECM	Economic growth and electricity consumption are cointegrated. EL → EG (five countries). EG is less dependent on EL (in three countries and independent in the remaining three countries, as a result of the variations in political and economic traits)
Chen et al. (2007)	1971—2001	10 Asian countries	ECM and Granger causality test	Short-run: EG → EL Evidence of unidirectional causality from GDP to electricity consumption. EL ↔ EC (when a panel procedure is implemented)
Narayan and Prasad (2008)	1970—2002 (USA) 1971—2002 (Mexico, Korea and the Slovak Republic) 1965—2002 (Hungary) 1960—2002 (Rest of OECD)	30 OECD countries	Bootstrapped causality test	EL → EG (in eight OECD countries) EL conversation policies do not affect EG in the remaining 22 countries
Hye and Riaz (2008)	1971—2007	Pakistan	Bounds test of cointegration and augmented Granger causality test	Short-run: EG ↔ EC Long-run: EG → EC EC does not lead to EG in the long-run, but an increase in energy prices impedes EG due to the cost of business.

TABLE 5.1 Economic growth—Energy consumption nexus.—cont'd

Authors	Study period	Countries	Estimation strategy	Findings
Yuan et al. (2008)	1963—2005	China	Johansen cointegration approach: VEC	Short-run: EL and OC → EG Confirmed a neutrality hypothesis for coal and total EC. Short-run: EG → EC, coal, and OL No evidence of causality from EG to EL.
Akinlo (2009)	1980—2006	Nigeria	ECM and variance decomposition analysis	EL → EG
Narayan and Smyth (2009)	1974—2002	Middle Eastern countries	Multivariate Granger causality test, FMOLS	Feedback effects between EL, EG, and exports.
Pao (2009)	1980—2007	Taiwan	A space modeling approach with cointegration and ECM	Evidence of cointegration between EG and EL. Both short-run and long-run: EG → EL
Payne (2009)	1949—2006	US	Toda-Yamamoto causality test	Evidence of a neutrality hypothesis: no causal relationship between renewable energy or nonrenewable energy and EG.
Acaravci and Ozturk (2010)	1990—2006	15 transition countries	Pedroni panel cointegration approach	No cointegration between EG and EC. EL does not have any effect on EG
Bartleet and Gounder (2010)	1960—2004	New Zealand	Multivariate analysis	Short-run: EG → EC without feedback effect. Evidence of a long-run relationship between energy prices, EG, and EC.

Continued

TABLE 5.1 Economic growth—Energy consumption nexus.—cont'd

Authors	Study period	Countries	Estimation strategy	Findings
Jamil and Ahmad (2010)	1960—2008	Pakistan	Johansen maximum likelihood approach, VECM	Real economic activity, such as output in commercial, agricultural, and manufacturing sectors → EL. Private expenditure and residential sectors increase EC.
Tsani (2010)	1960—2006	Greece	Modified Granger causality test by Toda and Yamamoto (1995)	EC → EG; EG ↔ disaggregate energy consumption indices such as residential and industrial energy consumption Evidence of a neutrality hypothesis exists in transport EC and EG.
Polemis et al. (2013)	1970—2011	Greece	Cointegration techniques and VECM	EL ↔ EG
Ciarreta and Zarraga (2010)	1971—2005	Spain	Toda and Yamamoto Granger causality test in a VECM in first difference	EG → EL; No evidence of nonlinear causality between both variables
Yoo and Kwak (2010)	1975—2006	7 South American countries	Hsiao version of the Standard Granger causality test, ECM	Causal relationship between EL and EG varies across the selected countries. Short-run: EL → EG (Argentina, Brazil, Chile, Columbia, and Ecuador); EL ↔ EG (Venezuela). No causality in Peru.
Eggoh et al. (2011)	1970—2006	21 African countries	Panel cointegration and causality tests	EC positively affects EG

TABLE 5.1 Economic growth—Energy consumption nexus.—cont'd

Authors	Study period	Countries	Estimation strategy	Findings
Wang et al. (2011)	1972–2006	China	ARDL bounds testing approach and a multivariate causality test	EC → EG Long-run and short-run: EC positively affects EG
Chaudhry et al. (2012)	1972–2012	Pakistan	ARDL, Granger causality test	EL increases EG but OC decreases EG
Shahbaz and Feridun (2012)	1971–2008	Pakistan	Toda-Yamamoto and Wald-test causality tests, ARDL	EG → EL, and not vice versa.
Shahbaz and Lean (2012)	1972–2009	Pakistan	ARDL, unrestricted ECM and VECM Granger causality	EG ↔ EC
Shahbaz et al. (2011)	1971–2009	Portugal	Bounds testing cointegration approach within an unrestricted ECM	Long-run: EG ↔ EC Short-run: EG → EL
Kouakou (2011)	1971–2008	Cote d'Ivoire	ECM	EG ↔ EC
Tang and Shahbaz (2013)	1972–2010	Pakistan	Johansen-Juselius cointegration test, VECM, Toda and Yamamoto (1995) and Dolado and Lutkepohl (1996) Granger causality test	EL → EG and across services and manufacturing sectors. No causality between real output in the agricultural sector and EL
Shahbaz et al. (2014)	1972QI and 2011QIV	Pakistan	ARDL	National GC and EG are complements

Continued

TABLE 5.1 Economic growth–Energy consumption nexus.—cont'd

Authors	Study period	Countries	Estimation strategy	Findings
Iyke (2015)	1971–2011	Nigeria	Trivariate VECM	Both long-run and short-run: EL → EG
Mirza and Kanwal (2017)	1971–2009	Pakistan	Johansen-Julius cointegration test, VECM	EC ↔ EG
Marques et al. (2017)	1971–2013	A panel of 43 countries	ARDL	Feedback hypothesis in the energy-growth nexus.
Saud et al. (2018)	1990–2014	Next-11 countries	Westerlund cointegration test, dynamic seemingly unrelated cointegration and Pairwise Demitrescu-Hurlin panel causality test	EC ↔ EG
Shahbaz et al. (2018)	1970–2015	BRICS countries	Brock-Dechert-Scheinkman test, NARDL	A positive shock in EG enhances EC and vice versa
Zafar et al. (2019)	1990–2015	Emerging economies	CUP-FM, CUP-BC, VECM	Renewable energy → EG; EG → nonrenewable energy

Note: ↔bidirectional causality; → unidirectional causality; ARDL, Autoregressive distributed lag; EC, Energy consumption; ECM, Error Correction Model; EG, Economic Growth; EL, Electricity consumption; GC, Gas consumption; GNP, Gross national product; OC, Oil consumption; PC, Petroleum consumption; VECM, Vector error correction model.

2.2 The globalization-economic growth nexus

Globalization goes beyond trade openness to encompass economic, political, and social dimensions. Because of this, the debate on the role of globalization to a country's development remains a controversial topic (see Fisher, 2003) since the effect varies across dimensions and a country's relative resources. For instance, theoretical growth studies have reported conflicting reviews on the role of globalization to development. On the one hand, globalization is perceived to stimulate economic growth (see Dollar, 1992; Dreher, 2006) through capital augmentation, improved factor productivity, effective resource allocation, and technological diffusion (see Mishkin, 2009; Samimi and Jenatabadi, 2014). In his seminal paper, Mishkin (2009) noted that globalization enables advanced technological transfer from developed to developing countries, which subsequently promotes division of labour to reap the increased benefits from comparative advantage of a country's production of different specialized activities (see Shahbaz, 2016). Dreher (2006) opined a positive effect of both social and economic globalization on growth with no effect of political globalization on growth.

On the other hand, globalization harms growth in countries with political instability and weak institutions (see Berg and Krueger, 2003; Borensztein et al., 1998; De Melo et al., 2008). For instance, some scholars argue that the growth effect of trade on growth is conditional to the structural progress of a country (see Calderón and Poggio, 2010), which shows the impact of a positive effect is not robust. Other studies link such evidence to the globalization index, and lack of consideration to prominent growth indicators weakening the positive effect of globalization on growth (see Edwards, 1998; Rodrik, 1997, 1998; Warner, 2003). From the Stolper-Samuelson Heckscher-Ohlin theorem, countries with relatively abundant factors would gain from freer trade while countries with relatively scarce factors would lose (see Potrafke, 2015). Thus, while the classical and neoclassical literature corroborates the benefits of globalization to the gains from trade, the concern accrued to the distributional effect on an economy. As argued by Potrafke (2015), the distributional effects from outsourcing, traded inputs and nontraded goods could influence social justice, when globalization affects the labour market to generate gender consequences; low-skilled women entering the labour market when low-income countries have a comparative advantage in the production of goods that is intensive in low-skilled labour. In a review, Crafts (2004) found that trade liberalization is good for growth, although, high institutional quality is required for a successful capital liberalization. Several studies have shown that the impact of globalization on economic growth varied across countries and regions, depending on the level of development and macroeconomic policies. Table 5.2 summaries the empirical literature on the globalization-economic growth nexus.

2.3 The globalization-energy consumption nexus

The impact of globalization factors such as trade openness, and FDI, among others, continues to influence energy demand through energy consumption (see Mahalik et al., 2017; Shahbaz et al., 2012; Saud et al., 2018). However, a dearth of literature has explored the relationship between globalization and energy consumption while the evidence in the existing literature remains unresolved due to the different dimensions of globalization, country characteristics, dataset, and econometric techniques employed for analysis. Table 5.3 provides a summary of the literature.

TABLE 5.2 Globalization-economic growth nexus.

Authors	Study period	Countries	Estimation strategy	Findings
Rousseau and Sylla (2003)	1850–1997	17 countries	Cross country OLS and Instrumental variables (IV) approach	Trade openness has a positive but insignificant impact on EG. EG has a positive and significant impact on Trade openness only from 1945 to 94 in the IV estimates
Ponzio (2005)	1710–1798	Mexico	Annual rates of estimated per capita growth	The increase in exports increased EG in Mexico, but this was not as expected
Dreher (2006)	1970–2000	123 countries	Arellano-Bond one-step GMM estimator	Globalization has a positive impact on EG
Afzal (2007)	1960–2006	Pakistan	Johansen cointegration technique	Dimensions of globalization such as trade openness and financial integration do not have any impact on economic growth in the short-run.
Villaverde and Maza (2011)	1970–2005	101 countries	OLS and GMM estimation techniques	Globalization is the main driver of EG to foster convergence in the income per capita.
Mudakkar et al. (2013)	1975–2011	SAARC countries	Granger causality tests and Toda-Yamamoto-Dolado-Lutkephol approach	Long-run –and-short-run: Feedback hypothesis between FDI and EG

TABLE 5.2 Globalization-economic growth nexus.—cont'd

Authors	Study period	Countries	Estimation strategy	Findings
Sakyi et al. (2012)	1970–2009	85 middle-income countries	Nonstationary heterogeneous panel cointegration techniques, CCEMG, FMOLS, and DOLS	Trade openness is the main driver of development but not of EG in middle-income countries. Trade openness is a cause and consequence of development in middle-income countries.
Samimi and Jenatabadi (2014)	1980–2008	33 OIC countries	GMM technique	An increase in globalization contributes to EG in countries with better-educated workers and a well-developed financial sector. The impact of globalization is influenced by a country's income level: low-income countries do not gain from globalization, but high- and-middle-income countries do.
Sakyi et al. (2015)	1970–2009	115 developing countries	Nonstationary heterogeneous panel cointegration techniques, CCEMG, FMOLS and DOLS, Granger causality test	Long-run: Positive bidirectional causality between income level and trade openness. Short-run: EG and trade openness move in the same direction

Continued

TABLE 5.2 Globalization-economic growth nexus.—cont'd

Authors	Study period	Countries	Estimation strategy	Findings
Valli and Saccone (2015)	1987–2009	33 sector level for China and 31 sector level for India	Vector Autoregressive model, Granger causality test	Evidence of feedback between economic growth and globalization over time. High levels of exports, imports, and FDI promote higher EG.
Olimpia and Stela (2017)	1990–2012	Romania	Least Squares and Pairwise Granger causality test	Positive relationship between overall globalization index and GDP per capita. Positive relationship between GDP growth rate and political and economic globalization.
Marques et al. (2017)	1971–2013	A panel of 43 countries	Autoregressive distributed lag (ARDL)	Globalization is a driver of EG in the long run.
Zerrin and Dumrul (2018)	1980–2015	Turkey	FMOLS cointegration test	EG positively affect economic and social globalization. Economic globalization negatively affects EG, insignificantly if KOF de facto and de jure are separated. Social globalization increases EG when KOF de facto is used but becomes negative when KOF de jure globalization is used.

TABLE 5.2 Globalization-economic growth nexus.—cont'd

Authors	Study period	Countries	Estimation strategy	Findings
Hassan et al. (2019)	1970–2014	Pakistan	ARDL	Globalization has a positive impact on EG. Additionally, there exists bidirectional causality between globalization and natural resources
Zafar et al. (2019)	1990–2014	OECD countries	Continuously updated FMOLS and Continuously updated Bias-corrected approaches	Evidence of a feedback hypothesis between globalization and EG
Zafar et al. (2019)	1990–2015	Emerging economies	Continuously updated FMOLS and Continuously updated Bias-corrected approaches, VECM, Pedroni and Westerlund panel cointegration test	For emerging economies, EG Granger causes trade openness in the long run only
Farooq et al. (2019)	1991–2017	47 OIC countries	System GMM panel estimation technique	Globalization adversely affects EG However, this effect varies across income groupings: In high-income OIC, globalization positively affects economic growth; In low-income OIC, this effect is negative.

TABLE 5.3 Globalization-energy consumption nexus.

Authors	Study period	Countries	Estimation strategy	Findings
Narayan and Smyth (2009)	1974–2002	Middle Eastern countries	Multivariate Granger causality test, FMOLS	Feedback effects between EL and exports.
Erkan et al. (2010)	1970–2006	Turkey	Cointegration, Granger causality test, the impulse response function	Unidirectional causality from EC to exports: shocks in EC positively affect exports.
Lean and Smyth (2010)	1970–2008	Malaysia	ARDL, Unrestricted ECM, TYDL Granger causality test.	No evidence of export-led nor handmaiden theories of trade and EC
Sadorsky (2011)	1980–2007	Eight middle eastern countries	VECM, Engle and Granger two-step approach	Short-run: Bidirectional feedback effect between EC and exports but unidirectional causality running from EC to imports. Long-run: trade and EC are cointegrated.
Sadorsky (2012)	1980–2007	Seven South American countries	FMOLS, Granger causality test	Short-run: unidirectional causality running from exports to EC but a bidirectional causality running from imports to EC Long-run: Both exports and imports increase EC, except the impact of exports is higher.
Mudakkar et al. (2013)	1975–2011	SAARC countries	Granger causality tests and Toda-Yamamoto-Dolado-Lutkephol approach	Feedback hypothesis between FDI and EC

TABLE 5.3 Globalization-energy consumption nexus.—cont'd

Authors	Study period	Countries	Estimation strategy	Findings
Shahbaz et al. (2013)	1971–2011	China	ARDL bounds testing approach, Granger causality analysis	International trade causes EC in the Chinese economy Bidirectional relationship between EC and trade.
Alt and Kum (2013)	1970–2010	Turkey	ARDL bounds testing approach, Unrestricted ECM	Long-term causality between electricity generation and export. Short-run: Feedback effect between electricity generation and export.
Shahbaz et al. (2016)	1971–2012	India	Bayer and Hanck cointegration test and ARDL	Long-run: an increase in globalization (social, economic, and overall globalization) leads to a reduction in energy demand
Marques et al. (2017)	1971–2013	A panel of 43 countries	Autoregressive distributed lag (ARDL)	Long-run: globalization is a driver of EC
Shahbaz et al. (2018)	1970–2014	25 developed countries	Panel cointegration and causality tests (Emirmahmutoglu and Kose panel causality test. Dumitrescu and Hurlin panel causality test.	Long-run relationship between EG, globalization, and EC Globalization augments EC in most countries but not in the USA and UK.

Continued

TABLE 5.3 Globalization-energy consumption nexus.—cont'd

Authors	Study period	Countries	Estimation strategy	Findings
Shahbaz et al. (2018)	1970–2015	BRICS countries	Brock-Dechert-Scheinkman test, NARDL bounds approach for cointegration	A positive shock in globalization affects EC positively A negative shock in globalization affects EC negatively
Saud et al. (2018)	1990–2014	Next-11 countries	Westerlund cointegration test, dynamic seemingly unrelated cointegration and Pairwise Demitrescu-Hurlin panel causality test	Globalization stimulates EC but this effect varies across individual countries.
Zafar et al. (2019)	1990–2014	OECD countries	Continuously updated FMOLS and Continuously updated Bias-corrected approaches	Unidirectional causality from globalization to EC

Scholars have argued that a critical factor to explain the lack of consensus in the energy-economic growth nexus is the role of globalization (see Shahbaz et al., 2016), which has not been explored in the literature, explicitly. This has necessitated the urgent need for further research on the role of globalization as a way to generate a win-win situation in developing countries (see Shahbaz et al., 2016). Thus, the literature, almost without exception, has not explored the moderating role of globalization in the energy consumption-economic growth nexus, particularly in the emerging economies literature. For instance, a study by Aïssa et al. (2014) employed a panel cointegration technique to investigate the relationship between renewable energy consumption, trade, and output in 11 African countries for the period 1980–2008. Their results showed that there exists bidirectional causality between output and exports as well as imports in the short-run and long-run. However, there was no evidence of causality running from output to energy consumption in both short- and-long-run. Although there was no evidence of causality running from trade or output to renewable energy in both periods, both trade and renewable energy had a significant positive impact on output.

Narayan and Smyth (2009) affirmed the feedback hypothesis in Middle Eastern countries, which showed that an increase in electricity consumption increased GDP while an increase in exports enhanced GDP. At the same time, an increase in GDP contributed to an increase in electricity consumption. Subsequently, Lean and Smyth (2010) conducted a multivariate Granger causality test between exports, prices, electricity, and growth in Malaysia using annual data from 1970 to 2008, to find a unidirectional causality from economic growth to electricity generation. Focusing on Australia, Shahbaz et al. (2015) employed Bayer and Hanck's (2013) newly developed cointegration test to show that economic growth was not emission-intensive but energy consumption was emission-intensive. Trade was found to be conducive in combating emissions in Australia. In a panel of 43 countries with data from 1971 to 2013, Marques et al. (2017) found that the energy-growth nexus is influenced by the heterogeneous impact of political, economic, and social globalization.

3. Methodology and data

3.1 Specification of empirical models and estimation technique

This study examines the effect of economic, social, and political globalization on economic growth and energy consumption in emerging economies. To explore the respective effect of the globalization indicators on economic growth and energy consumption, two main empirical models are specified.

3.1.1 Economic growth model

The first empirical model, which is economic growth model, follows Omri (2013), Omri and Kahouli (2014), and Shahbaz et al. (2013) to extend the Cobb–Douglas production function to probe the effect of globalization on economic growth. Thus, following these scholars, economic growth (*RGDPC*) is specified as a function of capital (*GCFC*), labour (*LABOUR*), energy consumption (*ENER*), and globalization indicators (*GB*). Therefore, the reduced-form of the log-linear empirical model for the estimation is given as:

$$
\begin{aligned}
lnRGDPC_{it} = \alpha_0 &+ \beta_1 lnGCFC_{it} + \beta_2 lnLABOUR_{it} + \beta_3 lnENER_{it} \\
&+ \beta_4 lnGB_{it} + \varepsilon_{it}
\end{aligned} \tag{5.1}
$$

Some scholars argue that the impact of globalization on economic growth is not always linear but can be nonlinear (Chang et al., 2009; Gu and Dong, 2011).

Based on this argument, we augment Eq. (5.1) with the quadratic term of globalization to capture its nonlinear effect.

$$lnRGDPC_{it} = \alpha_0 + \beta_1 lnGCFC_{it} + \beta_2 lnLABOUR_{it} + \beta_3 lnENER_{it} + \beta_4 lnGB_{it} + \beta_5 lnGB_{it}^2 + \varepsilon_{it}$$

(5.2)

3.1.2 Energy consumption model

Also, for the second empirical model, which is energy consumption model, this study follows Sadorsky (2010) and Shahbaz et al. (2016) to specify energy consumption (*ENER*) as a function of economic growth (*RGDPC*), financial development (*FIN*), and globalization indicators (*GB*). Therefore, the log-linear form of the empirical model for estimating the energy consumption function is specified as:

$$lnENER_{it} = \alpha_0 + \varnothing_1 lnRGDPC_{it} + \varnothing_2 lnFIN_{it} + \varnothing_3 lnGB_{it} + \varepsilon_{it} \quad (5.3)$$

Eq. (5.3) is also augmented with the quadratic term of globalization to capture the nonlinear effect of globalization in energy consumption. Therefore, Eq. (5.4) is used to examine the nonlinear effect of globalization on energy consumption.

$$lnENER_{it} = \alpha_0 + \varnothing_1 lnRGDPC_{it} + \varnothing_2 lnFIN_{it} + \varnothing_3 lnGB_{it} + \varnothing_4 lnGB_{it}^2 + \varepsilon_{it}$$

(5.4)

where $i = 1...23$ and $t = 1970...2015$; α_0 is the constant parameter; $\beta_1...\beta_5$ and $\varnothing_1...\varnothing_4$ is the coefficient to be estimated; ε_{it} is the stochastic error term; $lnRGDPC$ is the log of economic growth; $lnGCFC$ is the log of capital; $lnLABOUR$ is the log of labour; $lnENER$ is the log of energy consumption, $lnGB$ is the log of globalization indicators; $lnGB^2$ is the log of globalization indicators squared. This study estimates the above equations using the PMG estimator of the dynamic heterogeneous panel. We find this technique to be appropriate to permit us to examine the short-run and long-run impact of globalization on energy consumption and economic growth across dynamic heterogeneous economies. Alternatively, we employ the Pedroni (2001a,b) fully modified OLS and dynamic OLS to test the consistency of the empirical estimates.

3.2 Data and descriptive statistics

This study investigates the impact of economic, social, and political globalization on the economic growth-energy consumption nexus in a panel of 23 emerging economies for the period 1970−2015. According to Morgan Stanley Capital International (2018), there are 23 current emerging economies. These emerging countries used for this study are presented in Table 5.4. For this study, economic growth (*lnRGDPC*) is measured by GDP per capita (constant 2010 US$). Energy consumption (*lnENER*) is measured using energy use (kg of oil equivalent per capita). Labour (*lnLABOUR*) is measured by the total labour force, capital (*lnGCFC*) is measured by gross capital formation (constant 2010 US$) while financial development (*lnFIN*) is measured using domestic credit to the private sector. These data are obtained from World Development Indicators. The globalization indices: economic (*lnEGB*), social (*lnSGB*), and political (*lnPGB*) globalization are obtained from the KOF globalization index (Gygli et al., 2018).

4. Empirical results and discussions

4.1 Economic growth

The results for the economic growth model are presented in Table 5.5. In column 2 of Table 5.5, we find that economic globalization, which involves the flows of goods, services, and financial assets, stimulates economic growth, significantly influence economic growth in emerging economies in the long-run. Economic globalization encourages specialization in economic sectors with a dynamic comparative advantage and enhances productivity growth through capital augmentation, improved factor productivity, effective resource allocation, and technological diffusion (Acheampong, 2018; Lucas, 1988; Redding, 1999; Krugman et al., 2017; Mishkin, 2009; Potrafke, 2015; Stiglitz, 2003). Also, consistent with the argument of Samimi and Jenatabadi (2014), economic globalization, directly and indirectly, stimulates economic growth through complementary reform (see also Gurgul and Lach, 2014; Marques et al., 2017; Samimi and Jenatabadi, 2014; Santiago et al., 2020) and by increasing the demand for factors of production, which subsequently increase the use of energy and boost economic growth (Cole, 2006; Sadorsky, 2011, 2012).

The result reveals that political globalization has a significant positive effect on economic growth (see column 4). Thus, a percentage point increase in political globalization would lead to a percentage point increase in economic growth by about 2.46 in the long-run. This positive influence of political globalization on economic growth in the emerging economies is reflected by the strength of institutions and governance in these economies. Well-established institutions in the emerging economies improve the efficacy of the domestic institutions, thereby promoting economic growth. Our results that

TABLE 5.4 Descriptive statistics, 1970–2015.

Countries	RGDPC	GCFC	LABOUR	FIN	ENER	EGB	SGB	PGB
Brazil	8562.84	2.80E+11	8.20E+07	50.7792	1037.63	33.3223	43.3027	72.9696
Chile	7961.27	2.10E+10	6.80E+06	58.2284	1285.72	51.4988	50.8052	73.3587
China	1745.64	1.80E+12	7.40E+08	94.3186	986.582	28.306	26.1589	71.0268
Colombia	4752.41	3.20E+10	1.90E+07	32.0822	677.776	34.0527	45.1575	61.0822
Czech Republic	16,566.9	4.60E+10	5.20E+06	46.7103	4366.98	72.5396	73.6511	85.3447
Egypt, Arab Rep.	1660.44	1.90E+10	2.20E+07	30.4229	569.023	46.2885	33.6437	80.7101
Greece	21,085.8	4.70E+10	4.80E+06	88.3386	2006.72	55.9245	65.5475	78.9721
Hungary	11,551.2	2.60E+10	4.30E+06	38.7881	2497.07	61.8167	63.6304	76.021
India	759.469	2.30E+11	4.10E+08	28.8449	384.957	21.1744	27.6981	77.626
Indonesia	1958.05	1.80E+11	1.00E+08	31.8647	578.473	45.821	28.905	72.1775
Korea, Rep.	11,278.6	1.80E+11	2.30E+07	72.6324	2787.03	41.1912	59.3338	68.0173
Malaysia	5647.31	3.30E+10	1.00E+07	89.4598	1609.6	62.654	64.9491	67.4435
Mexico	8144.15	1.70E+11	4.20E+07	21.1988	1417.53	42.4197	47.5469	68.9744
Pakistan	748.461	1.80E+10	4.70E+07	23.8696	397.193	30.4481	23.7794	71.6214
Peru	3748.19	1.60E+10	1.30E+07	20.2331	547.649	48.7467	39.0434	69.7418
Philippines	1678.59	2.50E+10	3.20E+07	29.5745	455.8	44.708	36.3412	67.1864

Poland	9530.66	8.20E+10	1.80E+07	30.7119	2811.74	44.1865	56.0045	81.645
Qatar	64,624.6	3.60E+10	750,354	32.0787	15,441.9	72.8327	62.7397	36.5878
Russian Federation	8641.11	3.40E+11	7.40E+07	36.5878	4698.61	40.4548	56.127	84.3038
South Africa	6458.64	4.50E+10	1.70E+07	100.622	2486.78	44.551	49.911	60.4857
Thailand	2924.67	5.50E+10	3.60E+07	84.4912	958.046	45.9657	40.7637	65.4428
Turkey	7529.6	1.50E+11	2.30E+07	24.0967	1026.27	41.4268	47.1963	78.6683
United Arab Emirates	66,890	6.80E+10	3.20E+06	36.3843	8638.52	72.0036	60.431	49.3654

TABLE 5.5 Pooled Mean Group results (Dependent variable: gross domestic product per capita).

	1	2	3	4	5	6	7
Long-run							
lnGCFC	0.828[c]	0.473[c]	0.470[c]	0.493[c]	0.470[c]	0.586[c]	0.616[c]
	(0.046)	(0.024)	(0.029)	(0.041)	(0.023)	(0.053)	(0.046)
lnLABOUR	−0.922[c]	−0.349[c]	−0.643[c]	−0.634[c]	−0.143[b]	−0.714[c]	−0.785[c]
	(0.080)	(0.069)	(0.057)	(0.078)	(0.070)	(0.056)	(0.110)
lnENER	0.065	0.330[c]	0.200[c]	0.101	0.323[c]	0.313[c]	0.277[c]
	(0.113)	(0.044)	(0.075)	(0.084)	(0.041)	(0.086)	(0.101)
lnEGB		0.432[c]			−0.432		
		(0.031)			(0.737)		
lnSGB			1.546[c]			3.556[a]	
			(0.097)			(1.879)	
lnPGB				2.455[c]			−31.122[c]
				(0.253)			(3.709)
lnEGB²					0.108		
					(0.096)		
lnSGB²						−0.275	
						(0.241)	

lnPGB²							3.876^c
							(0.447)
Short-run							
ECT	−0.004	−0.069^c	−0.037	−0.042^b	−0.077^c	−0.003	−0.056^a
	(0.024)	(0.019)	(0.031)	(0.020)	(0.021)	(0.041)	(0.031)
D.lnGCFC	0.135^c	0.120^c	0.127^c	0.114^c	0.117^c	0.130^c	0.100^c
	(0.015)	(0.016)	(0.016)	(0.018)	(0.015)	(0.018)	(0.026)
D.lnLABOUR	−0.011	0.005	0.006	0.019	−0.014	−0.030	−0.014
	(0.060)	(0.069)	(0.076)	(0.077)	(0.069)	(0.062)	(0.093)
D.lnENER	0.163^c	0.113^c	0.178^c	0.185^c	0.106^c	0.159^c	0.171^c
	(0.036)	(0.043)	(0.042)	(0.041)	(0.040)	(0.037)	(0.038)
D.lnEGB		−0.016			−9.674		
		(0.124)			(13.620)		
D.lnSGB			−0.010			−15.860	
			(0.032)			(18.088)	
D.lnPGB				0.088			0.410
				(0.107)			(4.784)
D.lnEGB²					1.157		
					(1.558)		

Continued

TABLE 5.5 Pooled Mean Group results (Dependent variable: gross domestic product per capita).—cont'd

	1	2	3	4	5	6	7
D.lnSGB^2						1.876	
						(2.127)	
D.lnPGB^2							−0.019
							(0.535)
Constant	0.023	−0.070[b]	0.016[a]	−0.185[b]	−0.208[c]	−0.005	4.024[a]
	(0.081)	(0.030)	(0.009)	(0.086)	(0.072)	(0.245)	(2.234)
Observations	515	512	512	512	512	512	512

Standard errors in parentheses.
[a]P < .10.
[b]P < .05.
[c]P < .01.

political globalization promotes economic growth is consistent with Acemoglu et al. (2005) finding that political factors shape economic growth, although insignificant in the short-run. The short-run results thus confirm previous evidence by Gurgul and Lach (2014), Marques et al. (2017), and Santiago et al. (2020), which revealed that political globalization has a negligible effect on economic growth.

Similarly, social globalization increases economic growth in emerging economies in the long-run (see column 3), to suggest that a percentage point increase in social globalization would lead to a percentage point increase in economic growth by about 1.55. In line with existing studies, social globalization involves the free flow of information and communication due to internets, phones, and televisions, it promotes economic growth by reducing transaction costs (see Gurgul and Lach, 2014; Marques et al., 2017; Santiago et al., 2020; Potrafke, 2015). In the nonlinear effect models, the results suggest that social globalization has a positive effect on economic growth while its squared term alternate in sign and level of statistical significance, in the long run. This result indicates that an inverted U-shaped relationship exists between social and economic growth in emerging economies. Thus, at the initial stage of social globalization, economic growth increases, but after a certain threshold, its influence on economic growth declines in the long-run. We, however, find no robust evidence that such a relationship exists between economic and political globalization and economic growth.

From Table 5.5, the results suggest that energy consumption exerts a positive effect on economic growth in both the short-run and long-run. These evidence affirm the results of Omri (2013), Nasreen and Anwar (2014), Mutascu (2016), which revealed that energy consumption significantly increases economic growth. Therefore, the *growth hypothesis* is valid in emerging economies, mainly, in the presence of economic and social globalization (see column 2 and 3). Thus, economic and social globalization promote the transfers the technology which could improve energy efficiency and subsequently increase economic growth. This supports the argument that globalization increases increase the use of energy required to boost economic growth (Cole, 2006; Sadorsky, 2011, 2012).

Also, consistent with the neoclassical economic growth model, increasing capital stock boosts economic growth in emerging economies. Thus, investment in capital stock in emerging economies induces higher economic growth. The findings also suggest that labour force exerts a significant negative effect on economic growth in emerging economies, in the long-run. Thus, labour retards economic growth in emerging economies. Similar to the argument of Omri (2013), this result could be attributed to the brain-drain, unskilled, and low-productivity of the labour force in the emerging economies.

4.2 Energy consumption

Table 5.6 presents the estimates for the energy consumption model. From Table 5.6, the results indicate that economic, social, and political globalization have negative and significant direct effects on energy consumption in emerging economies in the long-run (see column 2 to 4), but broadly insignificant in the short-run. Thus, economic, social, and political integration of countries enables economies to reduce the magnitude of energy consumption through the transfer and adoption of advanced technologies that efficiently stimulate economic activities. Additionally, Jena and Grote (2008) argue that the negative impact of globalization activities on energy consumption is explained by the shift of countries production activities from the agricultural sector to the service sector, which involves less demand on natural resources or energy consumption and thus improving the quality of the environment. This result confirms the empirical results of Shahbaz et al. (2016) and Marques et al. (2017), which indicated that economic, social, and political globalization significantly reduce energy consumption.

Recent evidence suggests that a nonlinear relationship exists between energy consumption and globalization. To test the practicality of this hypothesis, we included the squared term of the globalization indicators in the estimation. The results suggests that a U-shaped relationship exists between the economic and social globalization and energy consumption in emerging economies (see column 5 and 6). Thus, at the initial stage of globalization, energy consumption declines, but after a certain threshold of globalization, energy consumption increases. This observation could be explained using the scale, composition, and technique effects of globalization (Cole, 2006; Ghani, 2012). Concerning political globalization, we find that the direct term is significantly positive while the squared term is significantly negative, suggesting an inverted U-shaped behaviour between political globalization and energy consumption. Thus, energy consumption decreases after a certain threshold of political globalization in the long-run.

The findings also show that economic growth significantly increases energy consumption. Energy usage is critical for production and consumption activities across countries and thus increase in energy consumption in emerging countries appears favorable for economic progress. This finding is in line with previous studies by Shahbaz et al. (2016), Sadorsky (2010), and Destek (2018) who reported that economic growth is associated with

TABLE 5.6 Pooled Mean Group results (Dependent variable: Energy consumption).

	1	2	3	4	5	6	7
Long-run							
lnRGDPC	0.618ᶜ	0.714ᶜ	1.026ᶜ	1.129ᶜ	0.666ᶜ	0.978ᶜ	1.037ᶜ
	(0.043)	(0.051)	(0.039)	(0.050)	(0.051)	(0.043)	(0.066)
lnFIN	−0.196ᶜ	0.047	−0.034	−0.044	0.063ᵇ	−0.072ᶜ	0.007
	(0.028)	(0.033)	(0.027)	(0.035)	(0.031)	(0.026)	(0.034)
lnEGB		−0.177ᶜ			−0.255		
		(0.065)			(0.381)		
lnSGB			−0.343ᶜ			−1.505ᶜ	
			(0.066)			(0.517)	
lnPGB				−0.699ᶜ			11.494ᶜ
				(0.108)			(2.918)
lnEGB²					0.020		
					(0.058)		
lnSGB²						0.175ᵇ	
						(0.073)	
lnPGB²							−1.438ᶜ
							(0.347)

Continued

TABLE 5.6 Pooled Mean Group results (Dependent variable: Energy consumption).—cont'd

	1	2	3	4	5	6	7
Short-run							
ECT	-0.090^b	-0.085^c	-0.101^c	-0.103^c	-0.080^c	-0.114^c	-0.104^c
	(0.041)	(0.017)	(0.019)	(0.026)	(0.017)	(0.021)	(0.021)
D.lnRGDPC	0.464^c	0.495^c	0.447^c	0.458^c	0.474^c	0.429^c	0.472^c
	(0.062)	(0.048)	(0.045)	(0.047)	(0.066)	(0.043)	(0.046)
D.lnFIN	0.012	-0.002	-0.001	0.003	-0.010	0.001	-0.005
	(0.015)	(0.015)	(0.016)	(0.016)	(0.016)	(0.017)	(0.017)
D.lnEGB		-0.095			-30.143		
		(0.073)			(29.586)		
D.lnSGB			0.027			-3.902	
			(0.068)			(5.566)	
D.lnPGB				0.003			31.077
				(0.064)			(26.121)
D.lnEGB^2					3.480		
					(3.417)		

						$D.lnSGB^2$	$lnPGB^2$
$D.lnSGB^2$						0.461 (0.653)	
$lnPGB^2$							−3.417 (2.907)
Constant	0.245[b] (0.101)	0.136[c] (0.026)	−0.034[a] (0.019)	0.038[c] (0.011)	0.153[c] (0.032)	0.240[c] (0.042)	−2.570[c] (0.539)
Observations	820	820	820	820	820	820	820

Standard errors in parentheses.
[a] $p < .10$.
[b] $p < .05$.
[c] $p < .01$.

increasing energy consumption in emerging economies. However, our findings contradict the results of Rafiq et al. (2016), which revealed that economic growth reduces energy demands in emerging economies. Besides, the impact of financial development on energy consumption in emerging economies is not robust in both the long-run and short-run periods.

4.3 Robustness checks

In this section, we employ an alternative estimator to examine the sensitivity of our results in Tables 5.5 and 5.6. Specifically, we present results estimated from the dynamic OLS (DOLS) and fully modified OLS (FMOLS) techniques. The empirical findings presented in Table 5.7, where we focus on the impact of globalization indices on economic growth reveals three things. First, we find that the effect of energy consumption on economic growth is consistent with the long-run results displayed in Table 5.5. Specifically, the results suggest that energy consumption exerts a significant positive effect on economic growth in both the FMOLS and DOLS estimates. Second, and consistent with our previous results, economic, social, and political globalization maintained their direct positive influence on economic growth. The results are also consistent with previous findings that globalization drives economic growth through technological spillover effects across economics (Stiglitz, 2003 Gurgul and Lach, 2014; Marques et al., 2017; Samimi and Jenatabadi, 2014; Santiago et al., 2020). Similarly, both the FMOLS and DOLS estimations suggest an inverted U-shaped behavior of the globalization measurements on economic growth in emerging economies.

Third, when the globalization factors are examined on energy consumption, we find that the FMOLS and DOLS estimates (see Table 5.8) produced almost similar evidence as those reported in Table 5.6. Thus, economic and social globalization inhibits energy consumption, but that of political globalization induces energy demands in the long-run. Moreover, we find that while globalization significantly influences energy consumption in emerging economies, it does so nonlinearly such that the initial stage of economic and social globalization decreases energy consumption, but the impact turns positive beyond a certain threshold of economic and social globalization, whereas an opposing behavior is found for political globalization, in the FMOLS anlaysis. This is explained by structural changes and international regulations that influence countries integration. Overall, our findings are less sensitive to alternative estimator and confirm the empirical results of Shahbaz et al. (2016) and Marques et al. (2017), which indicated that economic, social, and political globalization significantly reduces energy consumption.

TABLE 5.7 Fully modified ordinary least square and dynamic ordinary least square results (Dependent variable: gross domestic product per capita).

	1	2	3	4	5	6	7	8	9	10	11	12	13	14
	FMOLS							DOLS						
lnGCFC	0.230[c]	0.173[c]	0.208[c]	0.210[c]	0.177[c]	0.185[c]	0.209[c]	0.671[c]	0.463[c]	0.431[c]	0.530[c]	0.489[c]	0.386[c]	0.501[c]
	(0.027)	(0.024)	(0.022)	(0.026)	(0.024)	(0.019)	(0.023)	(0.085)	(0.045)	(0.051)	(0.052)	(0.048)	(0.033)	(0.046)
lnLABOUR	0.041	−0.020	−0.274[c]	−0.098	−0.027	−0.367[c]	−0.115[b]	−0.851[c]	−0.658[c]	−0.620[c]	−0.764[c]	−0.689[c]	−0.553[c]	−0.740[c]
	(0.056)	(0.049)	(0.055)	(0.061)	(0.049)	(0.051)	(0.053)	(0.263)	(0.130)	(0.188)	(0.156)	(0.130)	(0.122)	(0.141)
lnENER	0.787[c]	0.695[c]	0.558[c]	0.736[c]	0.694[c]	0.450[c]	0.521[c]	0.052	0.352[c]	0.238[a]	0.301[b]	0.369[c]	0.122	0.339[c]
	(0.060)	(0.054)	(0.555)	(0.058)	(0.054)	(0.052)	(0.057)	(0.199)	(0.113)	(0.129)	(0.133)	(0.125)	(0.085)	(0.117)
lnEGB		0.402[c]			−0.070				0.123			−4.666[c]		
		(0.054)			(0.551)				(0.081)			(1.561)		
lnSGB			0.568[c]			−2.418[c]				0.484[b]			−21.922[c]	
			(0.062)			(0.423)				(0.209)			(2.266)	
lnPGB				0.421[c]			−20.290[c]				−0.599[c]			−51.961[c]
				(0.105)			(2.488)				(0.127)			(4.798)
lnEGB²					0.063							0.564[c]		
					(0.073)							(0.197)		
lnSGB²						0.432[c]							2.812[c]	

Continued

TABLE 5.7 Fully modified ordinary least square and dynamic ordinary least square results (Dependent variable: gross domestic product per capita).—cont'd

	FMOLS								DOLS					
	1	2	3	4	5	6	7	8	9	10	11	12	13	14
lnPGB²							2.462[c]							5.782[c]
						(0.060)	(0.295)						(0.285)	(0.562)
Observations	515	512	512	512	512	512	512	108	90	90	90	90	90	90
Adjusted R^2	0.990	0.992	0.993	0.991	0.992	0.995	0.994	0.190	0.677	0.700	0.687	0.708	0.722	0.673

Standard errors in parentheses.
[a]$P < .10$.
[b]$P < .05$.
[c]$P < .01$.

TABLE 5.8 Fully modified ordinary least square and dynamic ordinary least square results (Dependent variable: Energy consumption).

	1	2	3	4	5	6	7	8	9	10	11	12	13	14
				FMOLS							DOLS			
lnRGDPC	0.654[c]	0.702[c]	0.695[c]	0.593[c]	0.730[c]	0.701[c]	0.696[c]	1.047[c]	1.154[c]	1.143[c]	1.135[c]	1.031[c]	1.450[c]	1.332[c]
	(0.030)	(0.039)	(0.042)	(0.036)	(0.039)	(0.039)	(0.042)	(0.057)	(0.049)	(0.062)	(0.078)	(0.050)	(0.064)	(0.066)
lnFIN	0.119[c]	0.137[c]	0.121[c]	0.076[b]	0.101[c]	0.057[b]	0.052[a]	-0.061[b]	-0.077[c]	-0.026	-0.101[c]	-0.068[c]	-0.065[c]	-0.140[c]
	(0.029)	(0.030)	(0.029)	(0.030)	(0.030)	(0.028)	(0.030)	(0.024)	(0.019)	(0.019)	(0.025)	(0.020)	(0.025)	(0.022)
lnEGB		-0.140[a]			-2.294[c]				-0.099[b]			2.421[c]		
		(0.075)			(0.543)				(0.049)			(0.887)		
lnSGB			-0.092			-2.573[c]				-0.199[b]			-5.404[c]	
			(0.060)			(0.402)				(0.093)			(1.791)	
lnPGB				0.285[c]			7.924[c]				-0.151			-1.561
				(0.085)			(1.965)				(0.132)			(3.872)
lnEGB²					0.299[c]							-0.327[c]		
					(0.074)							(0.123)		
lnSGB²						0.353[c]							0.619[c]	
						(0.057)							(0.233)	
lnPGB²							-0.934[c]							0.120
							(0.239)							(0.445)

Continued

TABLE 5.8 Fully modified ordinary least square and dynamic ordinary least square results (Dependent variable: Energy consumption).—cont'd

	1	2	3	4	5	6	7	8	9	10	11	12	13	14
				FMOLS							DOLS			
Observations	833	833	833	833	833	833	833	38	38	38	38	38	38	38
Adjusted R^2	0.971	0.971	0.971	0.972	0.973	0.974	0.974	0.993	0.995	0.993	0.994	0.996	0.993	0.993

Standard errors in parentheses.
[a]$p < .10.$
[b]$p < .05.$
[c]$p < .01.$

5. Conclusions and policy implications

This study employed the PMG estimation of a dynamic heterogeneous panel to examine the short-run and long-run impacts of economic, social, and political globalisation on the economic growth and energy consumption nexus in a panel of 23 emerging economies for the period 1970-2015. The findings that emerged from this study are presented as follows: First, for the economic growth model, the results revealed that energy consumption broadly contributes economic growth when economic and social globalisation are controlled. The findings further indicated that economic, social and political globalization directly promote economic growth in emerging economies. However, the nonlinear effect results showed that economic and political globalisation have a U-shaped relationship with economic growth, whereas that of social globalisation has an inverted U-shaped behaviour. Second, for the energy consumption model, we find that economic growth significantly increases energy consumption. The results also revealed that economic, social, and political globalisation adversely influence energy consumption when evaluated at standard levels of statistical significance. Notwithstanding, the globalisation factors affect energy consumption nonlinearly, with economic and social globalisation exhibiting a U-shaped relationship with energy consumption. Overall, the empirical findings lead to the following:

(i) *In the short-run, globalization improves economic growth but has a negligible influence on energy consumption.*

(ii) *In the long-run, globalization stimulates economic growth and reduces energy consumption.*

These findings offer some important policy implications for emerging economies. The findings reveal that economic, social, and political globalization enhance long-term economic growth while promoting energy efficiency in emerging economies. These findings suggest that policymakers should design and implement favorable policies that would integrate emerging economies with the rest of the world. Promoting economic, social, and political globalization would encourage the transfer of technological innovations into the productive sectors where technological innovations are limited. These technological innovations induced by globalization remain consequential for enhancing energy efficiency and economic growth in the long-term. Moreover, the integration of emerging economies with the rest of the world would provide them with an additional source of funding (through FDI), which is critical for boosting economic growth and energy efficiency. We also caution that, while economic globalization could stimulate growth, it also has a potential adverse impact on long-term growth through income inequality and unemployment (Krugman et al., 2017; Potrafke, 2015; Stiglitz, 2003). Therefore to harness the potential benefit of it, policymakers are required to implement desirable social interventions, which could act as a "safety net" for those who are hurt by economic globalization. Also, strengthening of existing political and economic institutions remains critical for globalization to enhance energy efficiency and long-term economic growth in the emerging economies. We also recommend that policymakers should incorporate globalization factors in the design and implementation of long-term energy efficiency and environmental conservation policies in emerging economies.

References

Acaravci, A., Erdogan, S., Akalin, G., 2015. The electricity consumption, real income, trade openness and foreign direct investment: the empirical evidence from Turkey. Int. J. Energy Econ. Pol. 5 (4), 1050—1057.

Acaravci, A., Ozturk, I., 2010. Electricity consumption-growth nexus: evidence from panel data for transition countries. Energy Econ. 32 (3), 604—608.

Acemoglu, D., Johnson, S., Robinson, J.A., 2005. Institutions as a fundamental cause of long-run growth. In: Aghion, P., Durlauf, S.N. (Eds.), Handbook of Economic Growth, vol. 1A, pp. 385—472.

Acheampong, A.O., 2018. Economic growth, CO_2 emissions and energy consumption: what causes what and where? Energy Econ. 74, 677—692. https://doi.org/10.1016/j.eneco.2018.07.022.

Adewuyi, A.O., Awodumi, O.B., 2017. Renewable and non-renewable energy-growth-emissions linkages: review of emerging trends with policy implications. Renew. Sustain. Energy Rev. 69, 275—291. https://doi.org/10.1016/j.rser.2016.11.178.

Afzal, M., 2007. The impact of globalisation on economic growth of Pakistan. Pakistan Dev. Rev. 723—734.

Aïssa, M.S.B., Jebli, M.B., Youssef, S.B., 2014. Output, renewable energy consumption and trade in Africa. Energy Pol. 66, 11—18.

Akinlo, A.E., 2009. Electricity consumption and economic growth in Nigeria: evidence from cointegration and co-feature analysis. J. Pol. Model. 31 (5), 681—693.

Alinsato, A.S., 2009. Electricity Consumption and GDP in an Electricity Community: Evidence from Bound Testing Cointegration and Granger-causality Tests. Munich Personal Research Archive. Working paper No. 20816.

Alt, H., Kum, M., 2013. Multivariate granger causality between electricity generation, exports, prices and economic growth in Turkey. Int. J. Energy Econ. Pol. 3, 41.

Altinay, G., Karagol, E., 2005. Electricity consumption and economic growth: evidence from Turkey. Energy Econ. 27 (6), 849—856.

Apergis, N., Ozturk, I., 2015. Testing environmental Kuznets curve hypothesis in Asian countries. Ecol. Indicat. 52, 16—22.

Apergis, N., Payne, J.E., 2014. Renewable energy, output, CO_2 emissions, and fossil fuel prices in Central America: evidence from a nonlinear panel smooth transition vector error correction model. Energy Econ. 42, 226—232. https://doi.org/10.1016/j.eneco.2014.01.003.

Aqeel, A., Butt, M.S., 2001. The relationship between energy consumption and economic growth in Pakistan. Asia Pac. Dev. J. 8 (2), 101—110.

Arslan, Y., Contreras, J., Patel, N., Shu, C., 2018. Globalisation and Deglobalisation in Emerging Market Economies: Facts and Trends. BIS Paper(100a).

Bartleet, M., Gounder, R., 2010. Energy consumption and economic growth in New Zealand: results of trivariate and multivariate models. Energy Pol. 38 (7), 3508—3517.

Bayer, C., Hanck, C., 2013. Combining non-cointegration tests. J. Time Ser. Anal. 34 (1), 83—95.

Belke, A., Dobnik, F., Dreger, C., 2011. Energy consumption and economic growth: new insights into the cointegration relationship. Energy Econ. 33 (5), 782—789.

Berg, A., Krueger, A.O., 2003. Trade, Growth, and Poverty: A Selective Survey. IMF Working Papers No.1047.

Borensztein, E., De Gregorio, J., Lee, J.W., 1998. How does foreign direct investment affect economic growth? J. Int. Econ. 45, 115—135.

Calderón, C., Poggio, V., 2010. Trade and Economic Growth Evidence on the Role of Complementarities for CAFTA-DR Countries. World Bank Policy Research. Working Paper No.5426.

Chang, R., Kaltani, L., Loayza, N.V., 2009. Openness can be good for growth: the role of policy complementarities. J. Dev. Econ. 90 (1), 33—49.

Chaudhry, I.S., Safdar, N., Farooq, F., 2012. Energy consumption and economic growth: empirical evidence from Pakistan. Pakistan J. Soc. Sci. 32 (2), 371—382.

Chen, S.T., Kuo, H.I., Chen, C.C., 2007. The relationship between GDP and electricity consumption in 10 Asian countries. Energy Pol. 35 (4), 2611—2621.

Ciarreta, A., Zarraga, A., 2010. Electricity consumption and economic growth in Spain. Appl. Econ. Lett. 17 (14), 1417—1421.

Cole, M.A., 2006. Does trade liberalization increase national energy use? Econ. Lett. 92 (1), 108—112. https://doi.org/10.1016/j.econlet.2006.01.018.

Crafts, N., 2004. Globalisation and economic growth: a historical perspective. World Econ. 27 (1), 45—58.

De Melo, J., Gourdon, J., Maystre, N., 2008. Openness, Inequality and Poverty: Endowments Matter. World Bank Policy Research Working. Paper No.3981.

Destek, M.A., 2018. Financial development and energy consumption nexus in emerging economies. Energy Sources B Energy Econ. Plann. 13 (1), 76—81. https://doi.org/10.1080/15567249.2017.1405106.

Dogan, E., 2014. Energy consumption and economic growth: evidence from low-income countries in sub-Saharan Africa. Int. J. Energy Econ. Policy 4 (2), 154.

Dogan, E, 2015. The relationship between economic growth and electricity consumption from renewable and non-renewable sources: a study of Turkey. Renew. Sustain. Energy Rev. 52, 534—546.

Dolado, J.J., Lütkepohl, H., 1996. Making Wald tests work for cointegrated VAR systems. Econom. Rev. 15 (4), 369—386.

Dollar, D., 1992. Outward-oriented developing economies really do grow more rapidly: evidence from 95 LDCs, 1976—1985. Econ. Dev. Cult. Change 40, 523—544.

Dreher, A., 2006. Does globalization affects growth? Empirical evidence from a new index. Appl. Econ. 6, 1091—1110.

Edwards, S., 1998. Openness, productivity and growth: what do we really know? Econ. J. 108 (447), 383—398.

Eggoh, J.C., Bangaké, C., Rault, C., 2011. Energy consumption and economic growth revisited in African countries. Energy Pol. 39 (11), 7408—7421.

Erkan, C., Mucuk, M., Uysal, D., 2010. The impact of energy consumption on exports: the Turkish case. Asian J. Bus. Manag. 2 (1), 17—23.

Farooq, F., Yusop, Z., Chaudhry, I.S., Iram, R., 2019. Assessing the impacts of globalization and gender parity on economic growth: empirical evidence from OIC countries. Environ. Sci. Pollut. Control Ser. 1—14.

Ferguson, R., Wilkinson, W., Hill, R., 2000. Electricity use and economic development. Energy Pol. 28 (13), 923—934.

Fischer, S., 2003. Globalization and its challenges. Am. Econ. Rev. 93, 1−30.

Ghani, G.M., 2012. Does trade liberalization effect energy consumption? Energy Pol. 43, 285−290. https://doi.org/10.1016/j.enpol.2012.01.005.

Ghosh, S., 2002. Electricity consumption and economic growth in India. Energy Pol. 30 (2), 125−129.

Gu, X., Dong, B., 2011. A theory of financial liberalization: why are developing countries so reluctant? World Econ. 34 (7), 1106−1123.

Gurgul, H., Lach, Ł., 2014. Globalization and economic growth: evidence from two decades of transition in CEE. Econ. Modell. 36, 99−107. https://doi.org/10.1016/j.econmod.2013.09.022.

Gygli, S., Haelg, F., Potrafke, N., Sturm, J.E., 2018. The KOF globalisation index-revisited.

Hassan, S.T., Xia, E., Huang, J., Khan, N.H., Iqbal, K., 2019. Natural resources, globalization, and economic growth: evidence from Pakistan. Environ. Sci. Pollut. Control Ser. 26 (15), 15527−15534.

Heckscher, E., 1919. The Effect of Foreign Trade on the Distribution of Income. Ekonomik Tidskrift 31. Reprinted in Readings in the Theory of International Trade. the American Economic Association, USA.

Huang, B.-N., Hwang, M.J., Yang, C.W., 2008. Causal relationship between energy consumption and GDP growth revisited: a dynamic panel data approach. Ecol. Econ. 67 (1), 41−54. https://doi.org/10.1016/j.ecolecon.2007.11.006.

Hye, Q.M.A., Riaz, S., 2008. Causality between energy consumption and economic growth: the case of Pakistan. Lahore J. Econ. 13 (2), 45−58.

International Energy Agency [IEA], 2018. Energy Efficiency 2018. IEA, Paris. https://www.iea.org/reports/energy-efficiency-2018.

Iyke, B.N., 2015. Electricity consumption and economic growth in Nigeria: a revisit of the energy-growth debate. Energy Econ. 51, 166−176.

Jafari, Y., Othman, J., Nor, A.H.S.M., 2012. Energy consumption, economic growth and environmental pollutants in Indonesia. J. Pol. Model. 34 (6), 879−889.

Jamil, F., Ahmad, E., 2010. The relationship between electricity consumption, electricity prices and GDP in Pakistan. Energy Pol. 38 (10), 6016−6025.

Jena, P.R., Grote, U., 2008. Growth-trade-environment nexus in India. Econ. Bull. 17, 1−17.

Karanfil, F., 2009. How many times again will we examine the energy-income nexus using a limited range of traditional econometric tools? Energy Pol. 37 (4), 1191−1194. https://doi.org/10.1016/j.enpol.2008.11.029.

Kouakou, A.K., 2011. Economic growth and electricity consumption in Cote d'Ivoire: evidence from time series analysis. Energy Pol. 39 (6), 3638−3644.

Kraft, J., Kraft, A., 1978. On the relationship between energy and GNP. J. Energy Dev. 401−403.

Krugman, P., Obstfeld, M., Melitz, M., 2017. International Economics: Theory and Policy. the latest ed. Addison-Wesley.

Lean, H.H., Smyth, R., 2010. Multivariate Granger causality between electricity generation, exports, prices and GDP in Malaysia. Energy 35 (9), 3640−3648.

Lucas, R.E., 1988. On the mechanics of economic development. J. Monetary Econ. 22 (1), 3−42.

Mahadevan, R., Asafu-Adjaye, J., 2007. Energy consumption, economic growth and prices: a reassessment using panel VECM for developed and developing countries. Energy Pol. 35 (4), 2481−2490. https://doi.org/10.1016/j.enpol.2006.08.019.

Mahalik, M.K., Babu, M.S., Loganathan, N., Shahbaz, M., 2017. Does financial development intensify energy consumption in Saudi Arabia? Renew. Sustain. Energy Rev. 75, 1022−1034. https://doi.org/10.1016/j.rser.2016.11.081.

Marques, L.M., Fuinhas, J.A., Marques, A.C., 2017. Augmented energy-growth nexus: economic, political and social globalisation impacts. Energy Proc. 136, 97−101. https://doi.org/10.1016/j.egypro.2017.10.293.

Menegaki, A., Tugcu, C., 2016. The sensitivity of growth, conservation, feedback & neutrality hypothesis to sustainable accounting. Energy Sustain. Dev. 34, 77−87.

Mirza, F.M., Kanwal, A., 2017. Energy consumption, carbon emissions and economic growth in Pakistan: dynamic causality analysis. Renew. Sustain. Energy Rev. 72, 1233−1240.

Mishkin, F.S., 2009. Globalization, macroeconomic performance, and monetary policy. J. Money Credit Bank. 41, 187−196.

Mudakkar, S.R., Zaman, K., Shakir, H., Arif, M., Naseem, I., Naz, L., 2013. Determinants of energy consumption function in SAARC countries: balancing the odds. Renew. Sustain. Energy Rev. 28, 566−574.

Mutascu, M., 2016. A bootstrap panel Granger causality analysis of energy consumption and economic growth in the G7 countries. Renew. Sustain. Energy Rev. 63, 166−171. https://doi.org/10.1016/j.rser.2016.05.055.

Narayan, P.K., Narayan, S., 2010. Carbon dioxide emissions and economic growth: panel data evidence from developing countries. Energy Pol. 38 (1), 661−666. https://doi.org/10.1016/j.enpol.2009.09.005.

Narayan, P.K., Prasad, A., 2008. Electricity consumption−real GDP causality nexus: evidence from a bootstrapped causality test for 30 OECD countries. Energy Pol. 36 (2), 910−918.

Narayan, P.K., Smyth, R., 2009. Multivariate Granger causality between electricity consumption, exports and GDP: evidence from a panel of Middle Eastern countries. Energy Pol. 37 (1), 229−236.

Nasreen, S., Anwar, S., 2014. Causal relationship between trade openness, economic growth and energy consumption: a panel data analysis of Asian countries. Energy Pol. 69, 82−91. https://doi.org/10.1016/j.enpol.2014.02.009.

Ohlin, B., 1933. Interregional and International Trade. Cambridge University Press, Harvard.

Olimpia, N., Stela, D., 2017. Impact of globalisation on economic growth in Romania: an empirical analysis of its economic, social and political dimensions. Studia Universitatis "Vasile Goldis" Arad−Economics Series 27 (1), 29−40.

Omri, A., 2013. CO2 emissions, energy consumption and economic growth nexus in MENA countries: evidence from simultaneous equations models. Energy Econ. 40, 657−664. https://doi.org/10.1016/j.eneco.2013.09.003.

Omri, A., 2014. An international literature survey on energy-economic growth nexus: evidence from country-specific studies. Renew. Sustain. Energy Rev. 38 (Supplement C), 951−959. https://doi.org/10.1016/j.rser.2014.07.084.

Omri, A., Kahouli, B., 2014. Causal relationships between energy consumption, foreign direct investment and economic growth: fresh evidence from dynamic simultaneous-equations models. Energy Pol. 67, 913−922. https://doi.org/10.1016/j.enpol.2013.11.067.

Ouedraogo, N.S., 2013. Energy consumption and economic growth: evidence from the economic community of West African States (ECOWAS). Energy Econ. 36 (Supplement C), 637−647. https://doi.org/10.1016/j.eneco.2012.11.011.

Ozturk, I., 2010. A literature survey on energy−growth nexus. Energy Pol. 38 (1), 340−349. https://doi.org/10.1016/j.enpol.2009.09.024.

Pao, H.T., 2009. Forecast of electricity consumption and economic growth in Taiwan by state space modeling. Energy 34 (11), 1779−1791.

Payne, J.E., 2009. On the dynamics of energy consumption and output in the US. Appl. Energy 86 (4), 575−577.

Payne, J.E., 2010. A survey of the electricity consumption-growth literature. Appl. Energy 87 (3), 723–731.

Pedroni, P., 2001a. Fully modified OLS for heterogeneous cointegrated panels. In: Nonstationary Panels, Panel Cointegration, and Dynamic Panels. Emerald Group Publishing Limited.

Pedroni, P., 2001b. Purchasing power parity tests in cointegrated panels. Rev. Econ. Stat. 83 (4), 727–731.

Pesaran, M.H., Shin, Y., Smith, R.J., 2001. Bounds testing approaches to the analysis of level relationships. J. Appl. Econom. 16 (3), 289–326.

Polemis, M.L., Dagoumas, A.S., 2013. The electricity consumption and economic growth nexus: evidence from Greece. Energy Pol. 62, 798–808.

Ponzio, C.A., 2005. Globalisation and economic growth in the third world: some evidence from eighteenth-century Mexico. J. Lat. Am. Stud. 37 (3), 437–467.

Potrafke, N., 2015. The evidence on globalisation. World Econ. 38 (3), 509–552.

Rafiq, S., Salim, R., Nielsen, I., 2016. Urbanisation, openness, emissions, and energy intensity: a study of increasingly urbanised emerging economies. Energy Econ. 56, 20–28. https://doi.org/10.1016/j.eneco.2016.02.007.

Redding, S., 1999. Dynamic comparative advantage and the welfare effects of trade. Oxf. Econ. Pap. 51 (1), 15–39.

Rodrik, D., 1997, October. Globalization, social conflict and economic growth. In: Conferencia de Raúl Prebisch. Ginebra. Versión revisada (en inglés) disponible en, vol. 24. http://www.ksg. harvard.edu/rodrik/global. pdf.

Rodrik, D., 1998. Who needs capital-account convertibility? In: Fischer, S. (Ed.), Should the IMF Pursue Capital Account Convertibility? Essays in International Finance. Department of Economics, Princeton University, Princeton, pp. 55–65.

Rousseau, P.L., Sylla, R., 2003. Financial systems, economic growth, and globalization. In: Globalization in Historical Perspective. University of Chicago Press, pp. 373–416.

Sadorsky, P., 2010. The impact of financial development on energy consumption in emerging economies. Energy Pol. 38 (5), 2528–2535. https://doi.org/10.1016/j.enpol.2009.12.048.

Sadorsky, P., 2011. Trade and energy consumption in the Middle East. Energy Econ. 33 (5), 739–749.

Sadorsky, P., 2012. Energy consumption, output and trade in South America. Energy Econ. 34 (2), 476–488.

Sakyi, D., Villaverde, J., Maza, A., Reddy Chittedi, K., 2012. Trade openness, growth and development: evidence from heterogeneous panel cointegration analysis for middle-income countries. Cuad. Econ. 31 (SPE57), 21–40.

Sadorsky, P., 2013. Do urbanization and industrialization affect energy intensity in developing countries? Energy Econ. 37, 52–59. https://doi.org/10.1016/j.eneco.2013.01.009.

Sakyi, D., Villaverde, J., Maza, A., 2015. Trade openness, income levels, and economic growth: the case of developing countries, 1970–2009. J. Int. Trade Econ. Dev. 24 (6), 860–882.

Samimi, P., Jenatabadi, H.S., 2014. Globalisation and economic growth: empirical evidence on the role of complementarities. PLoS One 9 (4).

Santiago, R., Fuinhas, J.A., Marques, A.C., 2020. The impact of globalisation and economic freedom on economic growth: the case of the Latin America and Caribbean countries. Econ. Change Restruct. 53 (1), 61–85. https://doi.org/10.1007/s10644-018-9239-4.

Saud, S., Baloch, M.A., Lodhi, R.N., 2018. The nexus between energy consumption and financial development: estimating the role of globalization in Next-11 countries. Environ. Sci. Pollut. Control Ser. 25 (19), 18651–18661.

Shahbaz, M., Feridun, M., 2012. Electricity consumption and economic growth empirical evidence from Pakistan. Qual. Quantity 46 (5), 1583−1599.

Shahbaz, M., Lean, H.H., 2012. The dynamics of electricity consumption and economic growth: a revisit study of their causality in Pakistan. Energy 39 (1), 146−153.

Shahbaz, M., Tang, C.F., Shabbir, M.S., 2011. Electricity consumption and economic growth nexus in Portugal using cointegration and causality approaches. Energy Pol. 39 (6), 3529−3536.

Shahbaz, M., Lean, H.H., Shabbir, M.S., 2012. Environmental Kuznets curve hypothesis in Pakistan: cointegration and Granger causality. Renew. Sustain. Energy Rev. 16 (5), 2947−2953.

Shahbaz, M., Khan, S., Tahir, M.I., 2013. The dynamic links between energy consumption, economic growth, financial development and trade in China: fresh evidence from multivariate framework analysis. Energy Econ. 40, 8−21. https://doi.org/10.1016/j.eneco.2013.06.006.

Shahbaz, M., Arouri, M., Teulon, F., 2014. Short-and long-run relationships between natural gas consumption and economic growth: evidence from Pakistan. Econ. Modell. 41, 219−226.

Shahbaz, M., Bhattacharya, M., Ahmed, K., 2015. Growth-globalisation-emissions nexus: the role of population in Australia. Monash Business School Discussion Paper 23 (15), p1−33.

Shahbaz, M., Mallick, H., Mahalik, M.K., Sadorsky, P., 2016. The role of globalisation on the recent evolution of energy demand in India: implications for sustainable development. Energy Econ. 55, 52−68. https://doi.org/10.1016/j.eneco.2016.01.013.

Shahbaz, M., Shahzad, S.J.H., Alam, S., Apergis, N., 2018a. Globalisation, economic growth and energy consumption in the BRICS region: the importance of asymmetries. J. Int. Trade Econ. Dev. 27 (8), 985−1009.

Shahbaz, M., Shahzad, S.J.H., Mahalik, M.K., Sadorsky, P., 2018b. How strong is the causal relationship between globalization and energy consumption in developed economies? A country-specific time-series and panel analysis. Appl. Econ. 50 (13), 1479−1494.

Shahbaz, M., Shahzad, S.J.H., Mahalik, M.K., et al., 2018c. Does globalisation worsen environmental quality in developed economies? Environ. Model. Assess. 23, 141−156. https://doi.org/10.1007/s10666-017-9574-2.

Shiu, A., Lam, P.L., 2004. Electricity consumption and economic growth in China. Energy Pol. 32 (1), 47−54.

Śmiech, S., Papież, M., 2014. Energy consumption and economic growth in the light of meeting the targets of energy policy in the EU: the bootstrap panel Granger causality approach. Energy Pol. 71, 118−129.

Squalli, J., 2007. Electricity consumption and economic growth: bounds and causality analyses of OPEC members. Energy Econ. 29 (6), 1192−1205.

Stiglitz, J.E., 2003. Globalisation and growth in emerging markets and the new economy. J. Pol. Model. 25 (5), 505−524. https://doi.org/10.1016/S0161-8938(03)00043-7.

Tang, C.F., Shahbaz, M., 2011. Revisiting the electricity consumption-growth nexus for Portugal: evidence from a multivariate framework analysis.

Tang, C.F., Shahbaz, M., 2013. Sectoral analysis of the causal relationship between electricity consumption and real output in Pakistan. Energy Pol. 60, 885−891.

Toda, H.Y., Yamamoto, T., 1995. Statistical inference in vector autoregressions with possibly integrated processes. J. Econom. 66 (1−2), 225−250.

Tsani, S.Z., 2010. Energy consumption and economic growth: a causality analysis for Greece. Energy Econ. 32 (3), 582−590.

Valli, V., Saccone, D., 2015. Structural change, globalization and economic growth in China and India. Eur. J. Comp. Econ. 12 (2), 133.

Villaverde, J., Maza, A., 2011. Globalisation, growth and convergence. World Econ. 34 (6), 952–971.

Wang, Y., Wang, Y., Zhou, J., Zhu, X., Lu, G., 2011. Energy consumption and economic growth in China: a multivariate causality test. Energy Pol. 39 (7), 4399–4406.

Warner, A., 2003. Once more into the breach: economic growth and integration. Center Glob. Dev. 12, 1.

Woetzel, J., Madgavkar, A., Seong, J., Manyika, J., Sneader, K., Tonby, O., et al., 2018. Outperformers: High-Growth Emerging Economies and the Companies that Propel Them. McKinsey Global Institute (MGI).

Wolde-Rufael, Y., 2006. Electricity consumption and economic growth: a time series experience for 17 African countries. Energy Pol. 34 (10), 1106–1114.

Yoo, S.H., 2006. The causal relationship between electricity consumption and economic growth in the ASEAN countries. Energy Pol. 34 (18), 3573–3582.

Yoo, S.H., Kwak, S.Y., 2010. Electricity consumption and economic growth in seven South American countries. Energy Pol. 38 (1), 181–188.

Yu, E.S., Choi, J.Y., 1985. The causal relationship between energy and GNP: an international comparison. J. Energy Dev. 249–272.

Yuan, J.H., Kang, J.G., Zhao, C.H., Hu, Z.G., 2008. Energy consumption and economic growth: evidence from China at both aggregated and disaggregated levels. Energy Econ. 30 (6), 3077–3094.

Zafar, M.W., Mirza, F.M., Zaidi, S.A.H., Hou, F., 2019a. The nexus of renewable and nonrenewable energy consumption, trade openness, and CO_2 emissions in the framework of EKC: evidence from emerging economies. Environ. Sci. Pollut. Control Ser. 26 (15), 15162–15173.

Zafar, M.W., Saud, S., Hou, F., 2019b. The impact of globalization and financial development on environmental quality: evidence from selected countries in the Organization for Economic Co-operation and Development (OECD). Environ. Sci. Pollut. Control Ser. 26 (13), 13246–13262.

Zerrin, K., Dumrul, Y., 2018. The impact of globalization on economic growth: empirical evidence from the Turkey. Int. J. Econ. Financ. Issues 8 (5), 115.

Chapter 6

Commodities spillover effect in the United States: insight from the housing, energy, and agricultural commodity markets

Andrew Adewale Alola[1,3], Uju Violet Alola[2]
[1]*Department of Economics and Finance, Istanbul Gelisim University, Istanbul, Turkey;*
[2]*Department of Tourism Guidance, Istanbul Gelisim University, Istanbul, Turkey;* [3]*Department of Economics, School of Accounting and Finance, University of Vaasa, Finland*

1. Introduction

Generally, connectedness is mainly associated with return market and credit risk, counterparty and gridlock risk, systemic risk, and the underpinning of fundamental microeconomic risks (Diebold and Yılmaz, 2014). The fact that shocks are transmitted around the global trading system has continued to be an interesting phenomenon. In reality, understanding this trend and deducing useful financial and economic inference, especially of the market correlation and spillovers is an associated bottleneck to researchers and investors. This is because the characteristic movements across markets are often associated with intranational (domestic) or international market linkages and connectedness. The transmission of financial market volatility or risk importantly informs an effective security pricing, sustainable asset allocation, and at establishing the limits of diversification (Steeley, 2006). An illustration of the international connectedness of the financial market was globally experienced during the Great Crisis of 2007–11. This global financial crisis (GFC) which forcefully began in the United States' subprime mortgage market (lasting about a year and half in the country at different stages) before unleashing strange distress in the financial and government institutions of the European Union countries between 2010 and 2011 (Diebold and Yilmaz, 2015a).

Energy-Growth Nexus in an Era of Globalization. https://doi.org/10.1016/B978-0-12-824440-1.00013-8
149

Having expressed the importance of understanding international (inter-country) connectedness of financial market movements, the similar notion is held for intranational real activity connectedness. Specifically, studying the connectedness of the major US financial institutions' stock return volatilities is as important as understanding the risks connectedness of the country's major markets, especially the risks of its macroeconomic markets. Being the largest gross domestic product measured at purchasing power parity conversion rates for more than a century, the US economy is currently the second largest behind China in the last 4 years (CIA, 2018). That is the reason the country's macroeconomic sectors have continued to largely determine the dynamics of the global markets. For instance, the country's economy is energy-intensive, so the dynamics of the US's energy market would potentially affects the global oil price. Despite the global campaign for energy transition amidst increasing concern of climate change challenges, energy consumption in the United States has remained high but with a declining rate of carbon emissions (British Petroleum, 2018). The demands for renewable energy sources (RES) in the United States in this circumstance and the continued exportation of agricultural commodities (especially grains: soybeans, wheat, and corn), also suggests the importance of the country's agricultural sector to its economy. Significantly, a similar position of importance is held for the US housing and real estate market, considering the association of the housing and real estate (mortgage) crisis with the GFC.

Considering the importance associated with the aforementioned sectorial markets of the United States economy, the current study investigates the (dynamic) returns and volatility spillovers of these US macroeconomic market (housing, energy, and agricultural commodities) components. For instance, the US's majorly cultivated agricultural crops or grains that were largely exported in 2017 are: wheat (worth $ 6.1 billion), corn (worth $ 9.1 billion), and soybeans (worth $ 216 billion) (USDA, 2018). The aforementioned agricultural commodities are also known RES. Hence, investigating the connectedness of the housing market, energy market (of fossil fuel and RES), and the agricultural commodity market shocks will potentially contribute a degree of novelty to the body of existing literature. And, to author's best knowledge, the described novel objective that is originally considered here is of unique note for certain reasons. First, in lieu of conventional methods of examining correlations of the dynamic markets (Kal et al., 2015; Balcilar et al., 2013, 2015; Basher and Sadorsky, 2016), the study employs the Diebold and Yilmaz (2012) for the measurement of specific spillover indexes that comprises of the gross and net spillover indexes, and directional interdependency (spillover) index. In addition to the spillover index computations, it compliments with the more efficient rolling-window analyses. Secondly, the study adopts RES component (renewable energy equity), oil prices of both West Texas Intermediate (WTI) and Brent crude, and the commodity prices of soybeans, corn, and wheat. As such, the investigation additionally provides the underpinning of the component-based shocks. Lastly, this investigation of microeconomic markets connectedness via returns and volatility spillovers also suggests potential market (s) useful for hedging and diversification opportunities in the United States.

The rest of the sections are in part. The next Section 2 contains a synopsis of the literature review. Data and methodology are presented in Section 3 while the results are discussed in Section 4. Section 5 offers concluding remarks that include policy implication of the study and proposal for future study.

2. Related literature: information transmission

2.1 The financial and commodity markets

In extant studies, the categorization of the literature includes the studies that address both identical financial assets and including inter-assets spillovers. For instance, there are studies specifically measuring the connectedness of financial markets (Diebold & Yilmaz, 2014, 2015a), stock markets (Baruník et al., 2016), and bond markets (Fernández-Rodríguez et al., 2015; Ahmad et al., 2018). Also, there are existing literature dealing with the spillovers of bond and equity or financial and bond markets (Salisu et al., 2018; Ahmad et al., 2018). In addition to financial and macroeconomic spillovers (Diebold and Yilmaz, 2015b; Barunik and Krehlik, 2016), studies have demonstrated macroeconomic connectedness with other market indicators. The macro-economic markets employed in the study of Diebold and Yilmaz (2015b) are the component assets of stock, bond, foreign exchange, and commodity with financial institutions. In the study, which adopts US and foreign countries financial institutions, the stock markets are noted for spreading the volatility shocks from the United States to other countries. However, there is gradual increase over time of the return spillover across stock markets. But during major crisis events, significant jumps are main determinant of the measures of volatility spillovers.

Similarly, Massacci (2016) demonstrated the spillover of the stock returns, macroeconomy, and global markets. The study measures tail connectedness by using the daily returns from the US size-sorted decile stock portfolios of large and small firms. And, during recessions, increase in large firms' tail risk is observed to be more than the small firms' risk. This perception of information transmission during recession could be aggravate the situation with the agricultural commodity market especially when there are unfavorable agricultural conditions and practices (Alola and Alola, 2018; Alola et al., 2019). Additionally, the degree of economic uncertainty resulting in the "boom" or "burst" in the prevailing economic activities such as in the tourism sector is potential catalyst for the spillover dynamics (Akadiri et al., 2019; Alola et al., 2019).

2.2 The energy market

Using the same time-varying but for specific commodity price shock and stock markets, Awartani and Maghyereh (2013) and Antonakakis et al. (2017) observed the dynamic spillovers connecting the shocks in the oil market and stock markets. Awartani and Maghyereh (2013) investigates the spillover effects between oil price shocks and the stock markets of the Gulf Cooperation Council countries for the period 2004−12. Using the Diebold and Yilmaz (2012) approach, the finding indicates that return and volatility spillovers are bidirectional and asymmetric in nature. In returns and volatilities, the evidence suggests that the oil market receives from other markets less than what it gives especially after the GFC in 2008.

In a similar dimension, Antonakakis et al. (2017) employs the oil supply-demand dynamics and the demand shocks of the sampled net oil-exporting and net-importing countries for the period 1995:09−2013:07. By extending the Diebold and Yilmaz (2014) approach, their investigation affirms that the returns and volatility of stock market spillovers are time-variant. Notably, the main transmitter of shocks to the stock markets is observed to be the aggregate demand shocks. But the periods of geopolitical unrest are characterized by the transmission of shocks to the stock markets by the supply related and demand shocks. Several other studies have also provided insight into the dynamics of oil prices (the energy market) in relation with the other markets (Raza et al., 2016; Mensi et al., 2017; Troster et al., 2018). For instance, Mensi et al. (2017) employed the variational mode decomposition approach with the time-varying and static symmetric and asymmetric copula functions to investigate the characteristic of the nexus of crude oil prices and the components of the major stock markets such as the S&P500, stoxx600, DJPI, and TSX indexes. By considering the bear, normal, and the bull markets in different investment horizons, the study further provided information on the upside and downside short-run and long-run spillovers between the stock and oil markets. Moreover, except for the S&P500 index, the study found tail dependence for the long-run nexus of the oil and stock markets. Similarly, Mensi et al. (2017) found a significant evidence of directional up and down risk asymmetric spillovers between the oil and stock markets in both short- and long-run horizons.

2.3 The housing market

Moreover, using a dynamic stochastic general equilibrium, Iacoviello and Neri (2010) empirically showed that the housing market spillovers are non-negligible, consumption-concentrated, and rarely a business investment, further studies have revealed the spillovers (of returns and volatilities) in the housing market (Tsai, 2015; Antonakakis and Floros, 2016; Lee and Lee, 2018). Interestingly, Iacoviello and Neri (2010) found that the housing demand, the housing technology shocks, and the monetary factors respectively explains

25%, 25%, and 20% of each volatility of the housing investment and the housing prices. However, the role of the monetary factors in the housing cycle over the past century is observed to be more substantial. Using the Diebold and Yilmaz (2012) of the generalized vector autoregression (VAR) approach, Tsai (2015) reveals that shock information is differently transmitted between the real estate and the stock markets during each of normal periods and financial crisis periods. It further shows that the housing market conveys information to the stock market during normal periods, while the financial crisis periods are characterized by the stock market net spillover of information.

Similarly, using the case of United Kingdom for the period 1997 M1–2015 M02, Antonakakis and Floros (2016) found the contagion from the housing and financial crisis to the real economy. In evidence, the study shows enormous spillover of shocks from the economic policy uncertainty, stock market, and the housing market to monetary policy stance, economic growth, and inflation. In relation to other studies, Antonakakis and Floros (2016) maintains that the observed volatility in the UK's economy are caused by shocks transmission which also varies over time. Additionally, Lee and Lee (2018) recently applied Diebold and Yilmaz (2012) to study the housing market volatility spillovers among G7 countries over the period 1970–2014. The evidence of degree of volatility interdependency over the business cycle especially with a surge during the GFC is similar to the aforementioned studies. The peculiarity of the Lee and Lee (2018) study is that it recognizes the United States and Italy as the respective major net transmitters of volatility shocks in the housing market to other countries during the GFC and the European debt crisis.

3. Data and empirical estimation

3.1 Data

We collect daily series of the agricultural commodity prices for soybeans, corn, and wheat (nonseasonally adjusted producer price index, index 1982 = 100), the US total market index of real estate investment trust (REIT, proxy for housing market and not seasonally adjusted), the renewable energy equity[1] (in the US dollars), crude oil prices (WTI and Brent crude, measured in US dollars per Barrel) over the daily period January 20, 2012 to August 2, 2018. Like the study of Tsai (2015), the stock market price index (by S&P500 and NASDAQ) are applied as control variables. While the dataset of the renewable energy equity is retrieved from the DataStream, other datasets are collected from the Federal Reserve Bank of St. Louis (FRED). After computing the level descriptive statistics of the series (see upper Table 6.1), we

---

1. Renewable energy equity is the aggregate equities of the renewable energy market from the Thomson Reuters DataStream.

TABLE 6.1 Descriptive statistics (Levels, returns, and volatility).

Variable	Mean	Maximum	Minimum	Standard dev	Skewness	Kurtosis	SSD	Observations
Wheat	5.54955	9.140000	3.260000	1.378616	0.677497	2.334214	3107.452	1636
Corn	4.43932	8.49000	2.790000	1.511770	1.136639	2.752206	3736.708	1636
Soybeans	11.42112	17.90000	7.840000	2.506784	0.617522	2.000673	10274.29	1636
Renewable	10.20077	18.20000	4.600000	2.605119	0.551461	3.641847	11096.16	1636
WTI	70.64891	110.6200	26.19000	24.08109	0.128501	1.441388	948134.6	1636
Brent	77.30312	128.1400	26.01000	28.56144	0.123797	1.404764	1333761	1636
REIT	8143.733	10343.63	5591.380	1370.860	−0.230504	1.625912	3.07E+09	1636
S&P 500	2003.387	2872.870	1278.040	402.6002	0.120974	2.294185	2.65E+08	1636
NASDAQ	4778.262	7932.240	2747.480	1316.900	0.436312	2.456764	2.84E+09	1636
Returns								
Wheat	0.092914	49.57662	−50.5939	8.688681	0.257892	7.116601	118599.8	1572
Corn	−0.06606	35.18677	−34.0258	5.765055	0.088556	12.28211	52213.53	1572
Soybeans	−0.11879	98.21836	−139.712	18.36424	−0.61013	10.10796	529812.2	1572
Renewable	0.764279	197.1507	−247.511	31.15916	0.182313	13.17848	1525273	1572
WTI	−0.28132	1412.484	−1207.30	223.8984	0.005479	5.271186	7875502	1572
Brent	−3.27479	921.7796	−1042.87	226.1803	−0.07585	4.305698	80368473	1572

REIT	951.6780	99400.06	−158265	29543.71	−0.47595	5.054160	1.37E+12	1572
S&P 500	321.1397	24071.83	−38745.8	5181.515	−0.75249	8.481974	4.22E+10	1572
NASDAQ	1148.415	87925.23	−105285	16698.09	−0.66559	7.626624	4.38E+11	1572
Volatility							
Wheat	73.71612	411.3041	13.62085	63.53579	2.511068	10.52967	6870627	1703
Corn	33.52751	389.3127	1.884149	53.66813	3.217958	15.66618	4902217	1703
Soybeans	342.4702	3860.470	32.98029	423.1509	3.961207	23.36834	3.05E+08	1703
Renewable	905.9236	3525.085	385.2254	547.0060	2.099559	7.334029	5.09E+08	1703
WTI	47247.05	145116.3	11920.52	22383.11	0.892338	3.556559	8.53E+11	1703
Brent	47632.13	126947.0	18402.53	19991.21	1.368626	4.992017	4.05E+20	1703
REIT	8.31E+08	2.97E+09	2.40E+08	4.88E+08	1.230504	1.625912	3.07E+09	1703
S&P 500	24585912	3.96E+08	5191366	27732579	6.042913	58.67918	1.31E+18	1703
NASDAQ	2.50E+08	3.01E+09	3881292	2.41E+08	4.414878	32.50099	9.86E+19	1703

A lag selection by SIC and AIC (maxlag = 1). SSD is the sum ofsquared deviation. Originally, the number of observations including missing dates is 1706.

compute the first difference of the returns of the series (r_t) (essentially to render the series stationary) of the natural logarithmic values of the series (S_t) for time t, such that

$$r_t = 100 * [\text{first difference } (\Delta) \text{of } \log(S_t)] \tag{6.1}$$

Again, we compute the volatility of the series from the return estimations from Eq. (6.1). In doing so, the series returns is regressed on the lag value of the series consecutively. Hence, the volatility series is obtained from EGARCH (p, q) where p and q are the lag values such that the common statistics of the returns and volatility series are also present in Table 6.1 (middle and bottom, respectively).

3.2 Empirical methodology

The current study employs the unique technique of Diebold and Yilmaz (2012) which is based on the generalized VAR model and insensitivity to variable ordering. In this approach, the total, directional, net, and net pairwise spillovers are the categories of spillover obtainable. Hence, a covariance stationary VAR (p) is considered

$$y_t = \sum_{i=1}^{p} \Phi_i y_{t-1} + \varepsilon_t \sim \left(0, \sum\right) \tag{6.2}$$

such that the moving average of the covariance stationary process of Eq. (6.2) above is

$$y_t = \sum_{i=0}^{\infty} A_i \varepsilon_{t-1} \tag{6.3}$$

where $y_t = (y_{1t}, y_{2t}, \ldots, y_{Nt})'$ is $N \times 1$ vector of the individual return and volatility series (vector endogenous variables), given that Φ is $N \times N$, ε is the vector of disturbance that are assumed to be independent (not necessarily identically) distributed over time, A (of Eq. 6.2) is assumed to follow the recursion $A_i = \Phi_1 \cdot A_{i-1} + \Phi_2 \cdot A_{i-2} + \ldots + \Phi_p \cdot A_{i-p}$, A_0 is the identity matrix (of $N \times N$ dimension), and $A_i = 0$ for all $i < 0$.

In the process of assessing the magnitude of the market spillovers as our priority instead of determining the causal effects of structural shocks, we adopt the conventional VAR framework such that the H-step-ahead forecast error variance contribution becomes

$$\theta_{ij}^g(H) = \frac{\sigma_{jj}^{-1} \sum\limits_{h=0}^{H-1} \left(e_i' A_h \sum e_j\right)^2}{\sum\limits_{h=0}^{H-1} \left(e_i' A_h \sum A_h e_i'\right)^2} \tag{6.4}$$

So that the variance matrix of the error vector is Σ, σ_{jj} is the standard deviation of the error term for variable j, e_i is the selection vector with $1 = ith$

element and $0 =$ otherwise. Then, the diagonally centralized elements (the own variance shares of shocks to variable y_i) is the fraction of the H-step-ahead error variance in forecasting y_i, given that $i = 1, 2, \ldots N$. Also, the off-diagonal (cross-variance shares or spillovers) are the fractions of the H-step-ahead error variances in forecasting y_i that are due to shocks to y_j, given that $j = 1, 2, \ldots N$ and i is not equal j. Furthermore, to use the full information, each entry of the variance decomposition matrix is normalized by taking the row sum such that

$$\theta_{ij}^g(H) = \frac{\theta_{ij}^g(H)}{\sum\limits_{j=1}^{N} \theta_{ij}^g(H)} \tag{6.5}$$

where $\sum\limits_{j=1}^{N} \theta_{ij}^g(H)$ (sum of the contributions to the variance of the forecast error) is not equal to 1, but $\sum\limits_{j=1}^{N} \widetilde{\theta}_{ij}^g(H) = 1$ and $\sum\limits_{i,j=1}^{N} \widetilde{\theta}_{ij}^g(H) = N$ by the construction.

In respect to the aforementioned estimations steps, the total spillover index (is directional spillover that specifically quantifies the contribution of spillovers (return and volatility shocks) among the examined commodity markets is provided as

$$S^g(H) = \frac{\sum\limits_{\substack{i,j=1 \\ i \neq j}}^{N} \widetilde{\theta}_{ij}^g(H)}{\sum\limits_{i,j=1}^{N} \widetilde{\theta}_{ij}^g(H)} \times 100 = \frac{\sum\limits_{\substack{i,j=1 \\ i \neq j}}^{N} \widetilde{\theta}_{ij}^g(H)}{N} \times 100 \tag{6.6}$$

Also, the total directional spillover exhibits two indicators: "to others" and "from others." While the directional spillover index from others is computed as

$$S_i^g(H) = \frac{\sum\limits_{\substack{J=1 \\ j \neq 1}}^{N} \widetilde{\theta}_{ij}^g(H)}{\sum\limits_{i,J=1}^{N} \widetilde{\theta}_{ij}^g(H)} \times 100 = \frac{\sum\limits_{\substack{J=1 \\ j \neq 1}}^{N} \widetilde{\theta}_{ij}^g(H)}{N} \times 100, \tag{6.7}$$

the directional spillover index to others is calculated as

$$S_i^g(H) = \frac{\sum\limits_{\substack{J=1 \\ j \neq 1}}^{N} \widetilde{\theta}_{ji}^g(H)}{\sum\limits_{i,J=1}^{N} \widetilde{\theta}_{ji}^g(H)} \times 100 = \frac{\sum\limits_{\substack{J=1 \\ j \neq 1}}^{N} \widetilde{\theta}_{ji}^g(H)}{N} \times 100 \tag{6.8}$$

Moreover, the difference between the "to others" and "from others" indicators is calculated using

$$S_i^g(H) = S_{.i}^g(H) - S_{i.}^g(H) \qquad (6.9)$$

So that, the net pairwise directional spillovers is also computed from

$$S_{ij}^g(H) = \left\{ \frac{\widetilde{\theta}_{ji}^g(H)}{\sum\limits_{i,k=1}^{N} \widetilde{\theta}_{ik}^g(H)} - \frac{\widetilde{\theta}_{ij}^g(H)}{\sum\limits_{j,k=1}^{N} \widetilde{\theta}_{jk}^g(H)} \right\} \times 100 = \left\{ \frac{\widetilde{\theta}_{ji}^g(H) - \widetilde{\theta}_{ij}^g(H)}{N} \right\} \times 100$$

$$(6.10)$$

Considering that Salisu et al. (2018) adopted the Diebold and Yilmaz (2012) to investigate the returns and volatility spillovers of the six global foreign exchange markets while several studies have also considered inter-market spillovers (Antonakakis and Floros, 2016; Lee and Lee, 2018), hence we advance the concept of aforesaid studies. In the current study, we measured the total spillover index, the contributions of spillovers of return and that of volatility shocks to the total forecast error variance. The study employs nine components of the US markets (stock, energy, housing, and agricultural commodity). As revealed, Tables 6.2A—6.2C are the categories (level, returns, and volatility series) of spillover indices.

4. Empirical results and discussion

In Table 6.1, the common statistics of the series for before computations, for returns, and volatility estimates are presented. The return estimates as contained in the middle section of Table 6.1 is for the entire period. In this part, the average returns of each series over the estimated period is the mean. Positive average returns are observed for *wheat, renewables, REIT, S&P500,* and *NASDAQ*, while *corn, soybeans, WTI,* and *crude Brent* exhibits negative average returns. The dynamics of the returns of all the series are displayed as Fig. 6.1 (for a—i). Form the observation from Table 6.1 as further illustrated in the aforesaid figures, the returns of all the series are skewed except the soybeans which apparently have a symmetric mean (skewness of −0.61013) compared to the mean of −0.11879. Also, the kurtosis statistics implies that the returns of the series are at least peaked (leptokurtic).

Again, the lower part of Table 6.1 depicts the descriptive statistics for the volatility series of all variables under the whole sample period. Giving Fig. 6.2 (of 1—9), there is significant evidence showing time events of high volatility are immediately followed by time events of relatively low volatility. As observed from the volatility statistics, all the significantly deviates from the mean, positively skewed, and the evidence is corroborated by the volatility figures.

TABLE 6.2A The directional spillover results at level.

Variable	1	2	3	4	5	6	7	8	9	Contribution from others	Net spillover
1	99.0	0.2	0.2	0.0	0.2	0.0	0.2	0.1	0.1	1.0	42.7
2	28.0	71.6	0.2	0.1	0.0	0.0	0.0	0.1	0.0	28.4	−5.4
3	14.6	19.8	65.3	0.1	0.2	0.1	0.0	0.0	0.0	34.7	−26.2
4	0.1	0.1	0.1	98.3	0.5	0.9	0.0	0.0	0.0	1.7	26.3
5	0.2	1.5	3.2	6.6	87.9	0.5	0.1	0.0	0.1	12.1	50.4
6	0.6	1.4	2.0	3.3	52.6	39.7	0.2	0.1	0.0	60.3	−58
7	0.0	0.1	0.8	3.1	0.7	0.1	94.9	0.1	0.2	5.1	46.7
8	0.2	0.1	1.3	7.8	5.7	0.3	28.9	55.5	0.3	44.5	9.2
9	0.0	0.0	0.8	6.8	2.7	0.5	22.3	53.2	13.7	86.3	−85.5
Contribution to others	43.7	23.0	8.5	28.0	62.5	2.3	51.8	53.7	0.8	274.2	
Contribution including own	142.7	94.6	75.6	126.2	150.3	42.1	146.7	109.1	14.5	Total[30.5%]	

Wheat = 1, Corn = 2, Soybeans = 3, renewable = 4, WTI = 5, Brent = 6, REIT = 7, S&P 500 = 8, NASDAQ = 9 Lag length by AIC selection = 2.

TABLE 6.2B The directional returns spillover.

Variable	1	2	3	4	5	6	7	8	9	Contribution from others	Net spillover
1	**99.4**	0.0	0.3	0.0	0.2	0.1	0.0	0.0	0.0	0.6	43.9
2	29.5	**70.2**	0.0	0.0	0.1	0.0	0.0	0.0	0.0	29.8	−13.1
3	13.6	15.0	**71.0**	0.1	0.1	0.0	0.0	0.0	0.0	29.0	−24.4
4	0.0	0.3	0.0	**99.5**	0.1	0.0	0.0	0.0	0.0	0.5	17.3
5	0.5	0.7	1.5	2.5	**94.7**	0.0	0.0	0.0	0.0	5.3	39.8
6	0.2	0.6	1.0	1.7	36.6	**59.6**	0.0	0.2	0.0	40.4	−40
7	0.3	0.0	0.3	2.6	0.8	0.1	**95.7**	0.2	0.1	4.3	40.9
8	0.2	0.0	0.9	5.9	4.9	0.2	26.2	**61.7**	0.1	38.3	22.9
9	0.1	0.0	0.6	5.0	2.3	0.1	18.9	60.8	**12.2**	87.8	−87.3
Contribution to others	44.5	16.7	4.6	17.8	45.1	0.4	45.2	61.2	0.5	236.0	
Contribution including own	143.9	86.9	75.6	117.3	139.8	60.0	140.9	122.9	12.6	**Total(26.2%)**	

Wheat = 1, Corn = 2, Soybeans = 3, renewable = 4, WTI = 5, Brent = 6, REIT = 7, S&P 500 = 8, NASDAQ = 9 Lag length by AIC selection = 1.

TABLE 6.2C The directional volatility spillover.

Variable	1	2	3	4	5	6	7	8	9	Contribution from others	Net spillover
1	**97.6**	0.6	0.0	0.1	0.1	0.8	0.4	0.1	0.3	2.4	26.6
2	18.3	**78.6**	0.8	0.1	0.7	1.0	0.3	0.0	0.1	21.4	9.3
3	7.5	27.8	**62.3**	0.0	1.3	0.4	0.3	0.1	0.2	37.7	−36
4	0.1	0.4	0.0	**98.8**	0.0	0.0	0.5	0.2	0.1	1.2	0.9
5	0.3	1.0	0.0	0.0	**97.6**	0.2	0.0	0.8	0.0	2.4	24.4
6	2.1	0.8	0.7	0.0	19.8	**76.0**	0.3	0.2	0.1	24.0	−21.4
7	0.6	0.0	0.0	0.4	2.2	0.1	**94.7**	1.9	0.1	5.3	32.1
8	0.0	0.0	0.0	0.8	1.7	0.1	19.9	**77.4**	0.0	22.6	52.1
9	0.0	0.1	0.0	0.7	1.0	0.0	15.6	71.4	**11.2**	88.8	−87.8
Contribution to others	29.0	30.7	1.7	2.1	26.8	2.6	37.4	74.7	1.0	205.9	
Contribution to others	126.6	109.3	63.9	100.8	124.3	78.6	132.1	152.1	12.2	**Total(22.9%)**	

Wheat = 1, Corn = 2, Soybeans = 3, renewable = 4, WTI = 5, Brent = 6, REIT = 7, S&P 500 = 8, NASDAQ = 9 Lag length selection by SIC (lag = 1).

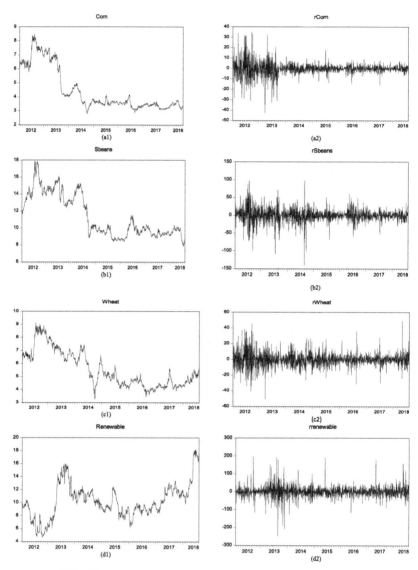

FIGURE 6.1 Plots (a—i) for series (left) and returns (right) of series.

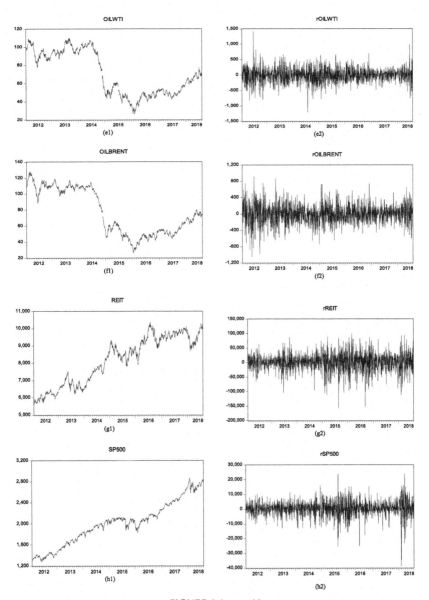

FIGURE 6.1 cont'd

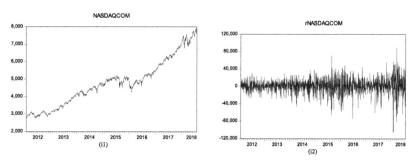

FIGURE 6.1 cont'd

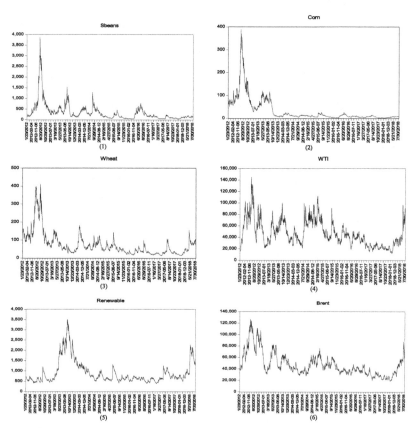

FIGURE 6.2 Plots showing volatility clustering (1−9) of the return volatility series.

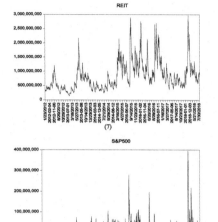

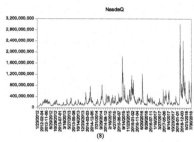

FIGURE 6.2 cont'd

4.1 Spillover indices

In estimating the total connectedness (spillover index) for the whole sample of the natural form of the series, the returns, and the volatility series, Eq. (6.6) (see above) is employed. The model selection is based on the Akaike information criterion (AIC) and Schwarz information criterion (SIC) of the VAR (1) model with 10-step-ahead forecasts horizon such that directional spillover indices are obtained as shown in Table 6.2A. In the indicated Table 6.2A, the connectedness of the natural series is 30% with high own-shares exhibited by wheat, corn, renewable, WTI, and REIT. A lower own-share is exhibited by Brent, NASDAQ, and moderately by S&P500.

Importantly, the total returns spillover over the whole period is 26.2% (see Table 6.2B; this implies that 73.8% of the variation is due to specific (idiosyncratic) shocks. Although it implies that return spillover among the examined markets (housing, energy, agricultural commodities, and stock) is rather low, large own contribution measures of approximately over 95% is observed for REIT, WTI, renewables, and wheat. Regarding directional spillover (expressed by Eq. 6.9 as "to others"— "from others"), wheat has the highest (positive) net return spillover (43.9%) to others and orderly followed by REIT, WTI, S&P500, and renewable. It implies that the variables (which are components of the examined markets) are net transmitters of returns. Similarly, negative return spillovers (net recipients) are experienced by NASDAQ, Brent, soybeans, and corn in highest-lowest order. This is obviously not without considering the individual directional from others. For instance, the S&P500 record the highest contribution (spillover) to the forecast error variance of the

NASDAQ returns with about 60.8%. This is followed by the spillover of WTI to Brent with 36.6%, wheat to corn with 29.5%, REIT to S&P500 and NASDAQ with 26.2% and 18.9%, respectively, corn to soybeans with 15.0%, wheat to soybeans with 13.6%. Also, corn (only agricultural commodity) record forecast error variance to renewable with 0.3%, but the commodities except corn record forecast error variance (although lower) to REIT, S&P500, and NASDAQ.

In a similar case, Table 6.2C presents the volatility spillovers over the entire sample period. The total volatility spillover over the estimated period is 22.9% so that 77.1% of the variation is due to idiosyncratic shocks. These values slightly differ from the returns spillovers earlier reported. And, the (net) directional risk spillovers "from" and "to" other market components are quite high and above the average of the directional return spillovers. It implies that lower magnitude of spillovers due to return may not translate to a lower magnitude of spillovers due to volatility (risk). The directional volatility spillover (expressed by Eq. 6.9 as "to others"—"from others") is highest in S&P500 (52.1%), and orderly followed by REIT (32.1%), wheat (26.1%), WTI (24.4%), corn (9.3%), and renewable (0.9%). And, the negative volatility spillover (net recipients) are experienced by NASDAQ (-87.8%), soybeans (-36%), and Brent (-21.4%). However, on the basis of components' risk spillovers, the NASDAQ seems to be more vulnerable to risk shocks of other markets (and component of other markets). This trend is empirically followed by soybeans (37.7%), Brent (24.0%), S&P500 (22.6%), corn (21.4%), REIT (5.3%), WTI (2.4%), wheat (2.4%), and renewable (1.2%). For the pairwise directional spillover, higher degree of net spillover is observed among the agricultural commodities. In overall, the net spillover of the intermarket components are low (ranging from 0.0% to 0.8%). Importantly, although the total returns and volatility spillover indices are lower than 50% average, there exists some significant level of interdependence among the examined components. Also, giving the small value of volatility spillover index, it hints that the return volatility for the market components is determined by exogenous factors that not examined in the VAR model.

4.2 Robustness tests

Considering the return and volatility spillover indices estimates presented earlier (in Tables 6.2B and 6.2C), the rolling-sample analysis with 200-week (and 100) windows and 10 steps horizons are subsequently provided as Fig. 6.3. In Fig. 6.3, it presents the 200-week windows (a) and 100-week windows (b) for return spillover index. Although the dynamic outlook of the two windows present the same movement, it peaks at about 50 in 2012 and downturn between 2017 and 2018 in 100-week windows for return spillover slightly differs from the 200-week windows. It implies that when more time is allowed, the return response would adjust as to avoid distress in the markets.

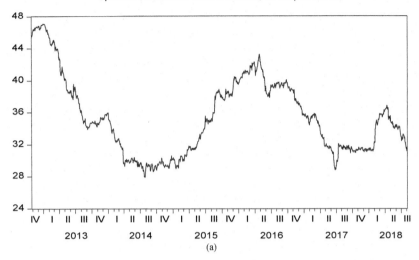

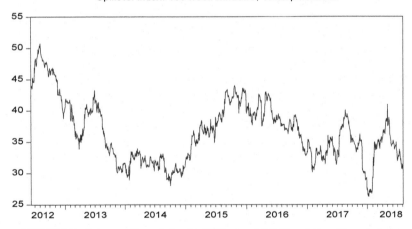

FIGURE 6.3 Returns spillover index for 200 (up) and 100 (down) for rolling window.

Moreover, the return volatility spillover index for 200-week windows (see Fig. 6.4) depict the existence of high volatility (period of high frequency immediately followed by period of low frequency) among the examined markets

Also, a robustness investigation with a resampled data (in-sample) for the period January 01, 2017—August 2, 2018 is employed to at least evaluate a new return spillover index as illustrated in Fig. 6.5 (Tsai, 2015). The sample period is considered because of the obvious policy shifts in the United States since the commencement of a new government on January 20, 2017. As evidently shown in Table 6.3, less of the variation is due to idiosyncratic shock

FIGURE 6.4   Return volatility spillover index for 200-week windows.

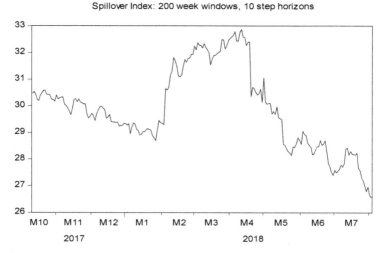

FIGURE 6.5   Returns spillover index for 100 for rolling window (An in-sample estimate).

in comparison with the return in the whole sample. Here, the total volatility spillover over the estimated resampled period is 30.5%. This implies that, although volatility spillover exists in the previous sample, the impact is explained in lower magnitude because of higher time lag.

Also, the net contribution to others by REIT which is higher in the whole sample estimate is intensely higher (72.4%). In this sense, the finding corroborates the evidence that the housing market (in the United Kingdom)

TABLE 6.3 The directional returns spillover (In-sample)

Variable	1	2	3	4	5	6	7	8	9	Contribution from others	Net spillover
1	**99.6**	0.0	0.0	0.0	0.1	0.0	0.2	0.0	0.0	0.4	53.4
2	33.9	**65.4**	0.1	0.0	0.4	0.0	0.0	0.2	0.0	34.6	−16.5
3	17.0	16.0	**66.2**	0.0	0.0	0.0	0.3	0.3	0.2	33.8	−30.4
4	0.2	0.2	0.2	**94.1**	3.1	0.5	0.4	0.0	1.3	5.9	10.4
5	0.8	0.8	0.7	2.0	**94.9**	0.0	0.7	0.0	0.0	5.1	35.6
6	0.3	0.5	0.1	1.9	33.4	**62.5**	0.1	1.0	0.2	37.5	−36.7
7	1.2	0.4	0.7	1.8	1.2	0.1	**94.2**	1.4	0.1	5.8	24.1
8	0.2	0.0	0.7	3.1	1.8	0.0	15.7	**78.3**	0.0	21.7	50.9
9	0.2	0.2	0.9	2.4	0.7	0.2	12.4	70.7	**12.4**	87.6	−85.8
Contribution to others	53.8	18.1	3.4	11.3	40.7	63.4	29.9	72.6	1.8	232.4	
Contribution to others	153.4	83.5	69.6	105.4	135.6	63.4	124.1	150.9	14.1	**Total(25.9%)**	

Wheat = 1, Corn = 2, Soybeans = 3, renewable = 4, WTI = 5, Brent = 6, REIT = 7, S&P 500 = 8, NASDAQ = 9 Lag length selection by SIC (lag = 1).

significantly determines the economic activity (Antonakakis and Floros, 2016). This is followed by wheat (39.6%), S&P500 (23.8%), and renewable (0.5%). Other variables, led by NASDAQ (followed by soybeans, corn, Brent, and WTI) are net receiver of volatility spillovers. Importantly, the pairwise volatility spillover among the market components is more significant in this estimate. For instance, the values above and below the diagonal indices are more significant, especially the interdependence of the agricultural commodities and renewable with the components of other markets.

5. Conclusions and policy direction

5.1 Conclusion

This investigation measures the magnitude of interdependence among the housing market, energy market, and agricultural commodities in the United States using the common components of the markets. Having employed the novelty of Diebold and Yilmaz (2012) over the daily period January 20, 2012—August 2, 2018, the findings revealed are characterized by the following empirical regularities. First, there is transmission of various types of shock among the housing market, energy market, stock market, and the agricultural commodities. Although slightly low, the intercomponent market transmission (especially the volatility connectedness) is significant. The estimate of the total returns and volatility spillover is significant evidence. Second, over the whole sample period, the total volatility spillover is lower than the return spillover. Hence, it translates that higher return spillover would not naturally indicate higher volatility spillover. Nevertheless, the spillovers show large deviation over period of time which implies a period of high information spillovers. As such, the period is characterized by increased correlation between the examined markets vis-à-vis the market components (Tsai, 2015). Third, some components of the market are observed to exhibit different contribution pattern. For instance, among the energy components, Brent receives more returns and volatility contributions more than it gives while renewable and WTI gives more than it receives. This could be because the country's economy is energy-driven which is largely of WTI and renewable-related and less of Brent. Lastly, using a smaller sample-size (January 01, 2012—August 2, 2018), we found that this relatively higher systemic risk currently existing among the examined markets.

5.2 Policy direction

The policymakers are expected to find the results of the current investigation very useful especially in the design of market framework of the United States. Importantly, part is the results from the robustness check which reveals higher total net volatility spillover where sample-size is limited to the time period of the current government of the United States. For instance, the relatively high net volatility contribution (spillover effect) from REIT to other markets should be a reminder of events preceding the GFC and its association with the United States' subprime mortgage market. Hence, real estate and the housing market boom should be prevented in a precautionary approach. This could be done by increasing the resilience of other markets possibly through economic diversification programs.

Further study similar to the current one is essential to examining the relevancy of connectedness among market components especially for regional perspectives.

References

Ahmad, W., Mishra, A.V., Daly, K.J., 2018. Financial connectedness of BRICS and global sovereign bond markets. Emerg. Mark. Rev. 37.

Akadiri, S.S., Alola, A.A., Uzuner, G., 2019. Economic policy uncertainty and tourism: evidence from the heterogeneous panel. Curr. Issues Tourism 1−8.

Alola, A.A., Alola, U.V., 2018. Agricultural land usage and tourism impact on renewable energy consumption among Coastline Mediterranean Countries. Energy Environ. 29 (8), 1438−1454.

Alola, A.A., Yalçiner, K., Alola, U.V., 2019. Renewables, food (in) security, and inflation regimes in the coastline Mediterranean countries (CMCs): the environmental pros and cons. Environ. Sci. Pollut. Control Ser. 1−11.

Alola, U.V., Cop, S., Adewale Alola, A., 2019. The spillover effects of tourism receipts, political risk, real exchange rate, and trade indicators in Turkey. Int. J. Tourism Res. 21.

Antonakakis, N., Chatziantoniou, I., Filis, G., 2017. Oil shocks and stock markets: dynamic connectedness under the prism of recent geopolitical and economic unrest. Int. Rev. Financ. Anal. 50, 1−26.

Antonakakis, N., Floros, C., 2016. Dynamic interdependencies among the housing market, stock market, policy uncertainty and the macroeconomy in the United Kingdom. Int. Rev. Financ. Anal. 44, 111−122.

Awartani, B., Maghyereh, A.I., 2013. Dynamic spillovers between oil and stock markets in the Gulf Cooperation Council countries. Energy Econ. 36, 28−42.

Balcilar, M., Demirer, R., Hammoudeh, S., 2013. Investor herds and regime-switching: evidence from Gulf Arab stock markets. J. Int. Financ. Mark. Inst. Money 23, 295−321.

Balcilar, M., Hammoudeh, S., Asaba, N.A.F., 2015. A regime-dependent assessment of the information transmission dynamics between oil prices, precious metal prices and exchange rates. Int. Rev. Econ. Finance 40, 72−89.

Barunik, J., Krehlik, T., 2016. Measuring the Frequency Dynamics of Financial and Macroeconomic Connectedness (No. 54). FinMaP-Working Paper.

Baruník, J., Kočenda, E., Vácha, L., 2016. Asymmetric connectedness on the US stock market: bad and good volatility spillovers. J. Financ. Mark. 27, 55−78.

Basher, S.A., Sadorsky, P., 2016. Hedging emerging market stock prices with oil, gold, VIX, and bonds: a comparison between DCC, ADCC and GO-GARCH. Energy Econ. 54, 235−247.

British Petroleum BP, 2018. BP Statistical Review of World Energy. https://www.bp.com/en/global/corporate/energy-economics/statistical-review-of-world-energy/downloads.html. (Accessed 15 October 2018).

Central Intelligence Agency CIA, 2018. https://www.cia.gov/library/publications/the-world-fact-book/geos/us.html. (Accessed 15 October 2018).

Diebold, F.X., Yilmaz, K., 2012. Better to give than to receive: predictive directional measurement of volatility spillovers. Int. J. Forecast. 28 (1), 57–66.

Diebold, F.X., Yılmaz, K., 2014. On the network topology of variance decompositions: measuring the connectedness of financial firms. J. Econom. 182 (1), 119–134.

Diebold, F.X., Yilmaz, K., 2015a. Trans-Atlantic equity volatility connectedness: US and European financial institutions, 2004–2014. J. Financ. Econom. 14 (1), 81–127.

Diebold, F.X., Yilmaz, K., 2015b. Financial and Macroeconomic Connectedness: A Network Approach to Measurement and Monitoring. Oxford University Press, USA.

Fernández-Rodríguez, F., Gómez-Puig, M., Sosvilla-Rivero, S., 2015. Volatility spillovers in EMU sovereign bond markets. Int. Rev. Econ. Finance 39, 337–352.

Iacoviello, M., Neri, S., 2010. Housing market spillovers: evidence from an estimated DSGE model. Am. Econ. J. Macroecon. 2 (2), 125–164.

Kal, S.H., Arslaner, F., Arslaner, N., 2015. The dynamic relationship between stock, bond and foreign exchange markets. Econ. Syst. 39 (4), 592–607.

Lee, H.S., Lee, W.S., 2018. Housing market volatility connectedness among G7 countries. Appl. Econ. Lett. 25 (3), 146–151.

Massacci, D., 2016. Tail risk dynamics in stock returns: links to the macroeconomy and global markets connectedness. Manage. Sci. 63 (9), 3072–3089.

Mensi, W., Hammoudeh, S., Shahzad, S.J.H., Shahbaz, M., 2017. Modeling systemic risk and dependence structure between oil and stock markets using a variational mode decomposition-based copula method. J. Bank. Finance 75, 258–279.

Raza, N., Shahzad, S.J.H., Tiwari, A.K., Shahbaz, M., 2016. Asymmetric impact of gold, oil prices and their volatilities on stock prices of emerging markets. Resour. Pol. 49, 290–301.

Salisu, A.A., Oyewole, O.J., Fasanya, I.O., 2018. Modelling return and volatility spillovers in global foreign exchange markets. J. Inf. Optim. Sci. 1–32.

Steeley, J.M., 2006. Volatility transmission between stock and bond markets. J. Int. Financ. Mark. Inst. Money 16 (1), 71–86.

Tsai, I.C., 2015. Dynamic information transfer in the United States housing and stock markets. N. Am. J. Econ. Finance 34, 215–230.

Troster, V., Shahbaz, M., Uddin, G.S., 2018. Renewable energy, oil prices, and economic activity: a Granger-causality in quantiles analysis. Energy Econ. 70, 440–452.

United States Department of Agriculture USDA, 2018. Foreign Agricultural Service. https://www.fas.usda.gov/data/top-us-agricultural-exports-2017. (Accessed 15 October 2018).

Further reading

Federal Reserve Bank of St. Louis, 2018. https://fred.stlouisfed.org/. (Accessed 25 September 2018).

Chapter 7

The effect of globalization on energy consumption: evidence from selected OECD countries

Burcu Ozcan[1], Ali Gokhan Yucel[2], Mehmet Temiz[1]

[1]*Faculty of Economics and Administrative Sciences, Department of Economics, Firat University, Elazig, Turkey;* [2]*Faculty of Economics and Administrative Sciences, Department of Economics, Erciyes University, Kayseri, Turkey*

1. Introduction

There are still ongoing discussions about the historical origins of the globalization era. For instance, the Industrial Revolution in the 18th century or the European colonialism in the late 15th century when Columbus discovered America was proposed as the breeding ground of globalization (Rennen and Martens, 2003). No matter how far it dates back, since the second half of the 20th century (1980s), the world has witnessed an increased breeze of globalization wind. In this respect, globalization became an inevitable process of the 20th century as a result of strong cultural, social, economic, and political interconnections across countries. We are aware of what is going on in the world and the globe via the information and communications technologies, which are the products of globalization, i.e., we are living in global villages instead of isolated villages. As stated by Shahbaz et al. (2016), globalization has opened up economies and supported their economic growth periods through the expansion of trade, investment activities, and technological inflows. However, in contrast to discussions about the origin of globalization, the literary discussion on the nexus of globalization and energy consumption was initiated with the birth of industrialization (Ahmed et al., 2016). Particularly, with the involvement of developing countries in the industrialization and urbanization processes, globalization has resulted in massive energy consumption in these countries (Shahbaz et al., 2019).

In this chapter, we focus on the effects of globalization on energy consumption based on some theoretical and empirical analyses. In this context, the present literature suggests two opposite views highlighting the effects of globalization on the energy sector. The first one argues that globalization can

Energy-Growth Nexus in an Era of Globalization. https://doi.org/10.1016/B978-0-12-824440-1.00015-1

173

reduce demand for energy via transfers of energy-efficient (saving) technologies and knowledge. In this sense, technology transfer through globalization changes knowledge on how to utilize energy in more efficient ways to reduce energy consumption, particularly the conventional energy consumption (Saud et al., 2018). The second view states that globalization may increase energy consumption in the host country if the country mainly aims at attracting more foreign firms to set up new businesses as a primary focus of globalization strategy (Shahbaz et al., 2016). At that time, the country allows foreign firms with dirty and obsolete technologies to enter the country, which, in turn, raised energy consumption and environmental pollution.

The remaining part of this chapter is in the following order: In Section 1.1, globalization concept with its dimensions is explained; In Section 1.2, the KOF Globalization Index is introduced; In Section 1.3, the effects of economic globalization on energy consumption are discussed (under the subtitles Section 1.3.1 "Effects of Trade Openness on Energy Consumption" and Section 1.3.2 "The Effects of FDI Flows on Energy Consumption"); In Section 2, a brief literature review is provided. A comprehensive econometric methodology is presented in Section 3. Section 4 introduces the data and reports the empirical findings. Finally, the study is concluded with important policy recommendations.

1.1 Concept of globalization and its dimensions

Globalization refers to a paradigm shift to a more connected world economy, broadly shaping economies and societies around the world (Etokakpan et al., 2020). It indicates a world order consisting of interconnected countries via trade and exchange of goods and services, capital flows, financial integration, technology transfer, and information spreading (Sabir and Gorus, 2019). In this sense, a profound and contemporary definition of globalization is stated by Rennen and Martens (2003, p. 143) as:

> We define contemporary globalisation as an intensification of cross-national cultural, economic, political, social and technological interactions that lead to the establishment of transnational structures and the global integration of cultural, economic, environmental, political and social processes on global, supranational, national, regional and local levels.

From an economic perspective, globalization is a move toward ensuring the economic, political, and sociocultural integration of the world by reducing the barriers to the exchange and increase of international capital and labor flows (Iheanacho, 2018). As such, the social, cultural, political, and economic interconnections have increased among countries as a result of globalization. Globalization is believed to have transformed the world from many fragmentary systems into a single system while connecting countries with the exchange of information and cross-border maintenance of produced technology (Azimova, 2019). Therefore, it is one of the building blocks of economic growth and welfare as it removes the cross-border restrictions on trade and capital flows across countries. It opens the borders for an economy to increase its trade and investment volumes and expand its economic activities (Danish et al., 2018). Besides, globalization encourages technical innovation, improves living and environmental conditions, and boosts total productivity as a result of increased economies of scale (Etokakpan et al., 2020). Moreover, globalization also improves the financial system by raising the functionality of financial markets through foreign direct investment (FDI) inflows. FDI increases investment opportunities leading to more business turnover and competition in the financial markets, which, in turn, enriches the operational efficiency of financial markets (Shahbaz et al., 2016).

Regarding the indicators of globalization, researchers have generally used a single indicator such as exports, imports, trade openness, FDI, etc. However, globalization is a multifaceted concept indicating much more than openness to trade and capital, and therefore it cannot be considered merely an economic, political, or technological process (Rennen and Martens, 2003). Besides its economic aspect, as stated by Gygli et al. (2019), globalization also includes citizens of different countries communicating with each other and exchanging their ideas and information as well as governments working together to struggle against global political issues. In particular, three concepts, namely economic globalization, social globalization, and political globalization, stand out as the dimensions of globalization. Of them, economic globalization is characterized by the long-distance flows of goods, capital, and services as well as information and apprehensions that accompany market exchanges; political globalization is characterized by the diffusion of government policies; and social globalization is expressed as the spread of ideas, information, images, and people (Dreher et al., 2010; Gygli et al., 2019). Social globalization builds a bridge between people through international mobility, personal contacts, and exposure to global media (Ahmed et al., 2019). The increased access to foreign media, the internet, and other sources have led people to more perceive environmental issues. As stated by Sabir and Gorus (2019), social globalization, by spreading information about best business norms and practices, helps to promote environmentally friendly energy sources in economic activities and improves environmental sustainability.

Besides its economic and social dimensions, globalization also has a political dimension since governments may support or restrict the private entrepreneurship with laws and incentives, e.g., the laissez-faire and free trade politics were mainly encouraged by the political and economic power of Great Britain by the mid-19th century (Rennen and Martens, 2003). Regarding the energy consumption and environmental quality, political globalization is of importance because it leads countries to participate in global environmental agreements and summits such as the Kyoto Protocol in 1997, the Paris agreement in 2015, etc.[1] All those environmental agreements and meetings have international standards for providing environmental safety and reducing fossil energy consumption. Paramati et al. (2016, 2017) suggest different ways by which political cooperation among countries can help the reduction of CO_2 and other greenhouse gases emissions: (1) Political globalization aims at alleviating both the national and global CO_2 emissions by providing strong politic cooperation across countries; (2) it supports energy production from clean energy sources, e.g., solar and wind plants rather than coal-based power plants, by providing a new infrastructure; (3) it deploys technology transfer and know-how in clean energy production and consumption across countries; and (4) it provides some technical and financial assistance to reduce CO_2 emissions and builds strong cooperation between developed and developing countries. Based on these explanations, it could be stated that globalization is a multidimensional concept and completely changes and transforms all living spaces of human beings.

1.2 The KOF globalization index

Under this subsection, the KOF Globalization Index is briefly explained because it is utilized as a proxy for globalization in the empirical section of the chapter. In the course of time, with the aim of combining all single indicators of globalization, Dreher (2006) first introduced the KOF Globalization Index in 2002 at the KOF Swiss Economic Institute. This index combines different aspects of globalization under the same umbrella and thereby represents globalization in a broad way. The KOF index divides globalization into three dimensions explained before, namely economic globalization, political globalization, and social globalization. Dreher et al. (2008) and Dreher (2006) measure the degree of economic globalization by two subindices. The first index measures the actual economic flows, i.e., trade, FDI, and portfolio investment (all in terms of percentage of GDP), while the second index measures restrictions on trade and capital through hidden import barriers, tariff rates, taxes on international trade, and an index of capital controls. Given a certain level of trade, higher revenues from taxes or tariffs indicate that country is less globalized (Dreher et al., 2010). All data on actual flows and restrictions are combined into two subindices that make up one overall economic globalization index. The degree of political globalization is determined by the number of embassies in a country, the number of international organizations of which the country is a member and the number of United Nations peace operations a country is involved in Dreher (2006). Besides, social globalization, the last dimension of globalization, is measured data from three categories

---

1. Interested reader may refer to Ulucak et al. (2019) for an excellent discussion on the environmental summits.

(see Dreher, 2006; Dreher et al., 2008, 2010): First, data on *personal contacts* reflect direct interactions among people living in different countries (e.g., outgoing telephone traffic, international tourism, average telephone cost of a call to the USA, government and workers' transfers received and paid, and the percentage of foreign population); second, data on *information flows* measure the potential flow of ideas and images (e.g., the numbers of Internet hosts and users, the numbers of telephone mainlines, the number of cable television, the number of daily newspapers, and the number of radios [all in per capita 1000]); and third, data on *cultural proximity* (the number of McDonald's restaurants by country).

In constructing the three subindices of globalization, Dreher (2006) transforms each variable in these dimensions to an index with a zero to ten scale, where higher values denote more globalization. The weights for the subindices are computed using principal components analysis, and 2000 is utilized as the base year. The 2002 KOF index consists of 23 variables and is available for 123 countries from 1970 to 2002 (see Dreher, 2006). Later on, Dreher et al. (2008) updated the original KOF index by using more recent data and making some methodological improvements. For instance, each variable reflecting dimensions of globalization has been transformed to an index on a scale of one to hundred, where hundred is the maximum value for a specific variable for the period 1970–2006, while one is the minimum value.

Finally, Gygli et al. (2019) revised the KOF index and composed the 2018 KOF Globalization Index by making a distinction between de facto and de jure measures throughout the dimensions of globalization, disentangling trade, and financial globalization within the economic dimension of globalization, using time-varying weighting of the variables, and increasing the number of variables from 23 to 43. The de facto globalization measures actual international flows and activities while the de jure globalization reflects policies and conditions that permit and foster flows and activities (Gygli et al., 2019). This index is available for 203 countries and spans from 1970 to 2017. In the computation of the 2018 KOF index, based on the aggregation of all 43 variables, the de facto and de jure subindices for five subdimensions of globalization (i.e., trade, financial, interpersonal, informational, and cultural globalization); the de facto and de jure indices for the three dimensions of globalization (economic, social, and political globalization); and finally one overall globalization index, which is the average of the de facto and the de jure indices, are computed. During the calculation process, each variable is transformed to an index with a scale from one to one hundred, where the observation with the highest value across all countries and years is scaled by one hundred (see Gygli et al., 2019). Table 7.1 tabulates the de facto and de jure variables for the (sub)dimensions of globalization with their weights in the related indices.

TABLE 7.1 Structure of the 2018 KOF globalization index.

Globalization index, de facto	Weights	Globalization index, de jure	Weights
1. Economic Globalization, de facto	33.3	**1. Economic Globalization,** de jure	33.3
1.1 Trade Globalization, de facto	50	**1.1 Trade Globalization,** de jure	50
Trade in goods	38.8	Trade regulations	26.8
Trade in services	44.7	Trade taxes	24.4
Trade partner diversity	16.5	Tariffs	25.6
		Trade agreements	23.2
1.2 Financial Globalization, de facto	50	**1.2 Financial Globalization,** de jure	50
Foreign direct investment	26.7	Investment restrictions	33.3
Portfolio investment	16.5	Capital account openness	38.5
International debt	27.6	International investment agreements	28.2
International reserves	2.1		
International income payments	27.1		
2. Social Globalization, de facto	33.3	**2. Social Globalization,** de jure	33.3
2.1 Interpersonal Globalization, de facto	33.3	**2.1 Interpersonal Globalization,** de jure	33.3
International voice traffic	20.8	Telephone subscriptions	39.9
Transfers	21.9	Freedom to visit	32.7
International tourism	21.0	International airports	27.4
International students	19.1		
Migration	17.2		
2. Informational Globalization, de facto	33.3	**2.2 Informational Globalization,** de jure	33.3
Used internet bandwidth	37.2	Television access	36.8
International patents	28.3	Internet access	42.6
High technology exports	34.5	Press freedom	20.6
2.3 Cultural Globalization, de facto	33.3	**2.3 Cultural Globalization,** de jure	33.3
Trade in cultural goods	28.1	Gender parity	24.7
Trade in personal services	24.6	Human capital	41.4
International trademarks	9.7	Civil liberties	33.9
McDonald's restaurant	21.6		
IKEA stores	16.0		

Continued

TABLE 7.1 Structure of the 2018 KOF globalization index.—cont'd

Globalization index, de facto	Weights	Globalization index, de jure	Weights
3. Political Globalization, de facto	33.3	3. Political Globalization, de jure	33.3
Embassies	36.5	International organizations	36.2
UN peace keeping missions	25.5	International treaties	33.4
International NGOs	37.8	Treaty partner diversity	30.4

Notes: Weights in percent for the year 2016. Weights for the individual variables are time variant. Overall indices for each aggregation level are calculated by the average of the respective de facto and de jure indices.
From Gygli, S., Haelg, F., Potrafke, N., Sturm, J.E., 2019. The KOF globalisation index — revisited. Rev. Int. Organ. 14, 543–574.

1.3 Effects of economic globalization on energy consumption

Since the strong globalization breeze in the 1980s, the role of energy in the economic growth process has gained much importance (Shahbaz et al., 2018a). In this regard, the acceleration of globalization has expanded economic activities via information sharing and technology transfer, both of which affect energy consumption (Saud et al., 2018). Globalization, through technological spillovers across borders as well as through its effects on production, consumption, and transportation of products, affects energy consumption (Shahbaz et al., 2019). However, the impact of globalization on energy consumption is not known a priori. On the one hand, globalization may lead to a reduction in the energy needs of production models by providing more advanced technology and knowledge transfers. In this sense, globalization creates novel technology, knowledge, and innovative production ways in different sectors of an economy. As stated by Saud et al. (2018), globalization causes a technology transfer that shifts knowledge on how energy could be used efficiently to cut energy demand, particularly from conventional energy sources. Further, globalization process that brings cleaner and more energy-efficient technologies along with knowledge transfer would not only minimize energy consumption from the conventional sources but also reduce demand for total energy (Danish et al., 2018).

Besides, if foreign firms install new businesses or modify their old ones by adapting sophisticated technologies, their energy consumption levels and overall production costs will diminish (Ahmed et al., 2019). Further, foreign firms may also create a demonstration effect if they lead domestic firms to switch over to similar types of new production methods, which are more energy-efficient (Gozgor et al., 2020; Shahbaz et al., 2016). On the other hand, only for the sake of high profits, if foreign firms do not consider energy saving and exploit natural resources of the host country with low-cost production techniques, globalization may raise energy consumption (Saud et al., 2018; Shahbaz et al., 2018c).

Based on the above-mentioned ideas, the effect of economic globalization on energy consumption is explained under the following two subtitles.

1.3.1 Effects of trade openness on energy consumption

As Koengkan (2018) states, trade openness may increase energy consumption if market liberalization creates more investments and industrialization boosting economic growth, while it may reduce energy consumption if market liberalization encourages the energy-efficient products and the research and development (R&D) intensive investments in the energy sector. In general, the effect of globalization, in terms of trade openness, on energy consumption is explained by the scale, composition, and technique effects. Of them, *the scale effect* suggests that globalization increases energy consumption and environmental pollution by extending the scale of economic activities and stimulating economic growth (Ahmed et al., 2019; Cetin et al., 2018; Cole, 2006; Dedeoğlu and Kaya, 2013). *The technique effect* states that developing countries imports more sophisticated technologies, technical know-how, and R&D from developed countries (Ahmed et al., 2019; Dinda, 2004; Etokakpan et al., 2020; Köksal et al., 2020). The implementation of these advanced technologies and knowledge into the production models consumes less energy and produces more output, which is usually referred to as the technique effect (Shahbaz et al, 2014, 2019). Lastly, the *composition effect* indicates a gradual shift from an agriculture-based economy to an industrial-based economy and finally to a more knowledge-based service economy (Ahmed et al., 2019; Shahbaz et al, 2018b, 2019), which results in less energy consumption. Considering all these effects, the relationship between energy consumption and globalization could be hypothesized as an inverted U-shaped curve, which is similar to the relationship between economic growth and environmental quality. In this respect, energy consumption climbs until a threshold level of globalization is obtained, and after that, it starts descending as the technique effect overwhelms the scale effect (Shahbaz et al., 2019).

Furthermore, the subcomponents of trade openness, i.e., exports and imports, may affect energy consumption, too. For instance, a rising level of exports causes more economic activity in the export-oriented sectors, which results in more demand for energy because the operation of the machinery and equipment utilized in the production, processing, and transportation of exported goods requires energy (Sadorsky, 2012). Without adequate energy supply, export volume will be adversely affected (Shahbaz et al., 2014). Regarding imports, imported goods may increase energy consumption in two ways (Sadorsky, 2011): First, a developed and functional transportation network necessary for the distribution of imported goods into a country requires energy. Second, since durable imported goods such as automobiles, air conditioners, refrigerators, etc., are significant users of energy, their imports increase demand for energy (Dedeoğlu and Kaya, 2013).

1.3.2 Effects of foreign direct investment flows on energy consumption

FDI inflows can affect energy consumption via many channels such as transfer and diffusion of technology, managerial skills, innovative methods, processes of doing things, spillover effects, and productivity gains (Saud et al., 2018). Shahbaz et al. (2019) state that restrictions on trade and FDI flows with tariffs, quotas, and some regulations that will decelerate globalization and negatively affect energy consumption. If FDI is considered as a proxy for globalization, we encounter two hypotheses, namely "the pollution haven hypothesis" (PHH) versus "the pollution halo hypothesis" which analyze the nexus of globalization and energy consumption. PHH underlines the positive effects of FDI inflows on environmental quality in developing countries. FDI is believed to reduce environmental degradation by transferring less energy-intensive and more environmentally friendly production methods from developed to developing countries (Solarin and Al-Mulali, 2018). From the perspective of a universal environmental standard, multinationals dealing with FDI tend to open out their clean technologies to their counterparts in the host country (Shahbaz et al., 2018d). Because foreign firms utilize more sophisticated and innovative production methods, the pollution halo hypothesis assumes that FDI reduces energy consumption. In contrast, the PHH is a theory suggesting that wealthy countries with high environmental regulations will push all their dirty industries while poor countries will pull them (Dinda, 2004). The PHH states that developed countries have strict environmental laws and regulations, while developing countries have lax regulations. In most cases, developing countries with weak regulations host for the profit-oriented foreign companies that try to skip the costly regulatory compliance in their home countries (Shahbaz et al., 2018b). Hence, developed countries transfer their energy (pollution)-intensive production technologies to developing countries via FDI flows (Antweiler et al., 2001; Koengkan et al., 2019). Thereby, developing countries become "pollution haven" if they lower environmental standards below their efficiency to attract foreign investments (Dinda, 2004).

Apart from the above-mentioned effects, economic globalization (either in the form of FDI flows or in the form of international trade) is of importance regarding renewable energy. Economic globalization can encourage renewable energy sources via the channels of FDI inflows and trade openness (Gozgor et al., 2020). For instance, an easier and cheaper access to foreign capital supports the deployment of technology in clean energy projects (Paramati et al., 2017). The adaptation of new technology in clean energy projects is both expensive and capital intensive. Therefore, developing countries mostly convert FDI inflows into clean energy projects that provide advanced and energy-efficient technologies significantly reducing CO_2 emissions (Paramati et al., 2016). Thereby, economic globalization reduces demand for conventional energy sources by boosting renewable energy investments and improves environmental quality by cutting carbon emissions level.

2. Literature review

A number of indicators representing globalization were used in the related literature. Quantities of trade (exports and/or imports), trade openness, trade liberalization, and FDI are frequently used indicators in the previous studies. Under this section, studies analyzing the nexus between globalization and energy consumption are classified based on their indicators.

First, a number of studies used some measures of trade volumes as an indicator (exports and/or imports) of globalization (Dedeoğlu and Kaya, 2013; Erkan et al., 2010; Katircioglu, 2013; Lean and Smyth, 2010a, 2010b; Li, 2010; Narayan and Smyth, 2009; Nnaji et al., 2013; Ozturk and Acaravci, 2013; Sadorsky, 2011, 2012; Sami, 2011; Shahbaz et al., 2013). Among these studies, Erkan et al. (2010) examined the relationship between energy consumption and exports for Turkey from 1970 to 2006. They found evidence of a Granger causality running from energy consumption to exports. Li (2010), using cointegration and Granger causality tests, observed that exports cause electricity consumption to increase in Shandong province of China. Another study for China was by Shahbaz et al. (2013), who investigated the relationship between energy consumption, economic growth, financial development, and international trade for the period 1971—2011. The authors confirmed the presence of a feedback relationship between trade and energy consumption. Lean and Smyth (2010a) investigated the relationship among electricity consumption, real GDP, exports, labor and capital in Malaysia based on the Granger causality test. They found a unidirectional causality running from exports to energy consumption. In another study, using the same methodology, Lean and Smyth (2010b) investigated the relationship between electricity generation, exports, economic growth, and price levels in Malaysia. Their findings revealed a unidirectional Granger causality running from electricity use to exports. Sami (2011) examined the relationship between exports, energy consumption, and output for the Japanese economy. The findings of the study confirmed causal relationship from both exports and GDP to energy consumption in the long-run. For the Singapore economy, Katircioglu (2013) obtained evidence of short- and long-term causalities from energy consumption to imports. Finally, for Nigeria, Nnaji et al. (2013) confirmed a causality running from energy consumption to exports.

Moreover, in this research line, there exist some panel data studies as well (see Narayan and Smyth, 2009; Sadorsky, 2011, 2012; Dedeoğlu and Kaya, 2013). Narayan and Smyth (2009) investigated the nexus of electricity consumption, exports, and economic growth for a panel of Middle Eastern countries. Results from the Granger causality test supported significant feedback effects among all variables. Using both exports and imports as indicators of globalization, Sadorsky (2011) found a causality running from exports to energy consumption in the short-run and a bidirectional causality between imports and energy consumption for a panel of eight Middle East countries.

In another study, Sadorsky (2012) confirmed a causal relationship between trade (exports and imports) and energy consumption in the long-run in a panel of seven South American countries. Finally, in their study including 25 OECD countries, Dedeoğlu and Kaya (2013) documented that there is a bidirectional causal relationship between energy consumption and both exports and imports.

Additionally, trade liberalization and trade openness are frequently used as proxies of globalization (Aïssa et al., 2014; Cole, 2006; Destek, 2015; Ghani, 2012; Gómez and Rodríguez, 2019; Koengkan, 2018; Kyophilavong et al., 2015; Nasreen and Anwar, 2014; Ozturk and Acaravci, 2013; Shahbaz et al., 2014). In an earlier study, Cole (2006) employed trade liberalization as a proxy of globalization in a panel of 32 developed and developing countries and found that trade liberalization increases energy consumption. However, Ghani (2012) found no relationship between trade liberalization and energy consumption for a panel of developing countries. Ozturk and Acaravci (2013) examined the relationship among financial development, trade, economic growth, energy consumption, and carbon emissions in Turkey over the years 1960–2007. Their results revealed that economic growth and trade openness increased energy consumption. By using trade openness as an indicator of globalization, Nasreen and Anwar (2014) revealed that trade openness has a positive effect on energy consumption in a panel of 15 Asian economies. Shahbaz et al. (2014) examined the relationship between energy consumption and trade openness for a panel consisting of 91 low-, middle-, and high-income countries from 1980 to 2010 based on several causality tests. Between trade openness and energy consumption, they confirmed a U-shaped relationship for the low- and middle-income countries, but an inverted U-shaped relationship for the high-income countries. However, Aissa et al. (2014) found no evidence of a causal relationship between renewable energy use and trade openness in 11 African countries, while Kyophilavong et al. (2015) found a bidirectional causality between trade openness and energy consumption in Thailand. Additionally, Destek (2015) revealed that economic growth and trade openness positively affect the energy consumption of the Turkish economy in the long-run. Finally, Koengkan (2018) found that trade openness positively affects energy consumption in the Andean community countries (Bolivia, Colombia, Peru, and Ecuador).

Few studies also used FDI as a proxy of globalization either alone or in combination with other indicators (Farhani and Solarin, 2017; Paramati et al., 2017; Sbia et al., 2014). Of them, Sbia et al. (2014) revealed that FDI and trade openness negatively affect energy demand in the United Arab Emirates (UAE). In their study covering 20 emerging market economies, Paramati et al. (2016) found that FDI inflows positively affect the clean energy consumption. In another study consisting of European Union (EU) members, G-20 and OECD countries, Paramati et al. (2017) found the same results as their previous study. Lastly, Farhani and Solarin (2017) found that financial development, FDI, and GDP negatively affect energy demand, while trade and capital positively affect it.

Finally, studies trying to explain the relationship between globalization and energy consumption using various globalization indicators have increased in recent years. The KOF Globalization Index developed by Dreher (2006) was utilized in the majority of these studies (see Can and Dogan, 2016; Dogan and Deger, 2016; Iheanacho, 2018; Danish et al., 2018; Koengkan, 2017; Shahbaz et al., 2016, 2018a,c,d, 2019). Among them, Shahbaz et al. (2016) investigated the relationship among globalization, economic growth, financial development, urbanization, and energy consumption for India from 1971 to 2012. Their results indicated that increases in globalization and financial development decrease energy consumption. Can and Dogan (2016) investigated the nexus between energy consumption and globalization in Turkey for the period 1970—2012. Their results revealed that globalization stimulates energy consumption in the long-run. Dogan and Deger (2016) analyzed the relationship between energy consumption, economic growth, and globalization for BRIC countries covering the period 2000—12 based on cointegration and causality tests. They found no causal linkage between energy consumption and globalization. Using an autoregressive distributed lag (ARDL) approach, Koengkan (2017) found that a 1% increase in globalization index causes a 0.44% rise in primary energy consumption in selected Latin America and Caribbean countries. Besides, Shahbaz et al. (2018d) examined the nexus between globalization, economic growth, and energy consumption using time series and panel data models for developed economies for the period 1970—2014. They found that globalization stimulates energy consumption for most of the countries under study. Only in the USA and the UK, globalization has a negative effect on energy consumption. By using the nonlinear ARDL (NARDL) bound testing approach, Shahbaz et al. (2018c) found that positive and negative shocks to globalization affect energy consumption in the same direction in BRICS countries. Shahbaz et al. (2018a) analyzed the relationship between globalization and energy consumption for the highly two globalized countries (i.e., Netherland and Ireland) with the quantile ARDL (QARDL) cointegration test. They found a positive correlation between variables in the long-run in both countries. Shahbaz et al. (2019) investigated the relationship among globalization and energy consumption for 86 economies with different income levels by cross-correlation estimation model. Their results revealed that globalization reduces energy consumption in the majority of countries in the long-run. Iheanacho (2018) examined the nexus of globalization, energy consumption, and economic growth for Nigeria during the period 1975—2011 through ARDL cointegration test. They supported a bidirectional relationship between globalization and energy consumption in the long-run. Danish et al. (2018) investigated the linkages between energy consumption, financial development and globalization in Next-11 countries for the period 1990—2014. Panel causality test results revealed that globalization has a positive effect on energy consumption.

In this research line, some studies employed the updated version of the KOF Globalization Index (see Ahmed et al., 2016; Azimova, 2019; Koengkan et al., 2019; Gozgor et al., 2020). Of them, Ahmed et al. (2016) examined the relationship between globalization, economic growth, financial development, trade, and energy consumption in China using modern time series analyses. They indicated that globalization affects energy consumption negatively in the short-run but positively in the long-run. Azimova (2019) examined the nexus between energy consumption, income, and globalization in a panel including 66 developing economies. They used the subindices of KOF Globalization Index. Results confirmed no linkage between economic and political globalization and energy consumption. Conversely, social globalization has a negative and significant effect on energy consumption. Koengkan et al. (2019) investigated the relationship between globalization, private capital stock, output level, and renewable energy in Latin American countries for the period 1980−2014 using panel ARDL approach. They used the KOF Globalization Index revised by Gygli et al. (2019). They obtained that globalization stimulates renewable energy consumption in 10 Latin American countries. Likewise, Gozgor et al. (2020) examined the linkage between renewable energy consumption and economic globalization in 30 OECD countries during the period 1970−2017. They used the economic globalization subindex in Dreher (2006) and its two updated versions. Their results confirmed that economic globalization stimulates renewable energy use. Table 7.2 provides a brief summary of the studies using the KOF index or its subindices while analyzing the relationship between globalization and energy consumption.

TABLE 7.2 A brief summary of studies using the KOF index (or its subindices).

Authors	Countries and time period	Methodology	Results
Shahbaz et al. (2016)	India 1971−2012	ARDL bounds test	Globalization negatively affects energy demand.
Ahmed et al. (2016)	China 1971−2013	ARDL bounds test Toda-Yamamoto causality test	Globalization has a positive effect on energy consumption in the long-run.
Can and Dogan (2016)	Turkey 1970−2012	Maki, Johansen Cointegration	Globalization stimulates energy consumption in the long-run.

Continued

TABLE 7.2 A brief summary of studies using the KOF index (or its subindices).—cont'd

Authors	Countries and time period	Methodology	Results
Dogan and Deger (2016)	BRIC countries 2000—12	Pedroni and Kao Cointegration, Granger causality	No causal relation between globalization and energy consumption.
Koengkan (2017)	Latin American and Caribbean countries 1991—2012	ARDL	Increases in the globalization index cause a rise in energy consumption.
Shahbaz et al. (2018d)	Developed economies 1970—2014	Panel cointegration and causality tests	Globalization stimulates energy consumption for most of the countries with the exceptions of the UK and USA.
Shahbaz et al. (2018c)	BRICS countries 1970—2015	NARDL	Shocks to globalization affect energy consumption in the same direction as shocks.
Shahbaz et al. (2018a)	Netherland and Ireland 1970Q1—2015Q4	QARDL	There is a positive correlation between globalization and energy consumption.
Shahbaz et al. (2019)	86 high, middle and low-income countries 1970—2015	cross-correlation estimates	Globalization decreases energy consumption in the majority of countries in the long-run.
Iheanacho (2018)	Nigeria 1975—2011	ARDL and VECM	Bidirectional relationship between globalization and energy consumption in the long-run.
Danish et al. (2018)	Next-11 countries 1990—2014	DSUR, Panel causality	Globalization has a positive effect on energy consumption.
Azimova (2019)	66 developing economies 1998—2014	VAR model	Globalization affects energy consumption positively in the long-run.

Continued

TABLE 7.2 A brief summary of studies using the KOF index (or its subindices).—cont'd

Authors	Countries and time period	Methodology	Results
Koengkan et al. (2019)	Latin America countries 1980–2014	Panel ARDL	Globalization stimulates renewable energy consumption
Gozgor et al. (2020)	30 OECD countries 1970–2015	Panel cointegration, FMOLS	Economic globalization stimulates renewable energy use

Notes: ARDL, NARDL, QARDL, VECM, DSUR, VAR, and FMOLS, respectively, refer to autoregressive distributed lag model, nonlinear autoregressive distributed lag model, quantile autoregressive distributed lag model, vector error correction model, dynamic seemingly unrelated regression, vector autoregressive model, and fully modified ordinary least squares.

3. Econometric methodology

A decisive factor for the success of empirical analyses lies in the selection of consistent and robust estimators. In this section, we review the econometric tests in panel data analysis and briefly introduce econometric tests that we adapt in our analysis.

3.1 Cross-sectional dependency test

O'Connell (1998) argues that if the cross-sectional dependence is ignored in panel data analysis, size and power properties of the tests weaken, which would bias the results. Similarly, Carrion-i-Silvestre et al. (2005) argue that the assumption of cross-sectional independency is hardly found in practice, especially in a globalized world where the shocks overpass the borders of the countries.

A generic cross-sectional dependency test in panel data analysis can be stated as below:

$$\Delta y_{i,t} = d_i + \delta_i y_{i,t-1} + \sum_{j=1}^{\rho_i} \lambda_{i,j} \Delta y_{i,t-j} + \mu_{i,t} \quad i = 1, \dots N; t = 1, \dots T \quad (7.1)$$

where N and T denote cross section dimension and time dimension of the panel, respectively. y is the variable of which the cross-sectional dependency is investigated, Δ is the difference operator. The deterministic component d_i is considered for constant and constant and trend, ρ is the lag length. The null and alternative hypotheses for the cross-sectional dependency are as below:

H_0: No cross-sectional dependency. $\left(\mathrm{Cov}\left(u_{it}, u_{jt}\right) = 0, \text{ for all } t \text{ and } i \right.$ $\left(i \neq j\right)\right)$

H_1: Cross-sectional dependency. $\left(\mathrm{Cov}\left(u_{it}, u_{jt}\right) \neq 0, \text{for at least one pair of } i \neq j\right)$

To test CD, Breusch and Pagan (1980) proposed a simple Lagrange multiplier (*LM*) test as shown below:

$$LM = T \sum_{i=1}^{N-1} \sum_{j=i+1}^{N} \widehat{\rho}_{ij}^2 \sim \chi_{\frac{N(N-1)}{2}}^2 \tag{7.2}$$

where $\widehat{\rho}_{ij}^2$ is the pair-wise correlations among the residuals from the ordinary least square (OLS) estimate as shown in Eq. (7.3):

$$\widehat{\rho}_{ij} = \widehat{\rho}_{ji} = \frac{\sum\limits_{t=1}^{T} \widehat{u}_{it}\widehat{u}_{jt}}{\left(\sum\limits_{t=1}^{T} \widehat{u}_{it}^2\right)^{1/2} \left(\sum\limits_{t=1}^{T} \widehat{u}_{jt}^2\right)^{1/2}} \tag{7.3}$$

Under the null hypothesis with a fixed N and $T \to \infty$, LM statistic is asymptotically distributed as chi-square with $N(N-1)/2$ degrees of freedom. LM test is applicable in cases where time dimension (T) is large relative to the cross section dimension (N).

To overcome the problem that the Breusch and Pagan's *LM* test is not applicable with large N, Pesaran (2004) developed a modified *LM* test which is applicable in the case of panel data models where first T then N increases and $T > N$:

$$CD_{LM} = \sqrt{\frac{1}{N(N-1)}} \sum_{i=1}^{N-1} \sum_{j=i+1}^{N} \left(T\widehat{\rho}_{ij}^2 - 1\right) \sim N(0, 1) \tag{7.4}$$

CD_{LM} test exhibits substantial size distortions if the cross section dimension is large relative to time dimension. In his same paper, Pesaran also proposed another CD test to be applied in the cases where $N > T$:

$$CD = \sqrt{\frac{2T}{N(N-1)}} \left(\sum_{i=1}^{N-1} \sum_{j=i+1}^{N} \widehat{\rho}_{ij}\right) \sim N(0, 1) \tag{7.5}$$

CD test in Eq. (7.5) lacks power in some situations in which the population average pair-wise correlations are zero, even though the underlying individual population pair-wise correlations are not zero. To overcome this issue, Pesaran et al. (2008) developed a bias-adjusted *LM* test constructed as follows:

$$LM_{\mathrm{adj}} = \sqrt{\left(\frac{2}{N(N-1)}\right)} \sum_{i=1}^{N-1} \sum_{j=i+1}^{N} \widehat{\rho}_{ij} \frac{(T-k)\widehat{\rho}_{ij}^2 - u_{Tij}}{\sqrt{v_{Tij}^2}} \sim N(0, 1) \tag{7.6}$$

where k is the number of explanatory variables, u_{Tij} is the exact mean, and v_{Tij}^2 is the variance of $(T-k)\widehat{\rho}_{ij}^2$.

3.2 Panel unit root test

In their seminal study, Granger and Newbold (1974) pointed out the *spurious* regression phenomenon when the stationarity of time series is neglected. Later on, in their influential paper, Nelson and Plosser (1982) showed that most economic variables are better characterized as nonstationary. Recognizing the deficiency in the literature, several panel unit root tests have emerged. The literature on panel unit root testing follows two strands as the first generation and the second generation tests (Baltagi and Pesaran, 2007; Pesaran et al., 2013). While the first generation panel unit root tests (e.g., Breitung, 2001; Choi, 2001; Hadri, 2000; Im et al., 2003; Levin et al., 2002; Maddala and Wu, 1999) ignore cross-sectional dependence between the error terms, the second generation panel unit root tests (e.g., Bai and Ng, 2004; Breuer et al., 2002; Hadri and Kurozumi, 2012; Nazlioglu and Karul, 2017; Pesaran, 2007; Reese and Westerlund, 2016; Vanessa Smith et al., 2004; Westerlund and Hosseinkouchack, 2016; Westerlund and Larsson, 2009) take account of possible cross-sectional dependence.

Among the various second generation panel unit root tests, cross-sectionally augmented ADF (*CADF*) test of Pesaran (2007) is one of the most popular tests in the literature (Westerlund and Hosseinkouchack, 2016). The unit root test proposed by Pesaran (2007) is based on standard augmented Dickey-Fuller (ADF) regressions augmented with the cross-sectional averages of lagged levels and first differences of the individual series.

Denoting N as the number of cross sections and t as time series, Pesaran (2007) regresses the below dynamic model:

$$\Delta y_{it} = \alpha_i + \beta_i y_{i,t-1} + c_i \bar{y}_{t-1} + d_i \Delta \bar{x} + \varepsilon_{it} \tag{7.7}$$

where $\bar{y}_{t-1} = \dfrac{\sum\limits_{i=1}^{N} y_{i,t-1}}{N}$ and $\Delta \bar{y}_t = \dfrac{\sum\limits_{i=1}^{N} \Delta y_{it}}{N}$.

Cross-sectional dependence is modeled through factor structure using cross-sectional averages of lagged levels and first differences of individual series.

The unit root hypothesis is expressed as $H_0 : \beta_i = 0$ for all i against the alternative hypothesis of $H_1 : \beta_i < 0$, $i = 1, 2 ..., N_1$

Based on *CADF* statistics, the unit root of *CIPS* is calculated as follows:

$$CIPS = \frac{1}{N} \sum_{i=1}^{N} CADF_i \tag{7.8}$$

where $CADF_i$ is the cross-sectional augmented DF statistic for each section obtained by the t-ratios of the ρ_i in the *CADF* regression. As standard central limit theorems do not apply to the *CIPS* statistic, Pesaran (2007) tabulated the critical values for three main specifications, without intercept and trend, with an intercept, with an intercept and trend.

3.3 Panel cointegration test

Long-run relationship among the nonstationary variables in panel data analysis is estimated through panel cointegration tests. Cointegration can simply be examined by applying unit root test on the residuals obtained from regressing one variable on a set of other variables. If the residuals do not contain unit root, then the series is said to be cointegrated. As in panel unit root tests, panel cointegration tests can be also grouped into two categories as first and second generation panel cointegration tests. While the first generation tests assume cross-sectional independency (Kao, 1999; Larsson et al., 2001; Pedroni, 1999; Westerlund, 2005, among others), second generation panel cointegration tests take into account cross-sectional dependence (Bai et al., 2009; Gengenbach et al., 2016; Westerlund, 2007, 2008; Westerlund and Edgerton, 2007, among others). In case of a correlation among the cross sections, first generation panel cointegration tests are subject to severe size distortion. Bai and Kao (2006) state that it is difficult to justify the assumption of cross-sectional independence in panel cointegration analysis.

Among the various panel cointegration tests proposed in the literature, Durbin-Hausman test suggested by Westerlund (2008) has several advantages over others (Ulucak et al., 2020): (1) the test takes into account of cross-sectional dependence by applying principal components method to the residuals, (2) as the asymptotic distribution of the test is standard normal, it could be applied with a large number of explanatory variables, (3) another feature of this test is that the regressor need not be nonstationary as opposed to many cointegration tests. In addition, Monte Carlo simulations indicate that the test has smaller size distortion and greater power compared to the other tests (Tong and Yu, 2018).

Two types of estimators are derived based on Durbin-Hausman principle. A common value for the autoregressive parameter is assumed in DH_p statistic, while autoregressive parameters are assumed to be heterogeneous in DH_g statistic. DH statistics are calculated as below (Westerlund, 2008, pp. 202−203):

$$DH_p = \widehat{S}_n \left(\widetilde{\phi} - \widehat{\phi} \right)^2 \sum_{i=1}^{n} \sum_{t=2}^{T} \widehat{e}_{it-1}^2 \text{ and } DH_g = \sum_{i=1}^{n} \widetilde{S}_i \left(\widetilde{\phi}_i - \widehat{\phi}_i \right)^2 \sum_{t=2}^{T} \widehat{e}_{it-1}^2$$

(7.9)

where $\widehat{S}_n = \widehat{\omega}_n^2 \big/ \left(\widehat{\sigma}_n^2 \right)^2$ and $\widehat{S}_i = \widehat{\omega}_i^2 \big/ \widehat{\sigma}_i^4$ are variance ratios. The Kernel estimator is obtained:

$$\widehat{\omega}_i^2 = \frac{1}{T-1} \sum_{j=-M_i}^{M_i} \left(1 - \frac{j}{M_i + 1} \right) \sum_{t=j+1}^{T} \widehat{v}_{it} \widehat{v}_{it-j}$$

(7.10)

where $\widehat{v}_{it}$ is the OLS residual and M_i is bandwidth parameter that determines how many autocovariances of $\widehat{v}_{it}$ to estimate in the kernel.

The null hypothesis of panel statistic is formulated as below:

$$H_0 : \phi_i = 1 \text{ for } i = 1, \ldots n$$

$$H_1^p : \phi_i = \phi \text{ and for } \phi < 1 \text{ for all } i$$

A common root for the autoregressive parameter is assumed under the null and alternative hypothesis. Therefore, rejection of null hypothesis will imply cointegration for all n cross sections.

The distinction between DH_g and DH_p statistics lies in the alternative hypothesis. For the group mean test, the alternative hypothesis is constructed as below:

$$H_1^g : \phi_i < 1 \text{ for at least some } i \text{ for at least some } i.$$

Therefore, as a common value for the autoregressive parameter is not assumed, rejecting the null hypothesis will imply cointegration for at least some of the cross sections. Both the DH_g and DH_p follow a standard normal distribution.

3.4 Panel cointegration estimators

To obtain long-term cointegration estimators, we apply continuously updated and fully modified (CupFM) and continuously updated and bias-corrected (CupBC) estimators proposed by Bai et al. (2009). This choice was motivated by three appealing features of these estimators: (1) these estimators control for the correlation among macroeconomic variables from different countries, (2) the estimators are robust to mixed $I(0)$ and $I(1)$ regressors, and (3) they are consistent due to correcting for serial correlation and endogeneity.

Bai et al. (2009) updated fully modified ordinary least squares (FMOLS) approach of Phillips and Hansen (1990) by considering unobserved factors formulated as below:

$$y_{it} = x_{it}' \beta_{it} + e_{it} \tag{7.11}$$

$$x_{it} = x_{it-1} + \varepsilon_{it}, \ e_{it} = \lambda_i' F_t + u_{it}$$

where $i = 1, \ldots, n$ represents cross section dimension and $t = 1, \ldots T$ represents time dimension. F_t and λ_i are unobserved common factors and country-specific factor loadings, respectively. If F_t and u_{it} are stationary, then, e_{it} would also be stationary. While CupBC estimates the asymptotic bias directly, CupFM modifies the data so that the limiting distribution does not depend on nuisance parameters. Both are continuously updated until the iteration reaches convergence. The estimators are formulated as follows:

$$\widehat{\beta}_{FM} = \left(\sum_{i=1}^{n} \sum_{t=1}^{T} x_{it} x_{it}' \right)^{-1} \sum_{i=1}^{n} \sum_{i=1}^{T} x_{it} y_{it} \tag{7.12}$$

4. Data and results

In their inspiring study, Shahbaz et al. (2016) recommend incorporating the degree of financial development of economies in multivariate regression analysis when investigating the impact of globalization on energy demand. The present study takes these suggestions as a starting point to investigate the effects of globalization on energy demand in a multivariate regression framework for OECD countries. We use annual data of 29 selected OECD countries covering the period of 1971−2017. We provided the list of countries in the Appendix. As the frequency of the series is annual, we do not deal with seasonality issues. Following Shahbaz et al. (2016) and Danish (2018), we estimate the following log-linear transformation of the variables:

$$EC_{it} = \beta_0 + \beta_{1i} EG_{it} + \beta_{2i} UR_{it} + \beta_{3i} \ln G_{it} + \beta_{4i} FD_{it} + \varepsilon_{it} \qquad (7.13)$$

where $\beta_1 \ldots \beta_4$ are the unknown parameters to be estimated, t indicates the time period of 1971−2017, and $i = 1,2 \ldots N$ represents cross sections in panel, consisting of 29 selected OECD countries[2] and ε is the idiosyncratic error term. We measure globalization using the revised version of the KOF Globalization Index, a composite index measuring globalization along the economic, social, and political dimensions. The inclusion of the revised KOF Globalization Index is important, as Gygli et al. (2019) propose that future research should use the new KOF Globalization Index to reexamine the consequences of globalization. We include private credit by banks to private sector as a proxy variable for financial development. We also use urbanization and economic growth as control variables to avoid specification bias. All the variables are transformed into natural logarithms to interpret the elasticities of the coefficient estimates. The definitions of the variables and data sources are given in Table 7.3.

We start our analysis by checking the cross-sectional dependence among the variables. We performed LM test proposed by Breusch and Pagan (1980), CD and CD_{LM} tests proposed by Pesaran (2004), and LM_{adj} test proposed by Pesaran et al. (2008). The results given in Table 7.4 show the strong existence of cross-sectional dependence in all variables.

An economic interpretation of the existence of CD is that a shock hitting a variable in any of the countries forming the panel will spread to other countries over time. As a result of increasing globalization and economic interdependence, countries are dependent on each other more than ever. As a result of this dependence, subprime mortgage crisis broke out in the USA in 2007 and rapidly spilled over other countries. The econometric implication of CD, on the other hand, is that when employing panel unit root and panel cointegration

---

2. Colombia recently became the 37th member of OECD on 28 April 2020. To form a balanced panel, we excluded eight OECD members (Czech Republic, Estonia, Hungary, Latvia, Lithuania, Poland, Slovak Republic, and Slovenia) due to data unavailability.

TABLE 7.3 Description of variables.

Variable	Symbol	Unit	Source
Energy consumption	EC	kg of oil equivalent per capita	World Development Indicators, WB
Economic growth	EG	GDP per capita in constant 2010 US$	World Development Indicators, WB
Urbanization	URB	Total urban population	World Development Indicators, WB
Globalization	G	KOF overall globalization index	Swiss Economic Institute
Financial development	FD	Private credit by banks, % of GDP	Global Financial Development, WB

TABLE 7.4 Cross-sectional dependency test results.

	LM	CD_{LM}	CD	LM_{adj}
lnEC	347.45*** (0.00)	5.96*** (0.00)	2.205** (0.00)	28.43*** (0.00)
lnEG	472.12*** (0.00)	10.26*** (0.00)	4.523*** (0.00)	36.17*** (0.00)
lnURB	75.23*** (0.00)	2.19** (0.01)	−1.982** (0.00)	16.83*** (0.00)
lnG	318.26*** (0.00)	8.29*** (0.00)	−1.664** (0.00)	18.88** (0.00)
lnFD	332.96*** (0.00)	3.17 (0.00)	−3.578 (0.00)	12.25 (0.00)

Note: *** and ** indicate statistical significance at 1% and 5% levels, respectively.

tests, estimators that account for CD should be employed. Otherwise, the estimates would be biased. For this purpose, we investigate the order of integration of the variables using cross-sectionally augmented IPS (CIPS) suggested by Pesaran (2007). The results of CIPS panel unit root test are presented in Table 7.5.

As shown in Table 7.5, the null hypothesis of unit root cannot be rejected for any of the variables at levels. However, the variables are stationary at their first differences. The fact that the variables contain unit root at their levels and stationary at their first differences provide us two important implications. Firstly, shocks to energy consumption, economic growth, urbanization, globalization, and financial development will have permanent effects. In other words, an initial shock hitting these variables will never die out without any intervention. Secondly, as the variables are integrated at the same order, there might exist a long-run cointegration relationship.

TABLE 7.5 Cross-sectionally augmented IPS panel unit root test results.

Variable	Constant		Constant and trend	
	Level	First difference	Level	First difference
lnEC	−1.731	−8.169***	−0.364	−6.316***
lnEG	−1.956	−5.574***	−0.465	−6.185***
lnURB	−2.128	−4.391***	−1.968	−3.158***
lnG	−1.645	−10.478***	0.188	−11.478***
lnFD	−2.082	−2.562***	−1.685	−2.729**

Notes: Maximum lag length is set to 2 and optimal lags were chosen based on Schwarz information criterion. *** and ** denote the rejection of the null hypothesis at the 1% and 5% significance levels, respectively. Rejection of the null hypothesis indicates stationarity at least one country. Critical values for the CIPS test are −2.30 at 1% and −2.16 at 5% for the constant case; −2.78 at 1% and −2.65 at 5% for the constant and trend case, respectively (see Tables 7.2B and C in Pesaran, 2007).

Having concluded that all of the variables are integrated of order one, we employ the Durbin-Hausman cointegration test proposed by Westerlund (2008) to observe the existence of long-run equilibrium. The results given in Table 7.6 indicate that the null hypothesis of no cointegration is strongly rejected for both DH_g and DH_p tests. This finding suggests that there is a long-run cointegration relationship among energy consumption, economic growth, urbanization, globalization, and financial development in the countries forming the panel.

In the final step of our analysis, we proceed with the estimation of the long-run coefficients in line with the Durbin-Hausman cointegration test results. In accordance with the results of CD tests, we performed CupFM and CupBC estimators developed by Bai et al. (2009) to obtain long-term cointegration estimators.

TABLE 7.6 Cointegration test results.

Test	Constant	Constant and trend
DHg	10.133*** (.00)	12.157*** (.00)
DHp	16.218*** (.00)	15.251*** (.00)

Notes: DH_g and DH_p statistics are Durbin-Hausman group and panel statistics, respectively. The maximum number of factors is set to 3. The bandwidth selection, Mi, corresponds to the largest integer less than $4(T/100)^{2/9}$ as suggested by Newey and West (1994). P-values are reported in parentheses. *** indicates the rejection of no cointegration null hypothesis at the 1% level of significance.

TABLE 7.7 Cointegration parameters.

	Dependent variable: EC	
Variable	CupFM	CupBC
lnEG	0.568*** [4.122]	0.617*** [3.982]
lnURB	0.106** [2.315]	0.117** [2.104]
lnG	0.124*** [5.128]	0.146*** [5.215]
lnFD	0.143** [1.976]	0.115** [2.053]

Notes: Values in brackets are t-statistics. *** and ** denote significance at 1% and 5% levels, respectively. CupFM refers to the continuously updated and fully modified estimator, and CupBC refers to the continuously updated and bias-corrected estimator.

The cointegration parameters are given in Table 7.7. Both the CupFM and CupBC estimators yielded similar results. All estimated coefficients have statistically significant impact on energy consumption. Among all the variables in the model, economic growth is by far the most influential variable on energy consumption with a coefficient of 0.568% for CupFM estimator and 0.617% for CupBC estimator. This is in line with the theoretical expectations as economic growth is a major contributor to energy consumption. With an increase in income, both households and firms demand more energy. In parallel with the increase in their incomes, households demand durable goods such as air conditioners, or fridges to increase their welfare. As for the firms, higher revenue encourages them to invest and produce more.

As for the relationship between urbanization and energy consumption, our findings suggest that urbanization is also positively linked to energy consumption. In other words, holding all other variables constant, a 1% increase in urbanization will cause energy consumption to increase 0.106%−0.117% percent. This result can be explained by two views. Urbanization increases income which changes the consumers' preferences in favor of energy-intense daily life products. Also, because of rapid urbanization, active workforce moves from primary sector (agriculture, forestry, mining, and fishing) to industry and service sectors which will also lead to an increase in energy demand. This finding is in line with the studies of Shahbaz et al. (2018d) for 25 developed countries, Liu et al. (2017) for China, and Kahouli (2017) for South Mediterranean countries.

Globalization has also a positive impact on energy consumption in OECD countries forming the panel. More specifically, a 1% increase in globalization index will lead to an average increase of 0.124%−0.146% in energy consumption. This result supports the view that globalization increase energy consumption in the host countries which aims at attracting more foreign firms to set up new businesses as a primary focus of globalization strategy.

In addition to investment opportunities, globalization brings trade liberalization which is another major reason behind the increase in energy demand. This finding is in line with the results of Shahbaz et al. (2018d) for 25 developed economies, Danish et al. (2018) for 11 countries, Ahmed et al. (2016) for China, Can and Dogan (2016) for Turkey, Koengkan (2017)) for Latin American and Caribbean countries, Shahbaz et al. (2018c) for BRICS countries, Shahbaz et al. (2018a) for Netherland and Ireland, Azimova (2019) for 66 developing countries, and Gozgor et al. (2020) for 30 OECD countries. However, our findings are not supported by Shahbaz et al. (2016) for India, Dogan and Deger (2016) for BRIC countries, and Shahbaz et al. (2019) for 86 countries.

Regarding the relationship between financial development and energy consumption, our findings reveal that a 1% increase in financial development is accompanied by an 0.115%–0.143% increase in energy consumption. The explanation of this finding is that access to financial instruments will be easier as the financial system develops. Therefore, households and firms will have easy access to credits and loans, which will finally lead to an increase in their energy demand to produce or consume more.

5. Conclusion and policy recommendations

Shahbaz et al. (2018d) argue that understanding the linkage between globalization and energy consumption in advanced economies is important as the shocks to globalization and energy consumption in these countries have substantial effects on not only the developed countries but also on developing countries. Based on their suggestion, we investigated the relationship among energy consumption, economic growth, urbanization, and globalization for 29 selected OECD countries using annual data for 1971–2017. We applied a battery of cross-sectional dependency tests, CIPS panel unit root test of Pesaran (2007), Durbin-Hausman cointegration test of Westerlund (2008) and Bai et al. (2009), CupFM second generation panel unit root test, second generation cointegration test, and obtained the long-term parameters using the estimators proposed by Bai et al. (2009).

Our findings reveal that there is a strong cross-sectional dependence in all variables in the countries forming the panel. This finding stems from rapid globalization and economic integration among the countries. Therefore, in the modern era of globalization, cross-sectional independence is a highly rigid assumption. As for the unit root tests, we failed to reject the null hypothesis of unit root. This implies that the effects of the shocks to the series will be permanent rather than temporary unless the policy makers intervene. Panel cointegration results suggest a long-run relationship among economic growth, urbanization, globalization, and financial development in the selected OECD countries. In the final step of our analysis, we estimated the long-run coefficients. Results show that economic growth is the major contributor to energy consumption with an elasticity rate of 0.568–0.617. We find that a 1%

increase in urbanization increases energy consumption by 0.568%—0.617%, while a 1% increase in financial development increases energy consumption by 0.143—0.115. As for the globalization, the results show that a 1% increase in globalization leads to an increase in energy consumption by 0.124%—0.146%.

Important policy implications are to be drawn from this study. Urbanization, a significant contributor to energy demand, has been continuously increasing. In 2009, for the first time in history, the number of people living in urban areas surpassed the number of people living in rural areas in. OECD states that 55% of the world's population lives in urban areas which is projected to reach 68% by 2050. This dramatic increase in the rate of urbanization will have a substantial effect on energy demand. To mitigate the effects of this increase, governments and firms should invest in cleaner and green technologies in daily life. Energy demand will continue to be a key factor in OECD countries in the modern era of globalization. Policymakers should not underestimate the role of globalization in energy demand for achieving sustainable and secure energy. Globalization should be used for transferring advanced technologies and knowledge to restrict energy demand. In addition, globalization can also be used as a tool to raise awareness on energy conservation through media and communication industries. As financial development is also another determinant of energy consumption, financial institutions should provide funds and credits to encourage firms adopting new technologies that require less energy.

Appendix: list of 29 OECD countries forming the panel

Australia, Austria, Belgium, Canada, Chile, Colombia, Denmark, Finland, France, Germany, Greece, Iceland, Ireland, Israel, Italy, Japan, Korea Republic, Luxembourg, Mexico, Netherlands, New Zealand, Norway, Portugal, Spain, Sweden, Switzerland, Turkey, United Kingdom, and the United States.

References

Ahmed, K., Bhattacharya, M., Qazi, A.Q., Long, W., 2016. Energy consumption in China and underlying factors in a changing landscape: empirical evidence since the reform period. Renew. Sustain. Energy Rev.

Ahmed, Z., Wang, Z., Mahmood, F., Hafeez, M., Ali, N., 2019. Does globalization increase the ecological footprint? Empirical evidence from Malaysia. Environ. Sci. Pollut. Res. 26, 18565—18582.

Aïssa, M.S.B., Jebli, M.B., Youssef, S.B., 2014. Output, renewable energy consumption and trade in Africa. Energy Policy 66, 11—18.

Antweiler, W., Copeland, B.R., Taylor, M.S., 2001. Is free trade good for the environment? Am. Econ. Rev. 91, 877—908.

Azimova, T., 2019. Globalization in reverse: the missing link in energy consumption. Glob. J. Hum. Soc. Sci. E Econ.

Bai, J., Kao, C., 2006. On the estimation and inference of a panel cointegration model with cross-sectional dependence. In: Baltagi, B.H. (Ed.), Contributions to Economic Analysis. Elsevier, pp. 3–30.

Bai, J., Kao, C., Ng, S., 2009. Panel cointegration with global stochastic trends. J. Econ. 149, 82–99.

Bai, J., Ng, S., 2004. A PANIC attack on unit roots and cointegration. Econometrica 72, 1127–1177.

Baltagi, B.H., Pesaran, M.H., 2007. Heterogeneity and cross section dependence in panel data models: theory and applications introduction. J. Appl. Econ. 22, 229–232.

Breitung, J., 2001. The local power of some unit root tests for panel data. In: Nonstationary Panels, Panel Cointegration, Dynamic Panels (Advances in Econometrics). JAI Press, pp. 161–177.

Breuer, J.B., McNown, R., Wallace, M., 2002. Series-specific unit root tests with panel data. Oxf. Bull. Econ. Stat. 64, 527–546.

Breusch, T.S., Pagan, A.R., 1980. The Lagrange multiplier test and its applications to model specification in econometrics. Rev. Econ. Stud. 47, 239.

Can, M., Dogan, B., 2016. The Relationship between globalization and energy consumption: Cointegration analysis in the case of Turkey. Journal of Finance 170, 59–70.

Carrion-i-Silvestre, J., Del Barrio-Castro, T., López-Bazo, E., 2005. Breaking the panels: an application to the GDP per capita. Econ. J. 8, 159–175.

Cetin, M., Ecevit, E., Yucel, A.G., 2018. The impact of economic growth, energy consumption, trade openness, and financial development on carbon emissions: empirical evidence from Turkey. Environ. Sci. Pollut. Res. 25, 36589–36603.

Choi, I., 2001. Unit root tests for panel data. J. Int. Money Financ. 20, 249–272.

Cole, M.A., 2006. Does trade liberalization increase national energy use? Econ. Lett. 92, 108–112.

Danish, K., Saud, S., Baloch, M.A., Lodhi, R.N., 2018. The nexus between energy consumption and financial development: estimating the role of globalization in Next-11 countries. Environ. Sci. Pollut. Res. 25, 18651–18661.

Dedeoğlu, D., Kaya, H., 2013. Energy use, exports, imports and GDP: new evidence from the OECD countries. Energy Pol. 57, 469–476.

Destek, M.A., 2015. Energy consumption, economic growth, financial development and trade openness in Turkey: Maki cointegration test. Bulletin of Energy 3, 162–168.

Dinda, S., 2004. Environmental Kuznets curve hypothesis: a survey. Ecol. Econ. 49, 431–455. https://doi.org/10.1016/j.ecolecon.2004.02.011.

Dogan, B., Deger, O., 2016. How globalization and economic growth affect energy consumption: Panel data analysis in the sample of BRIC countries. International Journal of Energy Economics and Policy 6, 806–813.

Dreher, A., 2006. Does globalization affect growth? Evidence from a new index of globalization. Appl. Econ. 38, 1091–1110.

Dreher, A., Gaston, N., Martens, P., 2010. Measuring globalization opening the black box. A critical analysis of globalization indices. J. Glob. Stud. 1, 129–147.

Dreher, A., Gaston, N., Martens, P., 2008. Measuring Globalisation, first ed. Springer-Verlag New York, New York, USA.

Erkan, C., Mucuk, M., Uysal, D., 2010. The impact of energy consumption on exports: The Turkish case. Asian Journal of Business Management 2, 17–23.

Etokakpan, M.U., Adedoyin, F.F., Vedat, Y., Bekun, F.V., 2020. Does globalization in Turkey induce increased energy consumption: insights into its environmental pros and cons. Environ. Sci. Pollut. Res.

Farhani, S., Solarin, S.A., 2017. Financial development and energy demand in the United States: New evidence from combined cointegration and asymmetric causality tests. Energy 134, 1029−1037.

Gengenbach, C., Urbain, J.-P., Westerlund, J., 2016. Error correction testing in panels with common stochastic trends. J. Appl. Econ. 31, 982−1004.

Ghani, G.M., 2012. Does trade liberalization effect energy consumption? Energy Policy 43, 285−290.

Gómez, M., Rodríguez, J.C., 2019. Energy consumption and financial development in NAFTA countries, 1971−2015. Applied Sciences 9, 302.

Gozgor, G., Mahalik, M.K., Demir, E., Padhan, H., 2020. The impact of economic globalization on renewable energy in the OECD countries. Energy Pol. 139, 111365.

Granger, C.W.J., Newbold, P., 1974. Spurious regressions in econometrics. J. Econ. 2, 111−120.

Gygli, S., Haelg, F., Potrafke, N., Sturm, J.E., 2019. The KOF globalisation index − revisited. Rev. Int. Organ. 14, 543−574.

Hadri, K., 2000. Testing for stationarity in heterogeneous panel data. Econ. J. 3, 148−161.

Hadri, K., Kurozumi, E., 2012. A simple panel stationarity test in the presence of serial correlation and a common factor. Econ. Lett. 115, 31−34.

Iheanacho, E., 2018. The role of globalisation on energy consumption in Nigeria. Implication for long run economic growth. ARDL and VECM analysis. Glob. J. Hum. Soc. Sci. 18, 10−28.

Im, K.S., Pesaran, M.H., Shin, Y., 2003. Testing for unit roots in heterogeneous panels. J. Econ. 115, 53−74.

Kahouli, B., 2017. The short and long run causality relationship among economic growth, energy consumption and financial development: evidence from South Mediterranean Countries (SMCs). Energy Econ. 68, 19−30.

Kao, C., 1999. Spurious regression and residual-based tests for cointegration in panel data. J. Econ. 90, 1−44.

Katircioglu, S.T., 2013. Interactions between energy and imports in Singapore: empirical evidence from conditional error correction models. Energy Policy 63, 514−520.

Koengkan, M., 2017. Is the globalization influencing the primary energy consumption? The case of Latin America and Caribbean countries. Cadernos UniFOA 12, 59−69.

Koengkan, M., 2018. The positive impact of trade openness on consumption of energy: fresh evidence from Andean community countries. Energy 158, 936−943.

Koengkan, M., Poveda, Y.E., Fuinhas, J.A., 2019. Globalisation as a motor of renewable energy development in Latin America countries. GeoJournal 0.

Köksal, C., Işik, M., Katircioglu, S., 2020. The role of shadow economies in ecological footprint quality: empirical evidence from Turkey. Environ. Sci. Pollut. Res. 27, 13457−13466.

Kyophilavong, P., Shahbaz, M., Anwar, S., Masood, S., 2015. The energy-growth nexus in Thailand: Does trade openness boost up energy consumption? Renew. Sustain. Energy Rev. 46, 265−274.

Larsson, R., Lyhagen, J., Löthgren, M., 2001. Likelihood-based cointegration tests in heterogeneous panels. Econ. J. 4, 109−142.

Lean, H.H., Smyth, R., 2010a. On the dynamics of aggregate output, electricity consumption and exports in Malaysia: evidence from multivariate Granger causality tests. Appl. Energy 87, 1963−1971.

Lean, H.H., Smyth, R., 2010b. Multivariate Granger causality between electricity generation, exports, prices and GDP in Malaysia. Energy 35, 3640−3648.

Levin, A., Lin, C.-F., Chu, J.C.-S., 2002. Unit root tests in panel data: asymptotic and finite-sample properties. J. Econ. 108, 1−24.

Li, L., 2010. An empirical analysis of relationship between export and energy consumption in Shandong Province. Int. J. Bus. Manag. 5, 214–216.

Liu, X., Zhang, S., Bae, J., 2017. The impact of renewable energy and agriculture on carbon dioxide emissions: investigating the environmental Kuznets curve in four selected ASEAN countries. J. Clean. Prod. 164, 1239–1247.

Maddala, G.S., Wu, S., 1999. A comparative study of unit root tests with panel data and a new simple test. Oxf. Bull. Econ. Stat. 61, 631–652.

Narayan, P.K., Smyth, R., 2009. Multivariate Granger causality between electricity consumption, exports and GDP: evidence from a panel of Middle Eastern countries. Energy Policy 37, 229–236.

Nasreen, S., Anwar, S., 2014. Causal relationship between trade openness, economic growth and energy consumption: A panel data analysis of Asian countries. Energy Policy 69, 82–91.

Nazlioglu, S., Karul, C., 2017. A panel stationarity test with gradual structural shifts: Re-investigate the international commodity price shocks. Econ. Model. 61, 181–192.

Nelson, C.R., Plosser, C.R., 1982. Trends and random walks in macroeconomic time series: some evidence and implications. J. Monetary Econ. 10, 139–162.

Newey, W.K., West, K.D., 1994. Automatic lag selection in covariance matrix estimation. Rev. Econ. Stud. 61 (4), 631–653.

Nnaji, C.E., Chukwu, J.O., Nnaji, M., 2013. Does domestic energy consumption contribute to exports? Empirical evidence from Nigeria. Int. J. Energy Econ. Policy 3, 297–306.

O'Connell, P.G.J., 1998. The overvaluation of purchasing power parity. J. Int. Econ. 44, 1–19.

Ozturk, I., Acaravci, A., 2013. The long-run and causal analysis of energy, growth, openness and financial development on carbon emissions in Turkey. Energy Econ. 36, 262–267.

Paramati, S.R., Apergis, N., Ummalla, M., 2017. Financing clean energy projects through domestic and foreign capital: the role of political cooperation among the EU, the G20 and OECD countries. Energy Econ. 61, 62–71.

Paramati, S.R., Ummalla, M., Apergis, N., 2016. The effect of foreign direct investment and stock market growth on clean energy use across a panel of emerging market economies. Energy Econ. 56, 29–41.

Pedroni, P., 1999. Critical values for cointegration tests in heterogeneous panels with multiple regressors. Oxf. Bull. Econ. Stat. 61, 653–670.

Pesaran, M.H., 2007. A simple panel unit root test in the presence of cross-section dependence. J. Appl. Econ. 22, 265–312.

Pesaran, M.H., 2004. General diagnostic tests for cross section dependence in panels (No. IZA DP No. 1240). In: Cambridge Working Papers in Economics.

Pesaran, M.H., Smith, V.L., Yamagata, T., 2013. Panel unit root tests in the presence of a multi-factor error structure. J. Econ. 175, 94–115.

Pesaran, M.H., Ullah, A., Yamagata, T., 2008. A bias-adjusted LM test of error cross-section independence. Econ. J. 11, 105–127.

Phillips, P.C.B., Hansen, B.E., 1990. Statistical inference in instrumental variables regression with I(1) Processes. Rev. Econ. Stud. 57, 125.

Reese, S., Westerlund, J., 2016. Panicca: panic on cross-section averages. J. Appl. Econ. 31, 961–981.

Rennen, W., Martens, P., 2003. The globalisation timeline. Integr. Assess. 4, 137–144.

Sabir, S., Gorus, M.S., 2019. The impact of globalization on ecological footprint: empirical evidence from the South Asian countries. Environ. Sci. Pollut. Res. 26, 33387–33398.

Sadorsky, P., 2012. Energy consumption, output and trade in South America. Energy Econ. 34, 476–488.

Sadorsky, P., 2011. Trade and energy consumption in the Middle East. Energy Econ. 33, 739–749.

Sami, J., 2011. Multivariate cointegration and causality between exports, electricity consumption and real income per capita: recent evidence from Japan. Int. J. Energy Econ. Policy 1, 59–68.

Saud, S., Danish, Chen, S., 2018. An empirical analysis of financial development and energy demand: establishing the role of globalization. Environ. Sci. Pollut. Res. 25, 24326–24337.

Sbia, R., Shahbaz, M., Hamdi, H., 2014. A contribution of foreign direct investment, clean energy, trade openness, carbon emissions and economic growth to energy demand in UAE. Econ. Model. 36, 191–197.

Shahbaz, M., Khan, S., Tahir, M.I., 2013. The dynamic links between energy consumption, economic growth, financial development and trade in China: Fresh evidence from multivariate framework analysis. Energy Econ. 40, 8–21.

Shahbaz, M., Lahiani, A., Abosedra, S., Hammoudeh, S., 2018a. The role of globalization in energy consumption: a quantile cointegrating regression approach. Energy Econ. 71, 161–170.

Shahbaz, M., Mahalik, M.K., Shahzad, S.J.H., Hammoudeh, S., 2019. Does the environmental Kuznets curve exist between globalization and energy consumption? Global evidence from the cross-correlation method. Int. J. Finance Econ. 24, 540–557.

Shahbaz, M., Mallick, H., Mahalik, M.K., Sadorsky, P., 2016. The role of globalization on the recent evolution of energy demand in India: implications for sustainable development. Energy Econ. 55, 52–68.

Shahbaz, M., Nasir, M.A., Roubaud, D., 2018b. Environmental degradation in France: the effects of FDI, financial development, and energy innovations. Energy Econ. 74, 843–857.

Shahbaz, M., Nasreen, S., Ling, C.H., Sbia, R., 2014. Causality between trade openness and energy consumption: what causes what in high, middle and low income countries. Energy Pol. 70, 126–143.

Shahbaz, M., Shahzad, S.J.H., Alam, S., Apergis, N., 2018c. Globalisation, economic growth and energy consumption in the BRICS region: the importance of asymmetries. J. Int. Trade Econ. Dev. 27, 985–1009.

Shahbaz, M., Shahzad, S.J.H., Mahalik, M.K., Sadorsky, P., 2018d. How strong is the causal relationship between globalization and energy consumption in developed economies? A country-specific time-series and panel analysis. Appl. Econ. 50, 1479–1494.

Solarin, S.A., Al-Mulali, U., 2018. Influence of foreign direct investment on indicators of environmental degradation. Environ. Sci. Pollut. Res. 25, 24845–24859.

Tong, T., Yu, T.E., 2018. Transportation and economic growth in China: a heterogeneous panel cointegration and causality analysis. J. Transport Geogr. 73, 120–130.

Ulucak, R., Yücel, A.G., İlkay, S.Ç., 2020. Dynamics of tourism demand in Turkey: panel data analysis using gravity model. Tour. Econ. 135481662090195.

Ulucak, R., Yücel, A.G., Koçak, E., 2019. The process of sustainability. In: Environmental Kuznets Curve (EKC). Elsevier, pp. 37–53.

Vanessa Smith, L., Leybourne, S., Kim, T.H., Newbold, P., 2004. More powerful panel data unit root tests with an application to mean reversion in real exchange rates. J. Appl. Econ. 19, 147–170.

Westerlund, J., 2008. Panel cointegration tests of the Fisher effect. J. Appl. Econ. 23, 193–233.

Westerlund, J., 2007. Estimating cointegrated panels with common factors and the forward rate unbiasedness hypothesis. J. Financ. Econ. 5, 491–522.

Westerlund, J., 2005. A panel CUSUM test of the null of cointegration. Oxf. Bull. Econ. Stat. 67, 231−262.

Westerlund, J., Edgerton, D.L., 2007. A panel bootstrap cointegration test. Econ. Lett. 97, 185−190.

Westerlund, J., Hosseinkouchack, M., 2016. Modified CADF and CIPS panel unit root statistics with standard chi-squared and normal limiting distributions. Oxf. Bull. Econ. Stat. 78, 347−364.

Westerlund, J., Larsson, R., 2009. A note on the pooling of individual PANIC unit root tests. Econ. Theor. 25, 1851.

Chapter 8

The electricity retail sales and economic policy uncertainty: the evidence from the electricity end-use, industrial sector, and transportation sector

Faik Bilgili[1], Pelin Gençoğlu[2], Sevda Kuşkaya[3], Fatma Ünlü[1]
[1]*Erciyes University, Faculty of Economics and Administrative Sciences, Kayseri, Turkey;* [2]*Erciyes University, Research and Application Center of Kayseri, Kayseri, Turkey;* [3]*Erciyes University, Justice Vocational College, Department of Law, Kayseri, Turkey*

1. Introduction

Recently, the concept of uncertainty has become one of the important issues in the literature. According to Rossi et al. (2016), there are two kinds of uncertainty. One of them is related to risk. It is defined by situations where one knows the odds of the unknown. This unknown indicates the probability distribution of the stochastic events. Second is that, even though rational agents can characterize the distributions, they may not assign the correct possibilities to future results. However, there is no consensus about the definition of uncertainty. The concept of uncertainty is quite important in all sciences. With the liberalization of capital movements since the 1980s, the process of integration of countries into global markets began to gain momentum. This process has led countries to intensively experience external shocks as well as internal shocks. These shocks have brought economic or/and political uncertainty to the forefront. The uncertainty reflects a lack of information about current events and policy actions could cause uncertainty about future outcomes for businesses, households, and the decisions of governments.

Economic/political uncertainty has a damaging impact on key macroeconomic variables, such as investment. It reduces the tendencies of the risk-taking of investors. So, as investments decrease, energy demand diminishes

Energy-Growth Nexus in an Era of Globalization. https://doi.org/10.1016/B978-0-12-824440-1.00016-3
203

due to reducing production. Given the strong link between economic development level of societies and energy use, the prediction of future energy demand plays a key role in determining an effective policy. However, the estimation of future energy demand is very complex and involves uncertainties depending on various factors such as technological development, sociocultural behavior, cost of energy, the sustainability of energy resources, enforced policies, etc. From this point of view, economic/political uncertainty has become one of the important elements of energy demand function which contains energy price, price of complementary goods, price of substitute goods, income, etc.

Many studies are investigating energy demand from different aspects. From the existing literature, it might be observed that there does not exist any studies investigating the relation between economic/political uncertainty and energy demand directly. The motivation of this paper lies in three points: (1) the economic/political uncertainty, which might be one of the main drivers of energy demand, has been neglected by most researchers and policymakers. In this context, although there exist many seminal studies investigating the energy demand from different aspects, there is a limited number of studies searching the nexus between energy demand and economic/political uncertainty. (2) The papers related to causality in the existing literature do not follow the structural vector autoregressive (SVAR) model as a method, hence they do not monitor the responses of energy demand to the structural impulses of uncertainty. (3) Eventually, this chapter by considering (1) and (2), aims at contributing to both theoretical and empirical literature. To this end, this work first establishes a theoretical demand function for energy (renewable and nonrenewable) in which political/economic uncertainty and geological risk are considered considerable determinants together with other basic theoretical determinants. Later, the short-run and long-run structural impacts of political/economic uncertainty and geological risk on energy demand will be estimated.

In this context, this study aims to contribute to analyze the relationship between economic/political uncertainty and energy demand by applying the SVAR model for the USA. In this paper, we will employ the variables such as electricity retail sales to represent energy demand in terms of end-use (EEU), industrial sector (IND) and transportation sector (TRANS), gross domestic product (GDP), energy price index (EPI), economic policy uncertainty index (EPUI), and World uncertainty index (WUI) in the USA for the quarterly period 1996−2019. In the literature related to energy demand, it is seen that it is preferred sectoral consumption as well as total energy consumption as variables. Energy consumption is classified by government agencies such as the US Energy Information Administration (EIA), taking into account the energy demand of subsectors such as residential, industrial, commercial, and transportation. This classification is also aggregated as residential and nonresidential in some studies. Additionally, as environmental impacts become more important, the source of the energy consumed (renewable or nonrenewable) has been another important classification title.

This book chapter will follow the sections of (Section 1) introduction, (Section 2) the literature review, (Section 3) the econometric methodology and data, (Section 4) empirical findings, and (Section 5) some relevant considerable remarks and policy implications.

2. Literature review

Recently, many studies are examining which factors affecting energy demand in the literature. Among these factors, political/economic risk and uncertainties come to the fore in this field. According to this, the literature review has been made by considering the studies both examining the impacts of political/economic risk and uncertainties on energy demand and defining the determinants of energy demand. These existing seminal papers in the literature are shown in Table 8.1, in terms of country or/and country groups of the research, period, methodology used, and the concluding remarks.

Table 8.1 includes studies that can be classified in the context of energy demand (energy consumption) and uncertainties in energy markets.

(i) *The studies examining the energy consumption-economic growth nexus:* The studies in this category can be classified as country-specific studies and multicountry studies. Considering the country-specific studies, one may say that the literature follows usually cointegration and causality analyses. Johansen (Yoo, 2005; Erdal et al., 2008; Akinlo, 2009; Ahamad and Islam, 2011; Abid and Sebri, 2012), ARDL (Halicioglu, 2007; Chandran et al., 2010; Ghosh, 2010; Alam et al., 2012; Islam et al., 2013; Pau and Fu, 2013; Sbia et al., 2014), and VAR (Holtedahl and Joutz, 2004; Sari and Soytas, 2004; Scarcioffolo and Etienne, 2018) have been generally preferred as cointegration analysis. All of these studies concluded that there exists cointegration between related variables. On the other side, some of these papers used together Granger causality analyses with their preferred methods. For instance, Yoo (2005), Erdal et al. (2008), Ahamad and Islam (2011), Abid and Sebri (2012), and Pao and Fu (2013) found bidirectional causality. Also, while Akinlo (2009) and Alam et al. (2012) determined unidirectional causality from energy consumption to economic growth, Halicioglu (2007) reached the reverse causality. Regarding multicountry studies, it can be seen that several estimation methods have used such as GMM (Al-Iriani, 2006; Omri, 2013; Tang and Abosedra, 2014; Komal and Abbas, 2015; Bhattacharya et al., 2017), panel FMOLS and DOLS (Salim and Rafiq, 2012; Koçak and Sarkgüneşi, 2017), and panel cointegration (Zeshan and Ahmed, 2013; Aissa et al., 2014; Khan et al., 2014; Ohler and Fetters, 2014, etc.). Besides the panel estimations, there are also the studies (Salim and Rafiq, 2012; Ohler and Fetters, 2014; Bercu et al., 2019, etc.) that used Granger causality. While Omri (2013), Ohler and Fetters (2014), Cai et al. (2018), and Bercu et al. (2019) found bidirectional causality, it was determined unidirectional causality from economic growth to energy consumption by Al-Iriani (2006).

TABLE 8.1 Literature review.

Author(s)	Country	Period	Methods	Result
Holtedahl and Joutz (2004)	Taiwan	1959–95	VAR	The impact of income and price on consumption are higher in the long-run.
Sari and Soytas (2004)	Turkey	1969–99	VAR	Different energy consumptions have different effects on income.
Wolde-Rufael (2005)	19 African countries	1971–2001	ARDL, Toda-Yamamoto causality	There exists a relationship between the gross domestic product (GDP) and energy consumption for eight countries and causality for 10 countries in the long-run.
Yoo (2005)	Korea	1970–2002	Johansen cointegration, Granger causality	Bidirectional causality exists between electricity consumption and GDP.
Al-Iriani (2006)	6 GCC countries	1970–2002	Panel cointegration, GMM	The causality runs from GDP to energy consumption.
Halicioglu (2007)	Turkey	1968–2005	ARDL, Granger causality	There exists cointegration between all variables and unidirectional causality from real GDP to electricity consumption, and from electricity price to electricity consumption.
Meijer et al. (2007)	Netherlands	1990–2006	Case study	The decision of whether an entrepreneur takes a specific action depends on the balance between perceived uncertainty and motivation.
Squalli (2007)	11 OPEC countries	1980–2003	ARDL, Toda-Yamamoto causality	The causality runs from GDP to energy consumption for Algeria, Iraq, Libya, and bidirectional causality for Iran, Qatar, Venezuela.

Erdal et al. (2008)	Turkey	1970–2006	Johansen cointegration, Pair-wise Granger causality	There exists bidirectional causality between energy consumption and GDP.
Akinlo (2009)	Nigeria	1980–2006	Johansen-Juselius cointegration, Granger causality	The causality runs from electricity consumption to GDP.
Barradale (2010)	USA	1999–2007	Comparative analysis by using 2006 Wind Industry Survey	Policy uncertainty exacerbates the decline in wind energy investments.
Chandran et al. (2010)	Malaysia	1971–2003	ARDL	There exist electricity consumption, real GDP and price share a long-run relationship.
Ghosh (2010)	India	1971–2006	ARDL	Unidirectional causality exists from economic growth to energy supply and energy supply to carbon emissions in the short-run.
Ahamad and Islam (2011)	Bangladesh	1971–2008	Johansen cointegration, Granger causality	The bidirectional causal Relationship between electricity consumption and economic growth
Abid and Sebri (2012)	Tunisia	1980–2007	Johansen cointegration, Granger causality	A bidirectional causal relationship between energy consumption and growth, while different directions of causality for different sectors.
Alam et al. (2012)	Bangladesh	1972–2006	ARDL, Granger causality	Unidirectional causality from energy consumption to economic growth
Fuinhas and Marques (2012)	Portugal, Italy, Greece, Spain, and Turkey	1965–2009	ARDL	Obtained findings support the feedback hypothesis.
Salim and Rafiq (2012)	6 Major Emerging Economies	1980–2006	FMOLS, DOLS, Granger causality	The main determinants are income and pollutant emission in Brazil, China, India, and Indonesia; however, it is income in the Philippines and Turkey.

Continued

TABLE 8.1 Literature review.—cont'd

Author(s)	Country	Period	Methods	Result
Sanquist et al. (2012)	The USA	2001, 2005	Factor analysis	It is explained by these factors approximately 40% of electricity consumption.
Islam et al. (2013)	Malaysia	1971–2009	ARDL	Energy consumption is influenced by economic growth and financial development.
Kavgic et al. (2013)	Serbia (Belgrade)	1946–2010	City-Scale Domestic Energy Model (BEDEM)	The scenarios predicting the energy consumption and carbon dioxide (CO_2) emissions of the housing stock indicate different high sensitivity levels.
Omri (2013)	14 MENA countries	1990–2011	GMM	There exists a bidirectional causal relationship between energy consumption and economic growth.
Pao and Fu (2013)	Brazil	1980–2010	ARDL, Granger causality	All kind of energy is cointegrated to GDP. There is bidirectional causality between economic growth and total renewable energy consumption.
Zeshan and Ahmed (2013)	5 South Asian countries	1980–2010	Panel cointegration	Energy consumption has a positive effect on output, whereas CO_2 emission has a negative effect.
Aissa et al. (2014)	11 African Countries	1980–2008	Panel cointegration	There are significant and positive effects of renewable energy consumption and trade on output.

Khan et al. (2014)	OECD, non-OECD and MENA countries	1975–2011	Panel cointegration, SUR	GDP per capita has a positive impact on energy consumption in countries, except high income OECD and non-OECD
Ohler and Fetters (2014)	20 OECD Countries	1990–2008	Panel data, Granger causality	There is a positive relationship between renewable electricity generation and GDP. There exist bidirectional causality between hydroelectricity and waste generation with GDP growth.
Sbia et al. (2014)	United Arab Emirates	1975:Q1–2011:Q4	ARDL, Granger causality	FDI, trade openness, and carbon emissions decline energy demand, but economic growth and clean energy have a positive impact.
Tang and Abosedra (2014)	24 MENA countries	2001–09	GMM	Energy consumption and tourism a have positive impact on economic growth while political instability affects economic growth negatively.
Eyre and Baruah (2015)	UK	2010–50	Quantified scenarios	Proposed scenarios point out that greater use of energy efficiency and biomass can also play a significant role.
Komal and Abbas (2015)	Pakistan	1972–2012	GMM	The positive and significant impact of economic growth and urbanization on energy consumption, but the impact of energy prices is a significant negative.
Kyophilavong et al. (2015)	Thailand	1971–2012	Bayer and Hanck cointegration	There exists a cointegration between the variables. Bidirectional causality exists between energy consumption and economic growth.
Irandoust (2016)	NORDIC countries	1975–2012	VAR	Unidirectional causality exists from technological innovation to renewable energy and from growth to renewable energy.

Continued

TABLE 8.1 Literature review.—cont'd

Author(s)	Country	Period	Methods	Result
Whitford (2016)	The USA	1972–2004	Granger causality	The changes in oil prices affect all types of hearings.
Bhattacharya et al. (2017)	85 developed and developing economies	1991–2012	System-GMM and FMOLS	Renewable energy deployment and institutions have promoted economic growth and reducing CO_2 emissions.
Koçak and Sarkgüneşi (2017)	Nine Black Sea and Balkan countries	1990–2012	Panel FMOLS and DOLS, Heterogeneous panel causality	It is a valid growth hypothesis for Bulgaria, Greece, Macedonia, Russia, and Ukraine; feedback hypothesis for Albania, Georgia, and Romania; neutrality hypothesis for Turkey.
Cai et al. (2018)	G-7 countries	1965–2015	Bootstrap, ARDL and causality test	There is no cointegration except, Germany and Japan. For Germany, there exist bidirectional causality between clean energy consumption and CO_2 emissions.
Scarcioffolo and Etienne (2018)	The USA	1994–2017	VAR, MGARCH	There are exists spillover effects between the two energy markets and the economic policy uncertainty.
Yu et al. (2018)	10 industrial sectors in the USA	02.01.1997 –18.04.2018	EGARCH-MIDAS, DCC-MIDAS	GEPU is positively related to the long-run volatility of financials and consumer discretionary industries; however, it is negatively related to information technology, materials, telecommunication services, and energy.
Bercu et al. (2019)	14 Central and Eastern European countries	1995–2017	Panel cointegration, Pairwise Granger causality	Good governance directly influences economic growth and indirectly through energy consumption. There exists bidirectional causality between electricity consumption and GDP.

Blas et al. (2019)	EU countries	1995–2009	MEDEAS-integrated assessment modeling	The changes in the improvement in energy efficiency and substitution of the final energy depend on the perception of scarcity of the different economic actors.
Hailemariam et al. (2019)	G-7 countries	1997:01 –2018:06	Nonparametric panel data	The impact of oil prices on economic policy uncertainty is time-varying.
Valadkhani et al. (2019)	60 Major Pollutant Countries	1965–2016	Panel breaking regression	There is no important effect on renewable energy in reducing CO_2 emissions, but some evidence indicates substitution toward hydroelectric in low-income countries.
Anton and Nucu (2020)	EU-28	1990–2015	Panel data analysis	Financial development has a positive effect on renewable energy consumption.

(ii) *The studies investigating the uncertainties in energy markets:* Meijer et al. (2007), Barradele (2010), Eyre and Baruah (2015), Scarcioffolo and Etienne (2018), and Hailemariam et al. (2019). One may assert that all of these studies reached that the uncertainties have negative impacts on energy markets for several aspects.

Briefly, while the studies related to energy demand focus on energy consumption and economic growth nexus, the studies examining political/economic risk and uncertainties mostly deal with fundamental macroeconomic indicators. Considering literature reviews as mentioned above, one might observe that there are limited numbers of studies directly examining the relationship between political/economic risk and uncertainties and energy demand. Additionally, although most of the studies applied cointegration and causality estimations, no studies have been coincided by using SVAR to determine the impact of uncertainties on energy consumption. However, this method allows us to monitor the responses of energy demand to the structural impulses of uncertainties. So, both short and long term structural effects political/economic risk and uncertainties on energy demand will be estimated in our studies by employing SVAR estimation. From all these points, the purpose of this paper is to contribute to the literature by filling these gaps.

3. Data and methodology: structural vector autoregressive model

Table 8.2 gives the data definition and the source of data. Eqs. (8.1−8.5) reveal the basic equations of SVAR models.

As given in the table, this work employs the data for Electricity End-Use (EEU), Electricity Retail Sales to the Industrial Sector (IND), Electricity Retail Sales to the Transportation Sector (TRANS), Country-level data of WUI (T6), Economic Policy Uncertainty Index (EPUI), Energy Price Index (EPI), GDP, and Population. T6 indicates the country-level data of the World Uncertainty Index (WUI) which contains the three-quarter weighted moving average of the WUI for 143 countries from 1996Q3 to 2019Q1. The three-quarter weighted moving average, computed as follows: 1996Q4 = (1996Q4*0.6) + (1996Q3*0.3) + (1996Q2*0.1)/3. WUI index measures overall uncertainty across the globe. The index is unbalanced GDP weighted average for 142 countries. The population data have been used to reach per capita GDP, per capita EEU, per capita IND, and per capita TRANS. Finally, this paper follows logarithmic forms of the variables. The quarterly data span from 1996:1−2019:2.

The paper employed SVAR approach that considers the impacts of unexpected shock(s) of the variable(s) on other variables in the vector autoregressive (VAR) system. The VAR model is a multivariate model that allows

TABLE 8.2 Data and source.

Data	Source
EEU = Electricity End Use, Total = Million Kilowatt-hours	EIA (2020), https://www.eia.gov/totalenergy/data/monthly/
ES Industrial = Electricity Retail Sales to the Industrial Sector = Million Kilowatt-hours	EIA (2020), https://www.eia.gov/totalenergy/data/monthly/
ES Transportation = Electricity Retail Sales to the Transportation Sector = Million Kilowatt-hours	EIA (2020), https://www.eia.gov/totalenergy/data/monthly/
T6: Country-level data of WUI	WUI (2020), https://worlduncertaintyindex.com/
EPUI = Economic Policy Uncertainty Index (policy-related economic uncertainty)	EPU (2020), https://www.policyuncertainty.com/
EPI = Energy Price Index (Consumer Price Index for All Urban Consumers: Energy in U.S. City Average, Index, 1982−84 = 100, Monthly, Seasonally Adjusted)	Federal Reserve Economic Data (2020), https://fred.stlouisfed.org
Real GDP = Seasonally Adjusted	Federal Reserve Economic Data (2020), https://fred.stlouisfed.org
Population	Federal Reserve Economic Data (2020), https://fred.stlouisfed.org

investigating the relationship between a set of economic variables. All parameters in the VAR system are determined endogenously. The SVAR can capture the dynamic behavior of all the variables in the model (Forni and Gambetti, 2014). By employing an SVAR approach, where all variables can be modeled as endogenous, we can account for all direct and induced indirect effects (Piroli et al., 2015; Camba-Mendez, 2012). The SVAR model allows the impulse-response functions to be obtained by introducing short- and long-term constraints into the model calculation process. In this context, the SVAR model is defined as follows:

$$\Delta_t = \Gamma_0 u_t + \Gamma_1 \Delta_{t-1} + \Gamma_2 \Delta_{t-2} + \Gamma_3 \Delta_{t-3} + \cdots + \Gamma_n \Delta_{t-n} = \sum_{i=0}^{\infty} \Gamma_i u_{t-i} \quad (8.1)$$

Here, Γ is the parameter of the structural parameters $(n \times n)$ and u_t is the parameter with normal distribution with zero mean and constant variance, showing structural shocks. In matrix form, it is expressed as:

$$\Delta_t = \Gamma(L) u_t \quad (8.2)$$

The restrictions are subject to change due to the assumptions of economic theories. For instance, in a SVAR model with X_1, X_2, X_3, X_4, and X_5 variables can be modeled in the matrix for as depicted in Eq. (8.3).

$$
\begin{pmatrix} X_t^1 \\ X_t^2 \\ X_t^3 \\ X_t^4 \\ X_t^5 \end{pmatrix} = \begin{pmatrix} 1 & A_{12}(L) & 0 & 0 & A_{15}(L) \\ 0 & 1 & 0 & 0 & A_{25}(L) \\ 0 & A_{32}(L) & 1 & A_{34}(L) & 0 \\ 0 & A_{42}(L) & A_{43}(L) & 1 & 0 \\ A_{51}(L) & A_{52}(L) & A_{53}(L) & A_{54}(L) & 1 \end{pmatrix} \begin{pmatrix} u_t^1 \\ u_t^2 \\ u_t^3 \\ u_t^4 \\ u_t^5 \end{pmatrix} \tag{8.3}
$$

and,

$$
\Gamma = \begin{pmatrix} \beta_{11} & \beta_{12} & \cdots & \beta_{15} \\ \beta_{21} & \beta_{22} & \cdots & \beta_{25} \\ \vdots & \vdots & \ddots & \vdots \\ \beta_{51} & \beta_{52} & \cdots & \beta_{55} \end{pmatrix} \tag{8.4}
$$

where L indicates the lag operator. Eq. (8.3) indicates that, for instance, at first row, X_1 does not respond to third and fourth variables in the system contemporaneously, it responds to them with lag. The variable X_1, however, reacts to X_2 and X_5 contemporaneously.

This work employs the variables in logarithmic forms and aims at monitoring the responses of energy demand to the shocks in other variables. In SVAR, the relevant long-run constraints take the form (Lee et al., 2012) as shown in Eq. (8.5).

$$
\begin{pmatrix} \Delta X_t^{LGDP} \\ \Delta X_t^{LWUI} \\ \Delta X_t^{LEPUI} \\ \Delta X_t^{LEPI} \\ \Delta X_t^{LED} \end{pmatrix} = \begin{pmatrix} B_{11}(L) & 0 & 0 & 0 & 0 \\ B_{21}(L) & B_{22}(L) & 0 & 0 & 0 \\ B_{31}(L) & B_{32}(L) & B_{33}(L) & 0 & 0 \\ B_{41}(L) & B_{42}(L) & B_{43}(L) & B_{44}(L) & 0 \\ B_{51}(L) & B_{52}(L) & B_{53}(L) & B_{54}(L) & B_{55}(L) \end{pmatrix} \begin{pmatrix} \varepsilon_t^{LGDP} \\ \varepsilon_t^{LWUI} \\ \varepsilon_t^{LEPUI} \\ \varepsilon_t^{LEPI} \\ \varepsilon_t^{LED} \end{pmatrix}
$$

$$\tag{8.5}$$

where LGDP, LWUI, LEPUI, LEPI, and LED denote log of per capita GDP, the log of country-level World uncertainty index, log of economic policy uncertainty index, log of the energy price index, and log of per capita retail electricity sales (energy demand), respectively. In Eq. (8.5), Δ is a difference operator. Hence, ΔX_t^{LGDP}, ΔX_t^{LWUI}, ΔX_t^{LEPUI}, ΔX_t^{LEPI}, and ΔX_t^{LED} depict the changes (differences) in LGDP, LWUI, LEPUI, LEPI, and LED, respectively.

First row of Eq. (8.5) indicates that LGDP does not respond to other variables contemporaneously, it responds to other variables with lag.

Third row indicates that the US economic policy uncertainty index (LEPUI) responds to the only LGDP, and LWUI, it does not respond to other variables contemporaneously but responds to other variables with lag. Fifth row states that log of per capita electricity retail sales (electricity demand) responds to all variables in the system contemporaneously. The next section will present the relevant outputs from the SVAR estimations.

4. Estimation output

In this section, we monitored two SVAR models. First group SVAR models contain four variables and second group SVAR models include five variables.

The outputs from four variable SVAR models are depicted in Tables 8.3–8.5. The results from five variable SVAR models are shown in Tables 8.6–8.8. Table 8.3 yields the estimations of SVAR with LPC_GDP, LWUI_T6, LEPUI, and LEEU. Table 8.3 indicates as follows:

(i) Per-capita electricity end-use (LEEU) has a positive income elasticity of 0.207. This output implies that electricity is a normal good. Normal good classification comprises two group commodities. Weak superior good (if income elasticity <1) and strong superior good (if income elasticity >1). It appears that electricity is a weak superior good.

(ii) As the country-level World uncertainty index (LWUI_T6) goes up by 1%, LEEU will be lower by 0.025%. The elasticity of retail electricity sales in terms of country-level (US) World uncertainty is −0.025. There exists a negative correlation between LEEU and LWUI as expected. The higher the uncertainty is, the lower the electricity demand will be.

(iii) As the economic policy uncertainty index (LEPUI) goes up by 1%, LEEU will decrease by 0.071%. The elasticity of retail electricity sales in terms of economic policy uncertainty is −0.071. There exists a negative association between LEEU and LEPUI as expected. The higher the economic policy uncertainty in the US is, the lower the retail electricity demand in the US will be.

Table 8.4 yields the estimations of SVAR with LPC_GDP, LWUI_T6, LEPUI, and LPC_IND, and explores as follows:

(i) The per capita electricity retail sales to the industrial sector (LPC_IND) responds to per capita GDP negatively (−0.408). The income elasticity of LPC_IND with respect to LGDP is −0.408. This output might yield that electricity in the industrial sector is an inferior good.

(ii) LPC_IND responds to the country-level World uncertainty negatively (−0.044). There exist a negative association between LPC_IND and LWUI_T6 as expected. As LWUI_T6 increases by 1%, LPC_IND will go down by 0.044%.

TABLE 8.3 Structural vector autoregressive estimates with recursive long run impulse-responses by F triangular matrix: 1997Q1–2019Q2 LPC_GDP, LWUI_T6, LEPUI, LEEU.

Model: e = Phi*Fu where E[uu'] = I

F=	C(1)	0	0	0
	C(2)	C(5)	0	0
	C(3)	C(6)	C(8)	0
	C(4)	C(7)	C(9)	C(10)

	Coefficient	Std. Error	z-Statistic	Prob.
C(1)	−0.994221	0.074105	−13.41638	.0000
C(2)	−6.906700	0.546238	−12.64411	.0000
C(3)	−2.237375	0.201321	−11.11349	.0000
C(4)	0.207093	0.018162	11.40266	.0000
C(5)	1.732781	0.129154	13.41641	.0000
C(6)	0.620047	0.102877	6.027042	.0000
C(7)	−0.024521	0.009394	−2.610347	.0090
C(8)	0.871958	0.064992	13.41641	.0000
C(9)	−0.070664	0.007561	−9.346139	.0000
C(10)	0.051460	0.003836	13.41641	.0000
Log-likelihood	457.2838			

Estimated S matrix:

0.000290	−0.001565	−0.003667	0.003671
−0.223183	0.542474	−0.515412	−0.267086
0.009242	0.164666	0.127506	0.073337
−0.022167	0.001283	0.005501	0.007323

Estimated F matrix:

−0.994221	0.000000	0.000000	0.000000
−6.906700	1.732781	0.000000	0.000000
−2.237375	0.620047	0.871958	0.000000
0.207093	−0.024521	−0.070664	0.051460

Notes: VAR Lag = 4 based on SC. The estimation method: Maximum likelihood via Newton–Raphson (analytic derivatives). Convergence is achieved after 32 iterations. Structural VAR is just-identified.

TABLE 8.4 Structural vector autoregressive estimates with recursive long run impulse-responses by F triangular matrix: 1997Q1–2019Q2 LPC_GDP, LWUI_T6, LEPUI, and LPC_IND.

		Model: e = Phi*Fu where E[uu′] = I		
F=	C(1)	0	0	0
	C(2)	C(5)	0	0
	C(3)	C(6)	C(8)	0
	C(4)	C(7)	C(9)	C(10)
		Coefficient	Std. Error	z-Statistic	Prob.
C(1)		0.480723	0.035634	13.49074	.0000
C(2)		2.916430	0.269358	10.82736	.0000
C(3)		0.825256	0.114276	7.221605	.0000
C(4)		−0.408018	0.032087	−12.71602	.0000
C(5)		1.532836	0.113621	13.49074	.0000
C(6)		0.522303	0.088420	5.907081	.0000
C(7)		−0.044327	0.010201	−4.345402	.0000
C(8)		0.758317	0.056210	13.49074	.0000
C(9)		−0.060943	0.008536	−7.139945	.0000
C(10)		0.069085	0.005121	13.49074	.0000
Log-likelihood		471.0963			
Estimated S matrix:					
0.004542		−0.001063	−0.001428	0.002280	
−0.204000		0.516676	−0.592266	0.131354	
−0.017843		0.163650	0.144678	0.025159	
−0.006867		−0.002505	0.001991	0.020326	
Estimated F matrix:					
0.480723		0.000000	0.000000	0.000000	
2.916430		1.532836	0.000000	0.000000	
0.825256		0.522303	0.758317	0.000000	
−0.408018		−0.044327	−0.060943	0.069085	

Notes: SVAR lag = 3 based on SC. Estimation method: Maximum likelihood via Newton–Raphson (analytic derivatives). Convergence is achieved after 21 iterations.

TABLE 8.5 Structural vector autoregressive estimates with recursive long run impulse-responses by F triangular matrix: 1997Q1–2019Q2 LPC_GDP, LWUI_T6, LEPUI, and LPC_TRANS.

Model: e = Phi*Fu where E[uu'] = I

F=	C(1)	0	0	0
	C(2)	C(5)	0	0
	C(3)	C(6)	C(8)	0
	C(4)	C(7)	C(9)	C(10)

	Coefficient	Std. Error	z-Statistic	Prob.
C(1)	0.406576	0.029812	13.63817	.0000
C(2)	1.676971	0.262975	6.376927	.0000
C(3)	0.450171	0.086169	5.224303	.0000
C(4)	0.373359	0.035816	10.42428	.0000
C(5)	2.241732	0.164372	13.63818	.0000
C(6)	0.425539	0.073225	5.811378	.0000
C(7)	−0.008498	0.023086	−0.368091	.7128
C(8)	0.638840	0.046842	13.63818	.0000
C(9)	−0.181525	0.018853	−9.628598	.0000
C(10)	0.128758	0.009441	13.63818	.0000
Log-likelihood	366.5859			

Estimated S matrix:

0.005466	0.001338	0.000201	0.001066
−0.215362	0.711446	−0.373199	−0.145039
−0.120101	0.078829	0.156430	0.072525
−0.024381	0.000530	−0.040258	0.050462

Estimated F matrix:

0.406576	0.000000	0.000000	0.000000
1.676971	2.241732	0.000000	0.000000
0.450171	0.425539	0.638840	0.000000
0.373359	−0.008498	−0.181525	0.128758

Notes: SVAR lag = 1 based on SC. Estimation method: Maximum likelihood via Newton–Raphson (analytic derivatives). Convergence is achieved after 21 iterations.

TABLE 8.6 Structural vector autoregressive estimates with recursive long run impulse-responses by F triangular matrix: 1997Q1–2019Q2 LPC_GDP, LWUI_T6, LEPUI, LEPI, and LPC_EEU.

Model: e = Phi*Fu where E[uu′] = I

F=	C(1)	0	0	0	0
	C(2)	C(5)	0	0	0
	C(3)	C(6)	C(9)	0	0
	C(4)	C(7)	C(10)	C(12)	0
	1	C(8)	C(11)	C(13)	C(14)
	Coefficient	Std. Error	z-Statistic	Prob.	
C(1)	1.498902	0.247088	6.066280	.0000	
C(2)	5.136753	4.905000	1.047248	.2950	
C(3)	−3.941089	1.299388	−3.033034	.0024	
C(4)	5.733520	1.426936	4.018064	.0001	
C(5)	2.355857	0.542508	4.342531	.0000	
C(6)	1.023961	1.527633	0.670293	.5027	
C(7)	−0.416956	0.842728	−0.494770	.6208	
C(8)	−0.124060	0.284182	−0.436553	.6624	
C(9)	2.067959	0.618842	3.341661	.0008	
C(10)	−1.173117	0.542817	−2.161163	.0307	
C(11)	−0.388148	0.107900	−3.597296	.0003	
C(12)	0.572620	0.041991	13.63681	.0000	
C(13)	−0.000565	0.017670	−0.031993	.9745	
C(14)	0.090254	0.016070	5.616206	.0000	
Log-likelihood	272.7530				
LR test for over-identification:					
Chi-square(1)	480.9198		Probability	.0000	
Estimated S matrix:					
0.025063	−0.001167	−0.006229	0.004504	0.000256	
2.792133	0.369905	−1.358546	−0.227501	0.043251	
−2.376264	0.339286	0.815800	−0.025148	−0.007592	

Continued

TABLE 8.6 Structural vector autoregressive estimates with recursive long run impulse-responses by F triangular matrix: 1997Q1–2019Q2 LPC_GDP, LWUI_T6, LEPUI, LEPI, and LPC_EEU.—cont'd

Model: e = Phi*Fu where E[uu'] = I

| 0.003758 | −0.009864 | −0.034885 | 0.036861 | 0.005904 |
| 0.991258 | −0.098177 | −0.351347 | −0.006672 | 0.121735 |

Estimated F matrix:

1.498902	0.000000	0.000000	0.000000	0.000000
5.136753	2.355857	0.000000	0.000000	0.000000
−3.941089	1.023961	2.067959	0.000000	0.000000
5.733520	−0.416956	−1.173117	0.572620	0.000000
1.000000	−0.124060	−0.388148	−0.000565	0.090254

Notes: SVAR lag = 1 based on SC. Estimation method: Maximum likelihood via Newton–Raphson (analytic derivatives). Convergence is achieved after 41 iterations.

TABLE 8.7 Structural vector autoregressive estimates with recursive long run impulse-responses by F triangular matrix: 1997Q1–2019Q2 LPC_GDP, LWUI_T6, LEPUI, LEPI, and LPC_IND.

Model: e = Phi*Fu where E[uu'] = I

F=	C(1)	0	0	0	0
	C(2)	C(5)	0	0	0
	C(3)	C(6)	C(9)	0	0
	C(4)	C(7)	C(10)	C(12)	0
	1	C(8)	C(11)	C(13)	C(14)

	Coefficient	Std. Error	z-Statistic	Prob.
C(1)	0.865951	0.087738	9.869794	.0000
C(2)	−11.66524	3.961646	−2.944545	.0032
C(3)	−3.647797	1.156231	−3.154904	.0016
C(4)	−2.866040	1.096140	−2.614666	.0089
C(5)	10.26738	2.629223	3.905100	.0001
C(6)	3.088685	0.782111	3.949164	.0001

Continued

TABLE 8.7 Structural vector autoregressive estimates with recursive long run impulse-responses by F triangular matrix: 1997Q1–2019Q2 LPC_GDP, LWUI_T6, LEPUI, LEPI, and LPC_IND.—cont'd

Model: e = Phi*Fu where E[uu'] = I			
C(7)	2.669699	0.723354	3.690724	.0002
C(8)	−1.034990	0.103063	−10.04230	.0000
C(9)	0.602925	0.087629	6.880448	.0000
C(10)	0.086647	0.229197	0.378044	.7054
C(11)	−0.110944	0.059870	−1.853074	.0639
C(12)	0.910183	0.178136	5.109485	.0000
C(13)	−0.161435	0.043264	−3.731364	.0002
C(14)	0.086233	0.010537	8.183648	.0000
Log-likelihood	321.6808			

LR test for over-identification:

Chi-square(1)	485.5641		Probability	.0000

Estimated S matrix:

−0.022879	0.017260	0.000482	0.005417	0.001741
−1.060117	0.687357	−0.577396	−0.430523	0.075624
−1.639844	1.022757	0.242220	0.000258	−0.040539
−0.286437	0.143457	0.012848	0.045115	0.014267
0.480127	−0.300944	−0.031917	−0.049193	0.053695

Estimated F matrix:

0.865951	0.000000	0.000000	0.000000	0.000000
−11.66524	10.26738	0.000000	0.000000	0.000000
−3.647797	3.088685	0.602925	0.000000	0.000000
−2.866040	2.669699	0.086647	0.910183	0.000000
1.000000	−1.034990	−0.110944	−0.161435	0.086233

Notes: SVAR lag = 1 based on SC. Estimation method: Maximum likelihood via Newton–Raphson (analytic derivatives). Convergence is achieved after 54 iterations.

TABLE 8.8 Structural vector autoregressive estimates with recursive long run impulse-responses by F triangular matrix: 1997Q1–2019Q2. LPC_GDP, LWUI_T6, LEPUI, LEPI, and LPC_TRANS.

Model: e = Phi*Fu where E[uu'] = I

F=	C(1)	0	0	0	0
	C(2)	C(5)	0	0	0
	C(3)	C(6)	C(9)	0	0
	C(4)	C(7)	C(10)	C(12)	0
	1	C(8)	C(11)	C(13)	C(14)

	Coefficient	Std. Error	z-Statistic	Prob.
C(1)	1.234501	0.075859	16.27362	.0000
C(2)	5.826423	0.666414	8.742949	.0000
C(3)	1.548862	0.251359	6.161945	.0000
C(4)	2.432432	0.092930	26.17483	.0000
C(5)	2.256007	0.165721	13.61329	.0000
C(6)	0.428208	0.082002	5.221902	.0000
C(7)	−0.086483	0.078776	−1.097839	.2723
C(8)	−0.024541	0.034317	−0.715143	.4745
C(9)	0.667771	0.055672	11.99477	.0000
C(10)	−0.378041	0.075351	−5.017086	.0000
C(11)	−0.238152	0.029225	−8.148851	.0000
C(12)	0.581458	0.045000	12.92124	.0000
C(13)	0.140659	0.015600	9.016427	.0000
C(14)	0.098733	0.007446	13.26042	.0000
Log-likelihood	476.8897			

LR test for over-identification:

Chi-square(1)	83.86338		Probability	.0000

Estimated S matrix:

0.009845	0.000590	−0.001758	0.004597	5.17E-05
−0.228240	0.742400	−0.314498	−0.204580	−0.100920
−0.172874	0.092590	0.192311	−0.043503	0.083004

Continued

TABLE 8.8 Structural vector autoregressive estimates with recursive long run impulse-responses by F triangular matrix: 1997Q1–2019Q2. LPC_GDP, LWUI_T6, LEPUI, LEPI, and LPC_TRANS.—cont'd

Model: e = Phi*Fu where E[uu'] = I				
−0.050263	0.002337	0.001242	0.039810	−0.007758
−0.023173	0.004273	−0.035872	5.10E-07	0.050998
Estimated F matrix:				
1.234501	0.000000	0.000000	0.000000	0.000000
5.826423	2.256007	0.000000	0.000000	0.000000
1.548862	0.428208	0.667771	0.000000	0.000000
2.432432	−0.086483	−0.378041	0.581458	0.000000
1.000000	−0.024541	−0.238152	0.140659	0.098733

Notes: SVAR lag = 1 based on SC. Estimation method: Maximum likelihood via Newton–Raphson (analytic derivatives). Convergence is achieved after 25 iterations.

(iii) The per capita electricity retail sales to the industrial sector responds to economic policy uncertainty negatively (−0.060). As LEPUI increases by 1%, LPC_IND will decrease by 0.060%.

Table 8.5 underlines the output from SVAR with LPC_GDP, LWUI_T6, LEPUI, and LPC_TRANS, and reveals as follows:

(i) The per capita electricity retail sales to the transportation sector (LPC_TRANS) has positive income elasticity (0.373). This output reveals that electricity in the transportation sector is a normal and weak superior-good/input.

(ii) LPC_TRANS has a negative association with the economic policy uncertainty index. The electricity demand of the transportation sector will diminish as the policy uncertainty index goes up. The elasticity of LPC_TRANS concerning LEPUI is −0.181. LPC_TRANS does not respond significantly to the country-level World uncertainty index.

The following estimations outputs have been obtained through SVAR with five variables.

Table 8.6 gives the estimation outputs from SVAR with the variables of LPC_GDP LWUI_T6 LEPUI LEPI LPC_EEU as follows:

(i) First row indicates theoretically that LGDP does not respond to other variables contemporaneously, it responds to other variables with lag.

(ii) Second row claims hypothetically that US uncertainty index (LWUI_T6) calculated by WUI responds to the only GDP and does not respond to all other variables contemporaneously, but responds to other variables with lag.

Empirical output (estimation from recursive long-run impulses) however finds an insignificant effect of LPC_GDP on LWUI_T6 (C2).

(iii) Third row states hypothetically that economic uncertainty index (LEPUI) responds to only GDP and LWUI_T6 and does not respond to all other variables contemporaneously, but responds to other variables with lag.

Empirical output (estimation from recursive long-run impulses) underlines that LEPUI is influenced only by LGDP by −3.941089 (C3) and is not affected by LWUI_T6 (C6).

As LPC-GDP goes up by 1%, LEPUI goes down by 3.941%. LEPUI has high LPC_GDP elasticity. GDP measures economic activities in a country. One might claim that the more the recorded economic activities are, the less the economic uncertainty will be.

(iv) Fourth row indicates theoretically that energy price index (LEPI) responds to LGDP, LWUI_T6, and LEPUI contemporaneously, but responds to LEEU with lag.

Empirical output (estimation from recursive long-run impulses) finds significant effects of LPC_GDP and LEPUI on LEPI. The coefficients of C4 and C10 yield the elasticity of LEPI for LPC_GDP (5.733520) and elasticity of LEPI with respect to LEPUI (−1.173117).

(v) Fifth row indicates theoretically that log of per capita energy demand (LEEU) responds to al variables in the system (LPC_GDP, LWUI_T6, LEPUI, and LEPI) contemporaneously. Fifth row first parameter being equal to 1 assumes that income elasticity of energy demand is unity. Fifth row second parameter (the effect of LWUI_T6 on LEEU) was found insignificant (C8). The fifth row, third parameter (the effect of LEPUI on LEEU) was found significant with the estimated parameter of −0.388148 (C11). It indicates that energy demand elasticity in terms of economic policy uncertainty is equal to −0.388. Fifth row fourth parameter (the effect of LEPI on LEEU) was found insignificant (C13).

Table 8.7 explores as follows:

(i) First row indicates theoretically that LPC_GDP does not respond to other variables contemporaneously, it responds to other variables with lag.

(ii) Second row claims theoretically that US country-level World uncertainty index (LWUI_T6) responds to the only LPC_GDP and does not respond to all other variables contemporaneously, but responds to other variables with lag.

Empirical output (estimation from recursive long-run impulses) reaches a negative significant effect of LPC_GDP on LWUI_LT6. As the per capita GDP increases by 1%, the US uncertainty index will reduce by 11.67% (C2).

(iii) Third row states hypothetically that economic uncertainty index (LEPUI) responses to only LPC_GDP and LWUI_T6 and does not respond to all other variables contemporaneously, but responds to other variables with lag.

Empirical output (estimation from recursive long-run impulses) shows that LEPUI is influenced by LPC_GDP and LWUI_T6 by −3.647797% and 3.088685%, respectively. Economic policy uncertainty has an income elasticity of −3.65 (C3) and country-level World uncertainty index elasticity of 3.089 (C6).

(iv) Fourth row indicates theoretically that energy price index (LEPI) responds to LPC_GDP, LWUI_T6, and LEPUI contemporaneously, but responds to LEEU with lag.

Empirical output (estimation from recursive long-run impulses) finds significant effects of LPPC_GDP and LWUI_T6 on LEPI. Log of the energy price index is influenced negatively (−2.87) by LPC_GDP, and positively (2.67) by LWUI_T6. These parameter estimations are depicted in Table 8.7 by C(4) and C(7), respectively.

(v) Fifth row indicates theoretically that log of per capita industrial energy demand (LPC_IND) responds to all variables in the system (LPC_GDP, LWUI_T6, LEPUI, and LEPI) contemporaneously. Fifth row first parameter being equal to 1 that assumes that income elasticity of industrial energy demand is equal to 1.

Fifth row second parameter (the effect of LWUI_T6 on PPC_IND) was found significant (C8). The elasticity of per capita industrial energy demand with respect to country-level World uncertainty index is 1.035. As LWUI_T6 increases by 1%, LPC_IND will decrease by 1.035%. Elasticity is almost unity.

Fifth row third parameter (the effect of LEPUI on LPC_IND) was found significant. As the policy uncertainty index goes up by 1%, the per capita industrial energy demand will decrease by 0.11%. It indicates that per capita industrial energy demand elasticity in terms of economic policy uncertainty is equal to −0.110.

Fifth row fourth parameter (the effect of LEPI on LPC_IND) was also found significant (C13). Price elasticity of per capita industrial energy demand (the elasticity of per capita industrial energy demand with regard in terms of energy price index) is equal to −0.161.

The outputs from Table 8.8 reveal as follows:

(i) First row indicates theoretically that LPC_GDP does not respond to other variables contemporaneously, it responds to other variables with lag.

(ii) Second row: The country-level World uncertainty index (LWUI_T6) will increase by 5.83% as per capita GDP (LPC_GDP) increases by 1%.

(iii) Third row: The percentage impacts of per capita GDP (LPC_GDP) and the country-level World uncertainty index (LWUI_T6) on economic uncertainty (LEPUI) are 1.55 and 0.43, respectively.

(iv) Fourth row: The percentage impacts of per capita GDP and economic policy uncertainty index on the energy price index are 2.43 and −0.24, respectively. Positive income elasticity of energy prices is greater than 1 whereas negative economic policy uncertainty index elasticity of energy prices is less than unity in absolute value.

(v) Fifth row: Per-capita energy demand of transportation responds to economic policy uncertainty negatively. The relevant elasticity is −0.24. It responds however positively to the energy price index. This outcome might yield that per capita energy for transportation seems to have snob good property.

5. Conclusion

By following four-variable VAR and five-variable VAR structural models, one can claim that in the United States:

(i) US country-level World uncertainty index and US economic policy index have negative impacts on electricity end-use and retail electricity sales to the industrial sector.

(ii) Retail electricity sales to the transportation sector are influenced negatively by economic policy uncertainty.

The results state that, although electricity is a necessary good/input in daily life and the real sectors, it will be affected adversely by an increase in uncertainties in the United States. This output might be confirmed by the statement that it is widely accepted that uncertainty diminishes the economic activities as evidenced in some microeconomic and macroeconomic models of energy consumption or energy investment.

For instance, Liu et al. (2020) reveal that economic policy uncertainty significantly inhibits traditional energy enterprises' investment in China. Bahmani-Oskooee and Nayeri (2020) claim that policy uncertainty has an asymmetric impact on consumer expenditure in all G7 countries. Adedoyin and Zakari (2020) investigate the nexus between energy consumption CO_2 emissions and economic policy uncertainty and reveal a positive impact of economic uncertainty on environmental quality (CO_2 emissions).

Possible future research might consider other types of uncertainties to observe the parameters of energy consumption. For instance, Pelletier et al. (2019) underline some energy consumption uncertainties (due to weather and road conditions, driver behavior, etc., rather than economic policy uncertainty) through the electric vehicle routing problem with energy consumption uncertainty and underline energy consumption uncertainties. Hence, energy efficiency and optimal use of energy are closely associated with several types of uncertainties to be measured (Wang et al., 2019; Moghaddas-Tafreshi et al., 2019). Considering the nexus between uncertainty and energy power consumption, and consumption of plug-in hybrid electric vehicles, the energy consumption optimization problems might be also solved under risk-averse and risk-neutral strategies as defined in Moghaddas-Tafreshi et al. (2019).

Overall, this paper yields the negative influence of uncertainties on energy consumption (retail electricity demand). Energy consumption has two main aspects to investigate. First, the impact of energy consumption on the environmental quality (Öztürk, 2010; Shahbaz and Balsalobre-Lorente, 2019; Bilgili et al., 2017, 2020; Kuşkaya and Bilgili 2019, 2020a,b; Balsalobre-Lorente et al., 2019; Menegaki and Tugcu, 2018; Bilgili et al., 2019; Ozcan et al., 2018; Bilgili and Ulucak, 2018) and, secondly, the influence of demand for energy on economic growth (Ozcan et al., 2020; Bulut and Menegaki, 2020; Bilgili and Ozturk; 2015; Özturk and Bilgili, 2015; Shahbaz et al., 2020; Balsalobre-Lorente et al., 2018; Bilgili et al., 2019b; Koçak and Sarkgüneşi, 2017; Menegaki, 2019).

Upon the outcomes of this research, one might suggest that policy makers follow the strategies to lower the policy uncertainties in the United States to boost energy consumption and hence economic growth through incentives, subsidies, and loan guarantee funds of sustainable energy policies to increase environmentally friendly renewable electricity power generation and consumption.

References

Abid, M., Sebri, M., 2012. Energy consumption-economic growth nexus: does the level of aggregation matter? Int. J. Energy Econ. Pol. 2 (2), 55–62.

Adedoyin, F.F., Zakari, A., 2020. Energy consumption, economic expansion, and CO_2 emission in the UK: the role of economic policy uncertainty. Sci. Total Environ. 738, 140014.

Ahamad, M.G., Islam, A.K.M.N., 2011. Electricity consumption and economic growth nexus in Bangladesh: revisited evidences. Energy Pol. 39, 6145–6150.

Aissa, M.S.B., et al., 2014. Output, renewable energy consumption and trade in Africa. Energy Pol. 66, 11–18.

Akinlo, A.E., 2009. Electricity consumption and economic growth in Nigeria: evidence from cointegration and co-feature analysis. J. Pol. Model. 31, 681–693.

Alam, M.J., Begum, I.A., Buysse, J., Huylenbroeck, G.V., 2012. Energy consumption, carbon emissions and economic growth nexus in Bangladesh: cointegration and dynamic causality analysis. Energy Pol. 45, 217–225.

Al-Iriani, M.A., 2006. Energy–GDP relationship revisited: an example from GCC countries using panel causality. Energy Pol. 34, 3342–3350.

Anton, S.G., Nucu, A.E.A., 2020. The effect of financial development on renewable energy consumption. A panel data approach. Renew. Energy 147, 330–338.

Bahmani-Oskooee, M., Nayeri, M.M., 2020. Policy uncertainty and consumption in G7 countries: an asymmetry analysis. Int. Econ. 163, 101–113.

Balsalobre-Lorente, D., Gokmenoglu, K.K., Taspinar, N., Cantos-Cantos, J.M., 2019. An approach to the pollution haven and pollution halo hypotheses in MINT countries. Environ. Sci. Pollut. Control Ser. 26 (22), 23010–23026.

Balsalobre-Lorente, D., Shahbaz, M., Roubaud, D., Farhani, S., 2018. How economic growth, renewable electricity and natural resources contribute to CO_2 emissions? Energy Pol. 113, 356−367.

Barradale, M.J., 2010. Impact of public policy uncertainty on renewable energy investment: wind power and the production tax credit. Energy Pol. 38, 7698−7709.

Bercu, A., Paraschiv, G., Lupu, D., 2019. Investigating the energy−economic growth−governance nexus: evidence from Central and Eastern European countries. Sustainability 11, 3355. https://doi.org/10.3390/su11123355.

Bhattacharya, M., et al., 2017. The dynamic impact of renewable energy and institutions on economic output and CO_2 emissions across regions. Renew. Energy 111, 157−167.

Bilgili, F., Kuşkaya, S., Ünlü, F., Gençoğlu, P., 2020. Does waste energy usage mitigate the CO_2 emissions? A time-frequency domain analysis. Environ. Sci. Pollut. Control Ser. 27 (5), 5056−5073.

Bilgili, F., 2015. Business cycle co-movements between renewables consumption and industrial production: a continuous wavelet coherence approach. Renew. Sustain. Energy Rev. 52, 325−332.

Bilgili, F., Ulucak, R., 2018. Is there deterministic, stochastic, and/or club convergence in ecological footprint indicator among G20 countries? Environ. Sci. Pollut. Control Ser. 25 (35), 35404−35419.

Bilgili, F., Ulucak, R., Koçak, E., 2019. Implications of environmental convergence: continental evidence based on ecological footprint. In: Balsalobre-Lorente, D., Shahbaz, M. (Eds.), Energy and Environmental Strategies in Era of Globalization, Handbook Springer Series. Springer eBook ISBN 978-3-030-06001-5, Hardcover ISBN 978-3-030-06000-8, Series ISSN 1865-3529.

Bilgili, F., Koçak, E., Bulut, Ü., Kuşkaya, S., 2017. Can biomass energy be an efficient policy tool for sustainable development? Renew. Sustain. Energy Rev. 71, 830−845.

Bilgili, F., Kuşkaya, S., Toguç, N., Muğaloğlu, E., Koçak, E., Bulut, Ü., Bağlıtaş, H.H., 2019. A revisited renewable consumption-growth nexus: a continuous wavelet approach through disaggregated data. Renew. Sustain. Energy Rev. 107, 1−19.

Blas, I., Miguel, L.J., Capellan-Perez, I., 2019. Modelling of sectoral energy demand through energy intensities in MEDEAS integrated assessment model. Energy Strategy Rev. 26, 100419.

Bulut, U., Menegaki, A., 2020. Solar energy-economic growth nexus in top 10 countries with the highest installed capacity. Energy Sources B Energy Econ. Plann. 1−14.

Cai, Y., et al., 2018. Nexus between clean energy consumption, economic growth and CO_2 emissions. J. Clean. Prod. 182, 1001−1011.

Camba-Mendez, G., 2012. Conditional forecasts on SVAR models using the Kalman filter. Econ. Lett. 115 (3), 376−378.

Chandran, V.G.R., Sharma, S., Madhavan, K., 2010. Electricity consumption−growth nexus: the case of Malaysia. Energy Pol. 38, 606−612.

Erdal, G., Erdal, H., Esengün, K., 2008. The causality between energy consumption and economic growth in Turkey. Energy Pol. 36, 3838−3842.

Eyre, N., Baruah, P., 2015. Uncertainties in future energy demand in UK residential heating. Energy Pol. 87, 641−653.

Forni, M., Gambetti, L., 2014. Sufficient information in structural VARs. J. Monet. Econ. 66, 124−136.

Fuinhas, J.A., Marques, A.C., 2012. Energy consumption and economic growth nexus in Portugal, Italy, Greece, Spain and Turkey: an ARDL bounds test approach (1965–2009). Energy Econ. 34, 511–517.

Ghosh, S., 2010. Examining carbon emissions economic growth nexus for India: a multivariate cointegration approach. Energy Pol. 38, 3008–3014.

Hailemariam, A., Smyth, R., Zhang, X., 2019. Oil prices and economic policy uncertainty: evidence from a nonparametric panel data model. Energy Econ. 83, 40–51.

Halicioglu, F., 2007. Residential electricity demand dynamics in Turkey. Energy Econ. 29, 199–210.

Holtedahl, P., Joutz, F.L., 2004. Residential electricity demand in Taiwan. Energy Econ. 26, 201–224.

Irandoust, M., 2016. The renewable energy-growth nexus with carbon emissions and technological innovation: evidence from the Nordic countries. Ecol. Indicat. 69, 118–125.

Islam, F., Shahbaz, M., Ahmed, A.U., Alam, M.M., 2013. Financial development and energy consumption nexus in Malaysia: a multivariate time series analysis. Econ. Modell. 30, 435–441.

Kavgica, M., Mumovic, D., Summerfield, A., Stevanovic, Z., Ecim-Djuric, O., 2013. Uncertainty and modeling energy consumption: sensitivity analysis for a city-scale domestic energy model. Energy Build. 60, 1–11.

Khan, M.A., Khan, M.Z., Zaman, K., Arif, M., 2014. Global estimates of energy-growth nexus: application of seemingly unrelated regressions. Renew. Sustain. Energy Rev. 29, 63–71.

Koçak, E., Sarkgüneşi, A., 2017. The renewable energy and economic growth nexus in Black Sea and Balkan countries. Energy Pol. 100, 51–57.

Komal, R., Abbas, F., 2015. Linking financial development,economic growth and energy consumption in Pakistan. Renew. Sustain. Energy Rev. 44, 211–220.

Kuşkaya, S., Bilgili, F., 2019. Yenilenebilir enerji kaynağı kullanımının çevre kirliliği üzerindeki etkisinin araştırılması: sürekli dalgacık uyumu modeli yaklaşımı. Anadolu Üniversitesi Sosyal Bilimler Dergisi. 19 (4), 39–60.

Kuşkaya, S., Bilgili, F., 2020b. Hidroelektrik enerji tüketiminin çevre üzerine etkisinin sürekli dalgacık uyumu modeli ile araştırılması: ABD örneği. Erciyes Üniversitesi Iktisadi ve Idari Bilimler Faküeltesi Dergisi 55, 263–284.

Kuşkaya, S., Bilgili, F., 2020a. The wind energy-greenhouse gas nexus: the wavelet-partial wavelet coherence model approach. J. Clean. Prod. 245, 118872.

Kyophilavong, P., Shahbaz, M., Anwar, S., Masood, S., 2015. The energy-growth nexus in Thailand: does trade openness boost up energy consumption? Renew. Sustain. Energy Rev. 46, 265–274.

Lee, H.L., Lim, E.S., Hwang, J., 2012. Panel SVAR model of women's employment, fertility, and economic growth: a comparative study of East Asian and EU Countries. J. Soc. Sci. 49, 386–389.

Liu, R., He, L., Liang, X., Yang, X., Xia, Y., 2020. Is there any difference in the impact of economic policy uncertainty on the investment of traditional and renewable energy enterprises? – a comparative study based on regulatory effects. J. Clean. Prod. 255, 120102.

Meijer, I.S.M., Hekkert, M.P., Koppenjan, J.F.M., 2007. The influence of perceived uncertainty on entrepreneurial action in emerging renewable energy technology; biomass gasification projects in The Netherlands. Energy Pol. 35, 5836–5854.

Menegaki, A.N., 2019. The ARDL method in the energy-growth nexus field; best implementation strategies. Economies 7 (4), 105.

Menegaki, A.N., Tugcu, C.T., 2018. Two versions of the index of sustainable economic welfare (ISEW) in the energy-growth nexus for selected Asian countries. Sustain. Prod. Consum. 14, 21−35.

Moghaddas-Tafreshi, S.M., Jafari, M., Mohseni, S., Kelly, S., 2019. Optimal operation of an energy hub considering the uncertainty associated with the power consumption of plug-in hybrid electric vehicles using information gap decision theory. Int. J. Electr. Power Energy Syst. 112, 92−108.

Ohler, A., Fetters, I., 2014. The causal relationship between renewable electricity generation and GDP growth: a study of energy sources. Energy Econ. 43, 125−139.

Omri, A., 2013. CO_2 emissions, energy consumption and economic growth nexus in MENA countries: evidence from simultaneous equations models. Energy Econ. 40, 657−664.

Ozcan, B., Apergis, N., Shahbaz, M., 2018. A revisit of the environmental Kuznets curve hypothesis for Turkey: new evidence from bootstrap rolling window causality. Environ. Sci. Pollut. Control Ser. 25 (32), 32381−32394.

Ozcan, B., Tzeremes, P.G., Tzeremes, N.G., 2020. Energy consumption, economic growth and environmental degradation in OECD countries. Econ. Modell. 84, 203−213.

Öztürk, İ., 2010. A literature survey on energy-growth nexus. Energy Policy 38 (1), 340−349.

Öztürk, İ., Bilgili, F., 2015. Economic growth and biomass consumption nexus: dynamic panel analysis for sub-Sahara African countries. Appl. Energy 137, 110−116.

Pao, H., Fu, H., 2013. Renewable energy, non-renewable energy and economic growth in Brazil. Renew. Sustain. Energy Rev. 25, 381−392.

Pelletier, S., Jabali, O., Laporte, G., 2019. The electric vehicle routing problem with energy consumption uncertainty. Transp. Res. B Methodol. 126, 225−255.

Piroli, G., Rajcaniova, M., Ciaian, P., Kancs, d'A., 2015. From a rise in B to a fall in C? SVAR analysis of environmental impact of biofuels. Renew. Sustain. Energy Rev., Elsevier 49, 921−930.

Rossi, B., Sekhposyany, T., Souprez, M., 2016. Understanding the Sources of Macroeconomic Uncertainty (09.09.2019). https://www.ecb.europa.eu/pub/conferences/shared/pdf/20160603_forecasting/Paper_10_Sekhposyan.pdf.

Salim, R.A., Rafiq, S., 2012. Why do some emerging economies proactively accelerate the adoption of renewable energy? Energy Econ. 34, 1051−1057.

Sanquist, T.F., Orr, H., Shui, B., Bittner, A.C., 2012. Lifestyle factors in U.S. residential electricity consumption. Energy Pol. 42, 354−364.

Sari, R., Soytas, U., 2004. Disaggregate energy consumption, employment and income in Turkey. Energy Econ. (26), 335−344.

Sbia, R., Shahbaz, M., Hamdi, H., 2014. A contribution of foreign direct investment, clean energy, trade openness, carbon emissions and economic growth to energy demand in UAE. Econ. Modell. 36, 191−197.

Scarcioffolo, A.R., Etienne, X.L., 2018. Does economic policy uncertainty affect energy market volatility and vice-versa?. In: Selected Paper Prepared for Presentation at the 2018 Agricultural Applied Economics Association Annual Meeting, Washington, D.C., August 5−August 7.

Shahbaz, M., Balsalobre, D., 2019. Energy and Environmental Strategies in the Era of Globalization. Springer.

Shahbaz, M., Raghutla, C., Chittedi, K.R., Jiao, Z., Vo, X.V., 2020. The effect of renewable energy consumption on economic growth: evidence from the renewable energy country attractive index. Energy 207, 118162.

Squalli, J., 2007. Electricity consumption and economic growth: bounds and causality analyses of OPEC members. Energy Econ. 29, 1192−1205.

Tang, C.F., Abosedra, S., 2014. The impacts of tourism, energy consumption and political instability on economic growth in the MENA countries. Energy Pol. 68, 458—464.

Valadkhani, A., et al., 2019. Effects of primary energy consumption on CO_2 emissions under optimal thresholds: evidence from sixty countries over the last half century. Energy Econ. 80, 680—690.

Wang, K., Li, X., Gao, L., Garg, A., 2019. Partial disassembly line balancing for energy consumption and profit under uncertainty. Robot. Comput. Integrated Manuf. 59, 235—251.

Whitford, A.B., 2016. Estimation of several political action effects of energy prices. Energy Pol. Res. 3 (1), 13—18.

Wolde-Rufael, Y., 2005. Energy demand and economic growth: the African experience. J. Pol. Model. 27, 891—903.

Yoo, S.H., 2005. Electricity consumption and economic growth: evidence from Korea. Energy Pol. 33, 1627—1632.

Yu, H., Fang, L., Sun, B., 2018. The role of global economic policy uncertainty in long-run volatilities and correlations of U.S. industry-level stock returns and crude oil. PLoS One 13 (2), e0192305. https://doi.org/10.1371/journal.pone.0192305.

Zeshan, M., Ahmed, V., 2013. Energy, environment and growth nexus in South Asia. Environ. Dev. Sustain. 15, 1465—1475.

Chapter 9

Economics of offshore renewable energy

Laura Castro-Santos[1], Almudena Filgueira-Vizoso[2],
Eugenio Baita-Saavedra[3], David Cordal-Iglesias[4]
[1]*Universidade da Coruña, Departamento de Enxeñaría Naval e Oceánica, Escola Politécnica
Superior, Ferrol, Spain;* [2]*Departamento de Química, Escola Politécnica Superior, Universidade da
Coruña, Ferrol, Spain;* [3]*Saitec, Leioa, Bilbao, Spain;* [4]*Universidade da Coruña, Escola
Politécnica Superior, Ferrol, Spain*

1. Introduction

Significant changes are taking place in the global energy sector due to several factors, such as: the increasing electrification, the great expansion of renewable energies, the problems caused in the production, and consumption of oil and the globalization of gas markets natural (International Energy Agency, 2018). The governments of the different countries are carrying out policies modeled on the new scenario of energy demand that is expected to increase by 25% until 2040, which requires more than two billion dollars per year of investment to get this new supply of energy. Renewable energies are presented as the new choice, representing almost two thirds of the aggregate global capacity until 2040, despite the fact that carbon remains the main source and gas continues to occupy a second place. This brings with it important environmental benefits allowing compliance with the agreements of the different environmental protocols, which are summarized in the Paris Agreement (European Union, 2015) that entered into force on November 4, 2016, and which establishes as a main objective to limit the increase in temperature at 1.5°C. For this, the International Renewable Energy Agency (International Renewable Energy Agency (IRENA), 2014) estimates that to meet the objectives of the agreement, it would be necessary to double the installation of clean sources in the next 15 years and an annual investment of 900,000 million dollars. The International Energy Agency (IEA) believes that carrying out this agreement *"will accelerate the transformation of the energy sector, as it will increase the speed of investments in clean technologies and energy efficiency."* The two areas where higher growth is expected are solar and wind. The IEA states that the objective is that the share of

Energy-Growth Nexus in an Era of Globalization. https://doi.org/10.1016/B978-0-12-824440-1.00005-9
233

renewables on total energy consumption in the world reaches 36% in 2030, which would double the fee on 2010.

With regard to renewable energy in the world, according to the Renewables 2019. Global Status Report (REN 21, 2019) states that renewable energies provide more than 26% of the world's electricity generation at the end of 2018. Within this type of wind energy, it is one of the main supports, reaching an installed capacity of 591 GW in the year, which compared to the installed power 10 years before (2008) of 121 GW represents an increase of 388% compared to 2008.

Regarding offshore wind power, seven countries in Europe and two in Asia connected a total of 4.5 GW in 2018, increasing the accumulated global capacity by 24%. As regards offshore wind energy, Europe remains the continent with the highest wind capacity in the world. In 2018 the installed power was increased by 2.6 GW, most of it is located in the North Sea, the United Kingdom increased its capacity by 1.3 GW, Germany almost 1 GW, Belgium 0.3 GW, and Spain installed its second offshore turbine. At the end of 2018, Europe housed around 79% of offshore wind power capacity (REN 21, 2019).

Within the marine renewable energies, we not only have wind energy. There are other energy sources such as wave energy or tidal energy that must also be taken into account because although there is much less installed power, they are still a new source of energy and its technological improvements can make them gain importance in the energy market, either as their own or combined sources of energy.

Most of the market for these energy sources are pilot projects with relatively small installations of less than 1 MW. The development of this type of energy sources focuses mainly on Europe and particularly on the coasts of Scotland.

2. Types of offshore energy technology

Within marine renewable energies, we can talk about (Falcão, 2010; Khan et al., 2017; O Rourke et al., 2010; Skilhagen et al., 2008; Sun et al., 2012):

- **Offshore wind energy:** Use the wind to produce energy. It is the most developed in the whole range of marine renewable energies.
- **Wave energy:** Wave energy, also known as wave energy, takes advantage, as its name implies, the energy of sea waves.
- **Tidal energy:** This uses the energy of the tides. The tides, that is, the rise and fall of sea waters, are produced by the gravitational actions of the sun and the moon.
- **Tidal or ocean thermal energy:** Tidal energy takes advantage of the difference in temperatures between the surface of the sea and the deep waters. The oceans are the major sources of solar energy accumulation, since their great thermal inertia allows temperatures to be more stable in them.

- **Saline gradient energy or osmotic power:** This type of energy is also known as blue energy, and takes advantage of the difference in salt concentration that exists between seawater and rivers. Energy is obtained through an osmosis process.
- **Current energy:** Take advantage of the kinetic energy contained in marine currents to produce energy.

The energies to be studied in this chapter are wind energy and wave energy and for this a classification will be made of them.

Within offshore wind energy, we can find two types of structures, fixed and floating. Fixed structures are used for depths less than 50 m and floating structures for depths greater than 50 m. Within the fixed structures, we can talk about monopile, tripile, tripod, jacket, gravity ... and within the floating ones, a classification will be made according to the material used for its construction distinguishing between steel and concrete structures. Most of the floating offshore wind farms are made with steel structures, but the maintenance problems that these structures pose can be supplemented with concrete structures. Today we find three possible concrete structures, which are the TELWIND (Project carried out by ESTEYCO), SATH Project (Project carried out by SAITEC), and WINDCRETE (Project carried out by the University of Barcelona) (Castro-Santos, 2015; Dankelmann et al., 2017; Molins i Borrell and Campos Hortigüela, 2016; Saitec, 2019; Sclavounos et al., 2010).

As for the energy of the waves, the main ways to obtain this energy are (Falnes and Løvseth, 1991):

- Through floats submerged at the bottom of the sea.
- Through articulated mobile devices: articulated device that tracks the movement of waves, such as the Pelamis device.
- Deposits that have an opening in the surface of the sea, that expels the air by the waves and moves a turbine generating electricity.
- Oscillating devices: they work with a hydraulic motor, hydraulic turbine, and a linear electric generator.
- Float: consists of buoys that run a generator and make a movement between the mast and the float.
- Rotation: device anchored to the seabed that operates pistons that transform energy.
- Wave collectors: devices that, by means of a ramp, enter the waves and start the hydraulic turbines.

3. Case of study

A particular case of study of two farms, one composed of wind energy and another farm composed of wave energy, will be studied to determine the most important aspects in economic terms of offshore wind and wave energy.

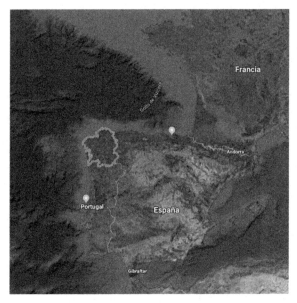

FIGURE 9.1 Location selected to develop the analysis (Google, 2019).

The location selected to do the comparison is the Galicia region (North-West of Spain) (see Fig. 9.1), the best area of the Iberian Peninsula in offshore wind and wave resource.

The platforms selected to do the analysis are the WindFloat for offshore wind and the Pelamis for wave energy (see Figs. 9.2 and 9.3).

In this context, the W2EC software (Castro-Santos and Filgueira-Vizoso, 2020) have been considered to calculate the main economic parameters of the farms. The inputs taken into account are common for the two farms to simplify the study. They are shown in Table 9.1. It is important to notice that the electric

FIGURE 9.2 Floating offshore wind platform selected to develop the analysis (Wikipedia, 2020a).

FIGURE 9.3 Floating offshore wave platform selected to develop the analysis (Wikipedia, 2020b).

TABLE 9.1 Inputs to develop the analysis.

Concept	Value	Units
Total power of the farm	500	MW
Life of the farm	20	years
Electric tariff	100	€/MWh
Capital cost	10%	–
Corporate tax	25%	–
Steel cost	524	€/ton
% Financing	50%	–
% Interest	7%	–

tariff has been considered as 100 €/MWh taking into account a medium value of the last years in Spain (although the present value is closest to 50 €/MWh), where the electricity cost varies a lot depending on the government laws.

On the other hand, the outputs that will be considered in this analysis are shown in Table 9.2. They have been calculated considering the method proposed by Castro-Santos et al. (2016)

TABLE 9.2 Outputs of the analysis.

Concept	Name	Units
Internal rate of return	IRR	%
Net present value	NPV	M€
Levelized cost of energy	LCOE	€/MWh

4. Results

Figs. 9.4 and 9.5 show the levelized cost of energy results for the case of the offshore wind farm and the wave energy farm, respectively. Their values go from 104 to 854 €/MWh for the WindFloat and from 829 to 11,722 €/MWh for the Pelamis structure. Therefore, the tariff considered is not adequate for a wave energy farm. Moreover, the best values for the offshore wind farm are close to the values of other traditional technologies such as onshore wind.

Figs. 9.6 and 9.7 show the internal rate of return (IRR) results for the case of the offshore wind farm and the wave energy farm, respectively. Their values go from −169% to 7% for the WindFloat and from −185% to −176% for the

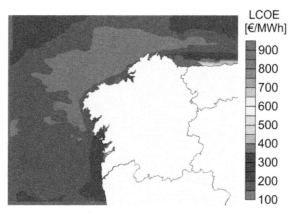

FIGURE 9.4   Levelized cost of energy (LCOE) for an offshore wind farm considering a Wind-Float platform.

FIGURE 9.5   Levelized cost of energy (LCOE) for an offshore wave farm considering a Pelamis platform.

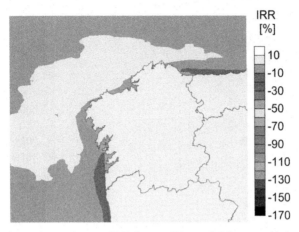

FIGURE 9.6   Internal rate of return (IRR) for an offshore wind farm considering a WindFloat platform.

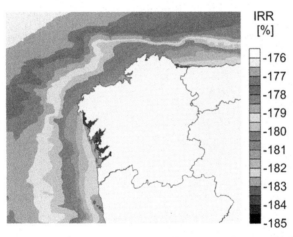

FIGURE 9.7   Internal rate of return (IRR) for an offshore wave farm considering a Pelamis platform.

Pelamis structure. Therefore, neither offshore wind farm nor wave energy farm are economically feasible in terms of IRR because their best values are lower than the capital cost considered.

Figs. 9.8 and 9.9 show the net present value (NPV) results for the case of the offshore wind farm and the wave energy farm, respectively. Their values go from −1488 to −31 M€ for the WindFloat and from −4162 to −2053 M€ for the Pelamis structure. Therefore, neither offshore wind farm nor wave energy farm are economically feasible in terms of NPV because their best values are lower than zero.

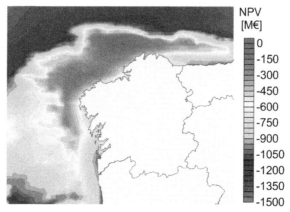

FIGURE 9.8   Net present value (NPV) for an offshore wind farm considering a WindFloat platform.

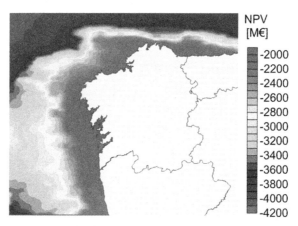

FIGURE 9.9   Net present value (NPV) for an offshore wave farm considering a Pelamis platform.

5. Conclusions

The present chapter has analyzed the importance of new ways of renewable energies in the world. The main types of offshore renewable energies are classified depending on different aspects such as bathymetry or distance to shore.

Moreover, the economics of offshore wind energy and wave energy have been studied considering a possible case of study of a 500 MW offshore wind farm and wave energy farm.

Results presented in the current work are related to an electric tariff of 100 €/MWh. It gives values of IRR and NPV not economically feasible for the floating platforms considered.

Therefore, it is important to know that the electric tariff considered has a great influence on the economic feasibility of an offshore renewable energy farm. In addition, the costs of these new technologies should be reduced in the future in order to improve their economic feasibility. Finally, governments should have this fact into account to design the laws that regulate the electric tariff to increase the competitivity of the offshore renewable energy.

References

Castro-Santos, L., Filgueira-Vizoso, A., 2020. A software for calculating the economic aspects of floating offshore renewable energies. Int. J. Environ. Res. Public Health 17. https://doi.org/10.3390/ijerph17010218.

Castro-Santos, L., Filgueira-Vizoso, A., Carral-Couce, L., Formoso, J.Á.F., 2016. Economic feasibility of floating offshore wind farms. Energy 112, 868–882. https://doi.org/10.1016/j.energy.2016.06.135.

Castro-Santos, L.D.C.V., 2015. Cost comparison of three floating offshore wind platforms. J. Coast. Res. 31, 1217–1221.

Dankelmann, S., Visser, B., Gupta, N., Serna, J., Couñago, B., Urruchi, Á., Fernández, J.L., Cortés, C., Guanche García, R., Jurado, A., 2017. TELWIND- Integrated Telescopic Tower Combined with an Evolved Spar Floating Substructure for Low-Cost Deep Water Offshore Wind and Next Generation of 10 MW+ Wind Turbines (WWW Document).

European Union, 2015. Acuerdo de París | Acción por el Clima [WWW Document]. URL. https://ec.europa.eu/clima/policies/international/negotiations/paris_es (Accessed 2.7.20).

Falcão, A.F. de O., 2010. Wave energy utilization: A review of the technologies. Renew. Sustain. Energy Rev. 14, 899–918. https://doi.org/10.1016/j.rser.2009.11.003.

Falnes, J., Løvseth, J., 1991. Ocean wave energy. Energy Policy 19, 768–775. https://doi.org/10.1016/0301-4215(91)90046-Q.

Google, 2019. Google Maps [WWW Document]. URL. https://maps.google.com.

International Energy Agency, 2018. World Energy Outlook 2018.

International Renewable Energy Agency (IRENA), 2014. Wave Energy Technology Brief, Future Energy: Improved, Sustainable and Clean Options for Our Planet. https://doi.org/10.1016/B978-0-08-099424-6.00017-X.

Khan, N., Kalair, A., Abas, N., Haider, A., 2017. Review of ocean tidal, wave and thermal energy technologies. Renew. Sustain. Energy Rev. https://doi.org/10.1016/j.rser.2017.01.079.

Molins i Borrell, C., Campos Hortigüela, A., 2016. WindCrete. Wind. Int. 12, 1–4.

O Rourke, F., Boyle, F., Reynolds, A., 2010. Tidal energy update 2009. Appl. Energy 87, 398–409. https://doi.org/10.1016/J.APENERGY.2009.08.014.

REN 21, 2019. Renewables 2019, Global Status Report (Presentation) 336.

Saitec, 2019. Technical Documentation Floating Offshore Wind Turbine SATH - 10 MW. O&M Strategy and Accesibility. Bilbao (Spain).

Sclavounos, P.D., Lee, S., DiPietro, J., 2010. Floating offshore wind turbines: tension leg platform and taught leg buoy concepts supporting 3–5 mw wind turbines. In: European Wind Energy Conference (EWEC). Warsaw, Poland, pp. 1–7.

Skilhagen, S.E., Dugstad, J.E., Aaberg, R.J., 2008. Osmotic power — power production based on the osmotic pressure difference between waters with varying salt gradients. Desalination 220, 476–482. https://doi.org/10.1016/J.DESAL.2007.02.045.

Sun, X., Huang, D., Wu, G., 2012. The current state of offshore wind energy technology development. Energy 41, 298–312. https://doi.org/10.1016/j.energy.2012.02.054.

Wikipedia, 2020a. Floating Wind Turbine [WWW Document]. URL. https://en.wikipedia.org/wiki/Floating_wind_turbine (Accessed 2.4.20).

Wikipedia, 2020b. Pelamis Bursts Out of a Wave [WWW Document]. URL. https://es.m.wikipedia.org/wiki/Archivo:Pelamis_bursts_out_of_a_wave.JPG (Accessed 1.8.20).

Chapter 10

Investigating the determinants of energy efficiency in emerging economies: the comparative roles of trade and financial globalization

Mehmet Akif Destek[1], Erkan Alsu[2], Cengizhan Karaca[3]

[1]*Department of Economics, Gaziantep University, Gaziantep, Turkey;* [2]*Department of Business Administration, Gaziantep University, Gaziantep, Turkey;* [3]*Department of Office Management and Executive Assistance, Gaziantep University, Gaziantep, Turkey*

1. Introduction

Energy has such a vital importance for the economies of developing and developing countries that today the development of a country is evaluated according to the height of energy consumption. The energy demand of the countries is increasing rapidly due to the increasing human population, urbanization, and modernization, and this demand is provided mostly from fossil fuels. As a matter of fact, fossil fuels such as oil, gas, and coal meet approximately 80% of global energy demand (IEA, 2018). Especially developing countries are struggling to meet this enormous amount of demand with limited fossil fuels and become an importer to meet this demand. As an expected result of this situation, the current account balance of the countries that import fossil fuels is affected negatively. Saving policies that reduce energy consumption, which are applied to reduce these negative effects, generally harm economic growth (Apergis and Tang, 2013; Yıldırım et al., 2014; Destek and Aslan, 2017; Apergis and Payne 2009; Tang et al., 2016). For this reason, most countries are seeking energy-saving policies that increase energy efficiency rather than policies that reduce energy consumption.

The concept of energy efficiency, which means obtaining more economic output by using less energy, is also important in terms of environmental quality targets. Because, using energy resources more effectively and efficiently, it also serves to the goal of preserving the current level of welfare with less

Energy-Growth Nexus in an Era of Globalization. https://doi.org/10.1016/B978-0-12-824440-1.00006-0
243

energy intensity and reducing greenhouse gas emissions. Although the responsibility of environmental destruction is largely in developed countries in the historical process, it is seen that developing economies will replace the developed economies in terms of pollution-enhancing activities if necessary measures are not taken (Sadorsky, 2010). However, if energy efficiency is provided, the use of fossil fuels, which play an active role in the production of developing countries, and which is the main energy source, can be taken under control. Moreover, ensuring energy efficiency will contribute to the current account balance by reducing the foreign dependency of countries.

Effective use of energy is largely possible with the technological and economic developments of countries. High-tech countries are increasing their energy use effectively thanks to research and development (R&D) projects aimed at increasing energy efficiency. Since developing countries have less developed technology compared to developed countries and these projects require high costs, they are inadequate in qualified R&D activities. Thus, developing countries have to make use of foreign resources in financing qualified R&D projects. The foreign financing resources to be obtained are provided both through foreign trade and through the funds to be obtained from the financial system of other countries. This reveals the importance of a strong trade and financial system and relational ties with other countries to ensure energy efficiency. The fact that developing countries have strong relations with high-tech countries provides technology transfer to these countries. When this information is considered in full, a strong financial system, trade liberalization, and technological innovations, which are important factors of globalization, are needed to ensure energy efficiency especially in developing countries (Boutabba, 2014; Shahbaz et al., 2013; Tan and Zhang 2010; Farhani and Solarin 2017).

As technological advances are largely associated with globalization, the relationship between commercial openness and financial liberalization, which is one of the important infrastructures of globalization, and energy efficiency or energy consumption is widely covered in the energy economy literature (Zhang 2011; Wang et al., 2011; Mahalik and Mallick 2014; Shahbaz et al. 2014, 2016; Cagno et al., 2015; Wei et al., 2016; Ahmed 2017; Sohag et al., 2015; Pan et al., 2017 Sbia et al., 2014; Cole 2006; Destek, 2018). Technological developments accelerate economic development by diversifying and increasing the volume and reveals energy efficiency through continuous improvement through learning by going (Fisher-Vanden et al., 2004; Zhou et al., 2010; Pan et al., 2017). On the other hand, thanks to financial development, consumers and businesses can easily find low-cost funds in a strong financial system. Accordingly, consumers contribute to the increase in energy consumption by increasing their spending (direct effect) and businesses by expanding their business volume (business effect). In addition, the strengthening of financial markets accelerates the use of financial instruments such as stocks and bonds, and paves the way for the strengthening of capital, the recovery of the economy, and the increase in energy consumption. In addition, countries need commercial liberalization due to economies of scale, efficient use of production factors, and technological effects. Commercial liberalization, on the other hand, is considered to have an impact on energy intensity, thus accelerating economic activities (Pan et al., 2017; Sadorsky, 2010, 2011; Coban and Topcu, 2013).

Considering the fact that the global fossil fuel demand has shifted from developed economies to developing economies and the increasing role of emerging economies in global environmental pollution in recent years, this study examines the determinants of energy efficiency for big emerging markets (BEMs): Argentina, Brazil, China, India, Indonesia, Mexico, Poland, South Africa, South Korea, and Turkey. In addition, the fact that countries with financing restrictions on projects that increase energy efficiency are largely high-income emerging economies is another reason for their selection in the study. The possible contributions of the study to the literature are as follows: (i) This is the first study to investigate the impact of globalization on energy efficiency in BEM. (ii) This study separates the impact of globalization with trade and financial globalization to compare the relative effects of different kind of globalization. (iii) The study employs second-generation panel data methodologies to consider the possible cross-sectional dependence (CSD) among emerging economies.

2. Literature review

In empirical literature, the effects of globalization on energy efficiency are mostly examined through the effects of the globalization index, and the distinction between trade globalization and financial globalization is not emphasized. Therefore, in order to be compatible with our study, we analyzed the studies in the literature in terms of trade and finance indicators, and reviewed them in two different tables.

In case of trade globalization, most of the studies focused on the impact of trade openness on energy consumption instead of on energy intensity/efficiency and the findings of these studies are mixed as a seen from Table 10.1. For instance, Nasreen and Anwar (2014) for 15 Asian countries; Ahmed (2017) for BRICS countries; Shahzad et al. (2017) for Pakistan; Koengkan (2018) for Andean community countries; Pan et al. (2019a,b) for Bangladesh; Alam and Murad (2020) for 25 OECD countries; and Zeren and Akkus (2020) for emerging economies found the energy consumption increasing effect of trade openness. On the other hand, Sbia et al. (2014) for United Arab Emirates and Shahbaz et al. (2016) for India concluded the evidence that trade openness reduces energy consumption. However, based on the fact that increasing or decreasing energy consumption does not reflect the level of energy efficiency, the limited number of studies on trade-energy efficiency should be observed. In this context, Pan et al. (2019a,b) attempts to fulfill this gap in literature to

TABLE 10.1 Summarized literature on trade globalization-energy nexus.

Author(s)	Country	Period	Methodology	Findings
Nasreen and Anwar (2014)	15 Asian countries	1980 −2011	FMOLS	TR increases EC
Shahbaz et al. (2014)	91 high, middle and low income countries	1980 −2010	PMG and MG	The impact of TR on EC exhibits an inverted U-shaped
Sbia et al. (2014)	UAE	1975Q1- 2011Q	ARDL	TR reduces EC
Shahbaz et al. (2016)	India	1971 −2012	ARDL	GLO reduces EC
Ahmed (2017)	BRICS countries	1991 −2013	Panel OLS	TR increases EC
Shahzad et al. (2017)	Pakistan	1971 −2011	ARDL	TR increases EC
Topcu and Payne (2018)	OECD countries	1990 −2015	CCE-MG and AMG	The impact of TR on EC exhibits an inverted U-shaped
Koengkan (2018)	Andean community countries	1971 −2014	GMM	TR increases EC
Shahbaz et al. (2018)	Netherlands and İreland	1970q1- 2015q4	QARDL	GLO increases EC
Pan et al. (2019a,b)	Bangladesh	1976 −2014	SVAR	TR increases EI
Gozgor et al. (2020)	30 OECD countries	1970 −2015	FMOLS	EG increases EC
Alam and Murad (2020)	25 OECD countries	1970 −2012	DOLS and FMOLS	TR increases EC
Zeren and Akkuş (2020)	Emerging Countries	1980 −2015	CCE-MG	TR increases EC

EC, energy consumption; EI, energy intensity; GLO, globalization index; TR, trade openness.

check the impact of trade openness on energy intensity for Bangladesh and found that trade openness increases the intensity.

Similar to the trade globalization, some studies consider the effect of financial globalization on energy consumption and used foreign direct investment (FDI) as an indicator of financial globalization. As a shown in Table 10.2, Omri and Kahouli (2014) for 65 countries and Abidin et al. (2015) for ASEAN countries concluded that FDI increases energy consumption. However, Doytch and Narayan (2016) for 74 countries confirmed the energy consumption reducing effect of FDI. Fortunately, there are also some studies to examine the impact of FDI on energy intensity/efficiency. For instance, Luo and Cheng (2013) concluded that increasing FDI leads to increase in energy efficiency in China. Similarly, Elliott et al. (2013) for 206 Chinese cities; Adom (2015) for Nigeria; Keho (2016) for African countries; Bu et al. (2019) for 13 cities in Jiangsu Province of China found that increasing FDI reduces energy intensity.

TABLE 10.2 Summarized literature on financial globalization-energy nexus.

Author(s)	Country	Period	Methodology	Findings
Omri and Kahouli (2014)	65 countries	1990 −2011	DSEM	FDI increases EC
Luo and Cheng (2013)	China	1997 −2018	DEA	FDI increases EE
Elliott et al. (2013)	206 Chinese cities	2005 −2008	Panel regression	FDI reduces EI
Abidin et al. (2015)	ASEAN countries	1980 −2014	ARDL	FDI increases EC
Adom (2015)	Nigeria	1971 −2011	FMOLS	FDI reduces EI
Keho (2016)	African countries	1970 −2011	ARDL	FDI reduces EI
Doytch and Narayan (2016)	74 countries	1985 −2012	GMM	FDI reduces EC
Bu et al. (2019)	13 cities in Jiangsu Province of China	2005 −2007	PCSE	FDI reduces EI
Cao et al. (2020)	BRICS and non-BRICS countries	1990 −2014	PSTR	FDI exerts insignificant impact on EI

EC, energy consumption; *EE*, energy efficiency; *EI*, energy intensity; *FDI*, foreign direct investment.

Overall, as a seen from above-reviewed literature, previous empirical studies which examined trade-energy efficiency nexus generally focused on the effect of trade openness. Similarly, the impact of financial globalization is mostly evaluated with focusing on the impact of FDI. However, both dimensions of globalization have many subcomponents thus the other components of both trade and financial globalization are ignored. To fulfill this gap, we used trade globalization index and financial globalization index which take into account the many dimensions of both globalization indices.

3. Materials and methodology

3.1 Data

Following the main hypothesis of our study, to compare the relative effects of trade and financial globalization on energy efficiency, we construct an empirical model with annual data from 1990 to 2015 in panel data form as follows:

$$EI_{it} = a_0 + a_1 Y_{it} + a_2 TEC_{it} + a_3 TG_{it} + a_4 FG_{it} + \varepsilon_{it} \qquad (10.1)$$

where EI is the energy intensity which means increasing values representing a decrease in energy efficiency and proxied as units of energy per unit of gross domestic product, Y is economic growth and proxied by real gross domestic product per capita in 2010 constant US dollars, TEC is technological development and proxied by patent numbers, TG is trade globalization and proxied by trade globalization index, and FG is financial globalization which is proxied by financial globalization index. In empirical analysis, all variables are used in natural logarithmic form. The dataset of energy intensity, economic growth, and technological development are retrieved from World Development Indicators of World Bank. In addition, trade and financial globalization indices are obtained from KOF Globalization Index.

There are some reasons for using the KOF Globalization Index developed by Dreher (2006) and revised by Gygli et al. (2018) as a trade and financial globalization indicator in the study. While examining the effects of trade globalization in the literature, the trade openness indicator is generally considered. In contrast, the trade globalization index consists of subcomponents such as trade in goods, trade in services, and trade partner diversity. Similarly, when empirical literature is observed, it is seen that financial globalization is generally represented by FDIs. The financial globalization index consists of components such as FDI, portfolio investment, International debt, International reserves, and international income payments. For these reasons, using these indices instead of the indicators commonly used in the literature will give us the effects of globalization more accurately.

3.2 Empirical procedure

In recent years, developments in panel data techniques have been differentiated as first and second-generation estimators, especially considering whether there is a correlation between cross sections in the panel. Therefore, in order for the policy recommendations made based on panel data estimates to be reliable and consistent, it is necessary to choose the right estimators and some pretests for the right estimator selection. In this context, the cross-sectional dependency test, which examines the validity of the dependency between the sections, should be used first. In this study, we used the CD test developed by Pesaran (2004) to check the CSD and using the cross-section independence as a null hypothesis. Another important point is to examine the stationary process of variables. It is also necessary to decide which unit root test should be used according to the results of the cross-section dependence tests. Therefore, under the unit root null hypothesis, this study employs cross-sectionally augmented IPS (CIPS) panel unit root test of Pesaran (2007). After observing the stationary properties of the variables, we use the Error Correction Mechanism (ECM)-based panel cointegration method (Westerlund, 2007) with the null hypothesis of there is no cointegration because using this cointegration test has several advantages. Its first advantage is that it allows CSD among the observed countries. Another advantage is that this methodology is one of the most appropriate tests for our empirical model, since error correction-based testing gives robust results while explanatory variables are weakly exogenous.

After determining the cointegration relationship between variables, this study employs second-generation panel cointegrating estimators such as common correlated effect (CCE) developed by Pesaran (2006) and augmented mean group (AMG) estimation of Eberhardt and Bond (2009) and Eberhardt and Teal (2011). Both estimation techniques are suitable for the case of CSD among countries in the panel.

4. Empirical findings

In the first step of empirical analysis, we should look at some preliminary tests to decide the most suitable methods for our empirical model. Therefore, we first examine the existence of CSD among countries with CD test of Pesaran (2004). As a seen in Table 10.3, CD test results show that the null hypothesis pointing to the absence of CSD is strongly rejected and CSD is validated. This result direct us to use second-generation panel data methodologies that allow CSD in the panel. In addition, it means that a policy shock in one of the BEM countries may be easily spill-over to the other countries. We also check another problematic issue that the homogeneity of the variables with delta and delta adjusted tests as also illustrated in Table 10.3. The finding from these tests reveal that the null of homogeneity is rejected for all variables and the presence of the heterogeneity is confirmed. Therefore, besides the CSD, we concluded that each country has also country-specific shocks.

In next step of empirical analysis, we examine the stationary properties of variables with CIPS unit root test and the results are presented in Table 10.3.

TABLE 10.3 The results of preliminary tests.

	EI	Y	TEC	TG	FG
CSD					
CD test	−3.003 [0.001]	−2.782 [0.003]	−2.647 [0.004]	−3.378 [0.000]	−3.169 [0.001]
Homogeneity					
Δ	10.197 [0.000]	20.256 [0.000]	14.542 [0.000]	12.073 [0.000]	6.175 [0.000]
Δadj	10.842 [0.000]	21.537 [0.000]	15.461 [0.000]	12.836 [0.000]	6.565 [0.000]
Stationarity					
Level	−1.759	−0.907	−2.279	−2.058	−1.892
First differences	−2.993***	−2.411*	−2.846**	−3.074***	−3.220***

The numbers in brackets are P-values. *, **, and *** indicates the statistical significance at 10%, 5%, and 1% level, respectively.

According to the results, all variables have unit root in the level form of variables. However, the null of unit root is rejected and all variables are stationary in the first differenced form. The finding that variables are integrated of order one I(1) give us a chance to examine the cointegration relationship between variables.

Following the unit root test results, the cointegration relationship between variables are observed with ECM-based cointegration test of Westerlund (2007), and the findings are shown in Table 10.4. It can be seen that the null hypothesis of there is no cointegration between variables are rejected by group tau and group alpha statistics while panel tau and panel alpha statistics point to

TABLE 10.4 The results of ECM-Based panel cointegration test.

	Statistic	P-value
G^τ	−4.454***	0.000
G^α	−1.747**	0.040
$P\tau$	−0.333	0.369
P^α	1.355	0.912

*, **, and *** indicates the statistical significance at 10%, 5%, and 1% level, respectively.

TABLE 10.5 Panel cointegration coefficients.

Variables	CCE		AMG	
	Coefficient	*P*-value	Coefficient	*P*-value
Y	−0.486***	0.001	−0.600***	0.000
TEC	−0.030**	0.027	−0.061**	0.048
TG	−0.120**	0.038	−0.154**	0.039
FG	0.046*	0.077	0.028*	0.092

*, **, and *** indicates statistically significance at 10%, 5%, and 1% level, respectively.

absence of cointegration. Westerlund (2007) states that if the heterogeneity assumption is valid in the panel, the group statistics show robust results whereas the panel statistics give consistent results in case of homogeneity. Therefore, we concluded that the cointegration is confirmed based on the finding from group statistics. Since the cointegration also means the long-run relationship between variables, we should detect the long-run effects of regressors on energy efficiency.

In final step of empirical analysis, the long-run coefficients of regressors are analyzed with CCE and AMG estimators and the results are shown in Table 10.5. The findings from both estimators reveal that increasing income reduces the energy intensity thus it can be said that economic growth leads to improve energy efficiency in observed countries. This finding is consistent with the studies of Pan et al. (2019a,b). In addition, technological development reduces energy intensity. As an expected result of energy efficiency increasing impact of technological development is also found by Huang et al. (2017) and Hille and Lambernd (2020). In case of globalization, we found that trade globalization increases energy efficiency while financial globalization reduces it.

Based on the finding of country-specific shocks, we also check the individual country results for CCE estimations as a seen in Table 10.6, the results show that economic growth increases energy efficiency in Argentina, China, India, Indonesia, Mexico, Poland, S. Africa, S. Korea, and Turkey. In addition, technological development increases energy efficiency in China, Indonesia, Mexico, Poland, and S. Korea. In case of globalization, it seems trade globalization increases the efficiency use of energy in Brazil, China, India, and Poland. However, trade globalization increases energy intensity in Indonesia. Moreover, financial globalization increases energy efficiency only in India, while it increases intensity in Brazil and China.

TABLE 10.6 Coefficients for cross-sections.

	Y	TEC	TG	FG
Argentina	−0.598*** [0.000]	0.006 [0.816]	−0.120 [0.268]	−0.014 [0.726]
Brazil	0.019 [0.985]	0.066 [0.308]	−0.203* [0.075]	0.080** [0.036]
China	−1.158*** [0.000]	−0.249*** [0.000]	−0.375*** [0.000]	0.272** [0.046]
India	−0.386*** [0.000]	−0.015 [0.708]	−0.071** [0.010]	−0.033* [0.079]
Indonesia	−0.750*** [0.000]	−0.034** [0.016]	0.169*** [0.000]	−0.090 [0.139]
Mexico	−0.465*** [0.001]	−0.043** [0.017]	−0.102 [0.235]	−0.166 [0.298]
Poland	−0.858*** [0.000]	−0.162** [0.012]	−0.688*** [0.005]	0.047 [0.721]
S. Africa	−0.822*** [0.000]	0.013 [0.342]	−0.118 [0.109]	0.003 [0.951]
S. Korea	−0.595*** [0.000]	−0.167*** [0.000]	0.063 [0.703]	0.048 [0.677]
Turkey	−0.369*** [0.008]	−0.025 [0.357]	−0.096 [0.340]	0.133 [0.129]

*, **, and *** indicates statistically significance at 10%, 5%, and 1% level, respectively.

5. Concluding remarks

This chapter explores the impact of different kinds of globalization (i.e., trade globalization and financial globalization) on energy efficiency by incorporating the impact of economic growth and technological development in BEMs. For this purpose, the annual period of 1990−2015 is analyzed with panel data techniques that allow CSD among emerging economies.

The results of the empirical analysis can be separated with panel case and country-specific case. In case of panel estimations, it is concluded that economic growth, technological development, and trade globalization increases energy efficiency. On the other hand, we found the evidence that financial globalization reduces the energy efficiency. The efficiency increasing effect of economic growth indicates that increasing wealth is returning as an investment in areas that increase energy efficiency rather than being directed to inefficient areas. Similarly, expected finding that technological development increases energy efficiency shows that energy efficiency enhancing R&D activities are successfully implemented and that technological innovations in other sectors create positive externality on energy sector. In perspective of trade, there may be some reasons of efficiency increasing effect of trade globalization. The first reason is that companies can purchase energy-saving modern machines in order to reduce energy costs with their export revenues. Secondly, due to the scale effect that occurs as a result of trade globalization, overproduction firms can produce more output with less energy consumption with the effect of learning method. Thirdly, due to the fact that trade globalization causes structural transformation in emerging economies, increasing share of the sectors where energy can be used more efficiently makes the emerging economies the exporting country in these sectors. The harmful effect of financial globalization on energy efficiency emerges largely as a counterpart to the financing opportunity provided by investors in developed countries to emerging economies. Policymakers in emerging economies do not want to overlook investors who have shifted their investments to emerging economies with relatively loose environmental regulations due to tight environmental regulations in developed countries. Foreign investors see these countries as cheap labor and cheap raw material resources and act with an understanding that maximizes profit instead of investing in energy-saving technologies in the host country.

In regard with policy implications, the policymakers in emerging economies are obliged to produce policies with the fact that economic growth, achieved without increasing energy efficiency and harmful to the environment, is only a temporary growth and that sustainable growth cannot be achieved with this understanding. Accordingly, they should make arrangements for transferring foreign investments to innovative activities that increase energy efficiency at certain rates. These investors should be supported by various tax exemptions and subsidies to investors who provide funding opportunities for these projects against the risk of withdrawal in the country's economy.

References

Abidin, I.S.Z., Haseeb, M., Azam, M., Islam, R., 2015. Foreign direct investment, financial development, international trade and energy consumption: panel data evidence from selected ASEAN countries. Int. J. Energy Econ. Pol. 5 (3).

Adom, P.K., 2015. Asymmetric impacts of the determinants of energy intensity in Nigeria. Energy Econ. 49, 570–580.

Ahmed, K., 2017. Revisiting the role of financial development for energy-growth-trade nexus in BRICS economies. Energy 128, 487–495.

Alam, M.M., Murad, M.W., 2020. The impacts of economic growth, trade openness and technological progress on renewable energy use in organization for economic co-operation and development countries. Renew. Energy 145, 382–390.

Apergis, N., Payne, J.E., 2009. Energy consumption and economic growth in Central America: evidence from a panel cointegration and error correction model. Energy Econ. 31 (2), 211–216.

Apergis, N., Tang, C.F., 2013. Is the energy-led growth hypothesis valid? New evidence from a sample of 85 countries. Energy Econ. 38, 24—31.

Boutabba, M.A., 2014. The impact of financial development, income, energy and trade on carbon emissions: evidence from the Indian economy. Econ. Modell. 40, 33—41.

Bu, M., Li, S., Jiang, L., 2019. Foreign direct investment and energy intensity in China: firm-level evidence. Energy Econ. 80, 366—376.

Cagno, E., Ramirez-Portilla, A., Trianni, A., 2015. Linking energy efficiency and innovation practices: empirical evidence from the foundry sector. Energy Policy 83, 240—256.

Cao, W., Chen, S., Huang, Z., 2020. Does foreign direct investment impact energy intensity? Evidence from developing countries. Math. Probl. Eng. Special Issue 2020. https://doi.org/10.1155/2020/5695684.

Çoban, S., Topcu, M., 2013. The nexus between financial development and energy consumption in the EU: a dynamic panel data analysis. Energy Econ. 39, 81—88.

Cole, M.A., 2006. Does trade liberalization increase national energy use? Econ. Lett. 92 (1), 108—112.

Destek, M.A., 2018. Financial development and energy consumption nexus in emerging economies. Energy SourcesPart B 13 (1), 76—81.

Destek, M.A., Aslan, A., 2017. Renewable and non-renewable energy consumption and economic growth in emerging economies: evidence from bootstrap panel causality. Renew. Energy 111, 757—763.

Doytch, N., Narayan, S., 2016. Does FDI influence renewable energy consumption? An analysis of sectoral FDI impact on renewable and non-renewable industrial energy consumption. Energy Econ. 54, 291—301.

Dreher, A., 2006. Does globalization affect growth? Evidence from a new index of globalization. Appl. Econ. 38 (10), 1091—1110.

Eberhardt, M., Bond, S., 2009. Cross-section Dependence in Nonstationary Panel Models: A Novel Estimator.

Eberhardt, M., Teal, F., 2011. Econometrics for grumblers: a new look at the literature on cross-country growth empirics. J. Econ. Surv. 25 (1), 109—155.

Elliott, R.J., Sun, P., Chen, S., 2013. Energy intensity and foreign direct investment: a Chinese city-level study. Energy Econ. 40, 484—494.

Farhani, S., Solarin, S.A., 2017. Financial development and energy demand in the United States: new evidence from combined cointegration and asymmetric causality tests. Energy 134, 1029—1037.

Fisher-Vanden, K., Jefferson, G.H., Liu, H., Tao, Q., 2004. What is driving China's decline in energy intensity? Resour. Energy Econ. 26 (1), 77—97.

Gozgor, G., Mahalik, M.K., Demir, E., Padhan, H., 2020. The impact of economic globalization on renewable energy in the OECD countries. Energy Policy 139, 111365.

Gygli, S., Haelg, F., Potrafke, N., Sturm, J.E., 2018. The KOF globalisation index-revisited.

Hille, E., Lambernd, B., 2020. The role of innovation in reducing South Korea's energy intensity: regional-data evidence on various energy carriers. J. Environ. Manag. 262, 110293.

Huang, J., Du, D., Tao, Q., 2017. An analysis of technological factors and energy intensity in China. Energy Policy 109, 1—9.

IEA, 2018. Global Energy Demand Grew by 2.1% in 2017, and Carbon Emissions Rose for the First Time since 2014). https://www.iea.org/news/global-energy-demand-grew-by-21-in-2017-and-carbon-emissions-rose-for-the-first-time-since-2014. (Accessed 15 May 2020).

Keho, Y., 2016. Do foreign direct investment and trade lead to lower energy intensity? Evidence from selected African countries. Int. J. Energy Econ. Pol. 6 (1).

Koengkan, M., 2018. The positive impact of trade openness on consumption of energy: fresh evidence from Andean community countries. Energy 158, 936−943.

Luo, J., Cheng, K., 2013. The influence of FDI on energy efficiency of China: an empirical analysis based on DEA method. In: Applied Mechanics and Materials, vol. 291. Trans Tech Publications Ltd, pp. 1217−1220.

Mahalik, M.K., Mallick, H., 2014. Energy consumption, economic growth and financial development: exploring the empirical linkages for India. J. Develop. Areas 139−159.

Nasreen, S., Anwar, S., 2014. Causal relationship between trade openness, economic growth and energy consumption: a panel data analysis of Asian countries. Energy Policy 69, 82−91.

Omri, A., Kahouli, B., 2014. Causal relationships between energy consumption, foreign direct investment and economic growth: fresh evidence from dynamic simultaneous-equations models. Energy Policy 67, 913−922.

Pan, X., Ai, B., Li, C., Pan, X., Yan, Y., 2019a. Dynamic relationship among environmental regulation, technological innovation and energy efficiency based on large scale provincial panel data in China. Technol. Forecast. Soc. Change 144, 428−435.

Pan, X., Uddin, M.K., Han, C., Pan, X., 2019b. Dynamics of financial development, trade openness, technological innovation and energy intensity: evidence from Bangladesh. Energy 171, 456−464.

Pesaran, M.H., 2004. General Diagnostic Tests for Cross Section Dependence in Panels. CESifo Working Paper 1229. IZA Discussion Paper, 1240.

Pesaran, M.H., 2007. A simple panel unit root test in the presence of cross-section dependence. J. Appl. Econom. 22 (2), 265−312.

Pesaran, M.H., 2006. Estimation and inference in large heterogeneous panels with a multifactor error structure. Econometrica 74 (4), 967−1012.

Sadorsky, P., 2010. The impact of financial development on energy consumption in emerging economies. Energy Policy 38 (5), 2528−2535.

Sadorsky, P., 2011. Financial development and energy consumption in Central and Eastern European frontier economies. Energy Policy 39 (2), 999−1006.

Sbia, R., Shahbaz, M., Hamdi, H., 2014. A contribution of foreign direct investment, clean energy, trade openness, carbon emissions and economic growth to energy demand in UAE. Econ. Modell. 36.

Shahbaz, M., Khan, S., Tahir, M.I., 2013. The dynamic links between energy consumption, economic growth, financial development and trade in China: fresh evidence from multivariate framework analysis. Energy Econ. 40, 8−21.

Shahbaz, M., Lahiani, A., Abosedra, S., Hammoudeh, S., 2018. The role of globalization in energy consumption: a quantile cointegrating regression approach. Energy Econ. 71, 161−170.

Shahbaz, M., Mallick, H., Mahalik, M.K., Sadorsky, P., 2016. The role of globalization on the recent evolution of energy demand in India: implications for sustainable development. Energy Econ. 55, 52−68.

Shahbaz, M., Nasreen, S., Ling, C.H., Sbia, R., 2014. Causality between trade openness and energy consumption: what causes what in high, middle and low income countries. Energy Policy 70, 126−143.

Shahzad, S.J.H., Kumar, R.R., Zakaria, M., Hurr, M., 2017. Carbon emission, energy consumption, trade openness and financial development in Pakistan: a revisit. Renew. Sustain. Energy Rev. 70, 185−192.

Sohag, K., Begum, R.A., Abdullah, S.M.S., Jaafar, M., 2015. Dynamics of energy use, technological innovation, economic growth and trade openness in Malaysia. Energy 90, 1497−1507.

Tan, Z.F., Zhang, J.L., 2010. Study on the dynamic relationship between energy efficiency and its influencing factors in China. China Popul. Resour. Environ. 4, 43−49.

Tang, C.F., Tan, B.W., Ozturk, I., 2016. Energy consumption and economic growth in Vietnam. Renew. Sustain. Energy Rev. 54, 1506−1514.

Topcu, M., Payne, J.E., 2018. Further evidence on the trade-energy consumption nexus in OECD countries. Energy Policy 117, 160−165.

Wang, Y., Wang, Y., Zhou, J., Zhu, X., Lu, G., 2011. Energy consumption and economic growth in China: a multivariate causality test. Energy Policy 39 (7), 4399−4406.

Wei, W.X., Chen, D., Hu, D., 2016. Study on the evolvement of technology development and energy efficiency—a case study of the past 30 years of development in Shanghai. Sustainability 8 (5), 457.

Westerlund, J., 2007. Testing for error correction in panel data. Oxf. Bull. Econ. Stat. 69 (6), 709−748.

Yıldırım, E., Sukruoglu, D., Aslan, A., 2014. Energy consumption and economic growth in the next 11 countries: the bootstrapped autoregressive metric causality approach. Energy Econ. 44, 14−21.

Zeren, F., Akkuş, H.T., 2020. The relationship between renewable energy consumption and trade openness: new evidence from emerging economies. Renew. Energy 147, 322−329.

Zhang, Y.J., 2011. The impact of financial development on carbon emissions: an empirical analysis in China. Energy Policy 39 (4), 2197−2203.

Zhou, N., Levine, M.D., Price, L., 2010. Overview of current energy-efficiency policies in China. Energy Policy 38 (11), 6439−6452.

Chapter 11

Is globalization a driver for energy efficiency and sustainable development?

Patrícia Hipólito Leal[1,2], António Cardoso Marques[1,2]
[1]University of Beira Interior, Management and Economics Department, Covilhã, Portugal;
[2]NECE-UBI, University of Beira Interior, Covilhã, Portugal

1. Introduction

Global warming is one of the greatest challenges facing mankind. Economically, environmental degradation has become one of the biggest issues confronting the planet. Resources are scarce, and the environment is considered within this scarcity. Therefore, the Paris Agreement and the Climate and Energy Framework Agreement reinforce the necessity to combat climate changes. In line with this, sustainable development is the main issue focused on by policymakers locally and globally (Gigliotti et al., 2019; Opoku, 2019). Sustainable development is frequently considered a process within the globalization context, constituted by three main components: namely economic, social, and environmental. Globalization is considered sustainable if it has a positive impact on the three main sustainable development components. Therefore, if globalization provokes negative effects, it may be considered unsustainable (Beumer et al., 2018). Since the 1990s, growing globalization has been noted, and since then, this phenomenon has affected economies, the environment, and human life. As asserted by Intriligator (2004), globalization is considered crucial to define economic development and the future of the planet. With increasing and enhanced globalization, the world's output is continuously growing, such as noted by Rahman (2020). Economic growth depends on energy, and society's excessive dependence on energy arises from the necessity to ensure the supply of growing energy demand and to sustain economic development. Consequently, carbon dioxide (CO_2) emissions are directly related to economic growth due to the fact that output growth needs energy consumption (Rahman, 2017).

Energy-Growth Nexus in an Era of Globalization. https://doi.org/10.1016/B978-0-12-824440-1.00004-7
257

Characterized by the integration of the national economy with the world economy in aspects of trade, capital, socioeconomic, and politics, globalization is viewed as a driving force of economic growth (Mishkin, 2009). Consequently, it could undermine the environment and induce environmental degradation (Shahbaz et al., 2018; You and Lv, 2018). On the one hand, the phenomenon of globalization stimulates environmental degradation by inducing trade openness and enabling the relocation of pollutant industries. The relocation of industries occurs from strict environmentally regulated countries to permissive environmentally regulated ones, that is, mainly from developed to developing countries. Consequently, this phenomenon could provoke significant environmental damages. According to the message of the Kyoto Protocol in 1997, and the Paris Agreement in 2015, both developed and developing countries have to mitigate CO_2 emissions, and consequently could have to sacrifice high growth of income or the ambition of high income growth in order to promote environmental quality (Rahman, 2017). Diverse environmentalists consider that globalization promotes the rising global demand for goods and services, which provokes an increase in economic activities and production, and consequently leads to environmental damage (Rahman, 2020). On the other hand, globalization can induce environmental quality by allowing clean and environmentally friendly technology import.

One of the foremost common goals of countries worldwide is to promote economic growth that ensures a clean and suitable environment. The difficulties lie in how to grow economies in a sustainable way. In order to accomplish sustainable development, understanding how to promote energy efficiency and its key benefits, consequently promoting environmental enhancement without compromising economic growth, is crucial. The effect of globalization on energy consumption and the environment is not consensual in the literature. Therefore, considering globalization as a tool to disseminate information and knowledge, able to transcend technological barriers and overcome national borders, the question arises: Could globalization be a driver of energy efficiency and sustainable development? Through the recent improvements in globalization measurement by Gygli et al. (2019), globalization can be analyzed through three dimensions, namely economic, social, and political, and each dimension divided into the variables that represent the policies that allow and control the flows and the activities, and into the variables that represent the flows and activities, the measure de jure and de facto, respectively. This measurement of globalization makes it possible to obtain specific results for each dimension and measure, which in turn allows the development of appropriate and assertive instruments and policies to suitably combat against environmental degradation and promote energy efficiency and sustainable development. Considering the necessity and urgency to act against climate change, it is crucial to implement measures and policies that stimulate the beneficial effects of globalization on the environment.

This chapter firstly aims to present the background of the main issues under analysis and discussed, namely globalization, energy efficiency, and sustainable development. Therefore, a comprehensive debate of the literature of these research topics is provided, which becomes a useful tool for the researchers that intend to study these issues. Besides that, this chapter innovates through employing empirical procedures to assess the effect of each dimension and measure of globalization on energy efficiency and sustainable growth, aiming to give an answer to the research question: Is globalization a driver of energy efficiency and sustainable development? Taking into account the gap in the literature of the analysis of the new measures of globalization, de jure and de facto, this chapter brings an innovative contribution to the current literature, filling the gap of the analysis of globalization de jure and de facto on sustainable development and energy efficiency. Sustainable development is assessed through the Index for Sustainable Economic Welfare (ISEW), which is considered an aggregate welfare measure that encompasses economic values that integrate distribution inequality of income, social impacts, environmental damage and quality, and elements that are beneficial to welfare (O'Mahony et al., 2018). Therefore, this research is able to stress effects, such as the transfer of know-how and good policy practices. Considering that the time span covers more than two economic cycles, the use of a methodology that permits the analysis of the distinction of effects between the short- and the long-run is required. Therefore, to answer the proposed research question, this research uses the autoregressive distributed lag (ARDL) model.

Henceforth, this chapter is structured as follows. In the next section, a comprehensive debate about the relevant literature of globalization, energy efficiency, and sustainable development is presented. Subsequently, data and empirical methods used in this research are explained in Section 3. Furthermore, Section 4 presents the obtained results. Lastly, Sections 5 and 6 consist of the discussion and conclusions of the main results, respectively.

2. Debate

To present a comprehensive literature review, in an organized way, this section is divided into three subsections. The first subsection presents a summary of the globalization literature. The second concentrates on a summary of the literature of energy efficiency. Lastly, the third one focusses on a summary of the sustainable development literature.

2.1 Globalization

First introduced by Dreher (2006), and later improved by Dreher et al. (2008), the KOF globalization index becomes the most used globalization measurement in the literature (Gygli et al., 2019) among diverse globalization indexes proposed over time such as Global Index (Raab et al., 2008); New Globalization Index (Vujakovic, 2010); and Maastricht Globalization Index (Figge and Martens, 2014). Therefore, the KOF globalization index is widely used in the current literature (Bilgili et al., 2020; Gozgor et al., 2020; Guan et al., 2020; Leal and Marques, 2020; Liu et al., 2020; Sethi et al., 2020; Suki et al., 2020; Ulucak et al., 2020; Wang et al., 2020; Woo and Jun, 2020). Recently, Gygli et al. (2019) updated the KOF globalization index and proposed the revisited KOF globalization index. The updated version includes new variables in the KOF globalization index structure, capturing diverse interaction activities, and introduces two new measures, de jure and de facto.

The phenomenon of globalization consists of economic, social, and political aspects, and it has a direct effect on production-consumption, environment, and human behaviors (Bilgili et al., 2020). Since the 1990s, the fast increasing globalization phenomenon has affected human life and economies through economic, political, and social aspects. Consequently, globalization represents diverse challenges in the widely varying areas of economics, energy, and environmental performance. Various authors have addressed the research questions of the role of globalization in the environment (Ahmed et al., 2021; Aluko et al., 2021; Bilgili et al., 2020; Leal and Marques, 2019a, 2020; Liu et al., 2020; Suki et al., 2020; Sun et al., 2020; Wang et al., 2020; Yameogo et al., 2021). A summary of the literature that examines the relationship between globalization and environmental performance is provided by Bilgili et al. (2020). Nevertheless, there is no consensus if globalization is beneficial or harmful to the environment. Bilgili et al. (2020) reveal a negative causality that runs from financial globalization, political globalization, and trade globalization to ecological footprint. In turn, Suki et al. (2020) reveal that economic and overall globalization undermine the environment, while political and social globalization are beneficial.

On the one hand, globalization is harmful to the environment, reducing environmental quality, through stimulating trade openness and promoting the relocation of pollutant industries from developed countries to developing (Bilgili et al., 2020). Globalization induces an increase in economic activities, such as trade and transportation, which consequently provokes more resource use and pollution (Copeland and Taylor, 2013). Furthermore, countries with an abundance of natural resources could overuse their resources in order to increase exports and consequently deteriorate the environmental quality (Bilgili et al., 2020). On the other hand, globalization could promote environmental quality through the importation of clean and green technology or by improving global ecological awareness. Furthermore, through the dissemination of technological progress, which is useful for decreasing the use of resources and consequently producing less pollution, and overcoming trade barriers, it is expected that globalization enhances countries' welfare (Shahbaz et al., 2018). The sustainable use of natural resources is a challenge with increasing globalization. The phenomenon of globalization provokes a split of production from consumption and consequently leads to the global redistribution of natural resources (Bolwig et al., 2010). According to Bilgili et al. (2020), globalization has a crucial influence on the shape of efficient material use and consequently to determinate environmental sustainability. Ulucak et al. (2020) conclude that higher globalization levels induce reduced material consumption and promote sustainable resource management.

By importing technology directed related to energy consumption and production activities, globalization has the ability to decrease the energy demand of economies (Baek et al., 2009). Therefore, economic globalization improves the level of technology and consequently increases the productivity of energy sources and decreases the energy cost. Motivated by the findings provided by Shahbaz et al. (2016) regarding the effect of economic globalization on the evolution of the energy demand in India, Gozgor et al. (2020) considered that this relationship should be addressed by developed countries. Therefore, Gozgor et al. (2020) focused on the effect of economic globalization on renewable energy consumption in the Organization for Economic Co-operation and Development (OECD) countries, concluding that a higher level of economic globalization induces the use of renewable energy. These authors provide a summary of energy-globalization nexus literature.

2.2 Energy efficiency

Introduced into the literature by Fisher (1921) with the objective of decomposing energy intensity into structural shift and energy efficiency (Metcalf, 2008), the Fisher ideal index is, according to the literature, commonly preferred and used (Boyd and Roop, 2004). This index perfectly decomposes the change in energy intensity into efficiency and activity changes, as well as the Log Mean Divisia index (LMDI). The Fisher ideal index and the LMDI derivate from the index decomposition analysis (IDA). The IDA includes Laspeyres index methods and Divisia index methods. The Fisher ideal index is included in the Laspeyres index methods (Ang and Liu, 2001), while the LMDI in the Divisia index methods. According to Dargahi and Khameneh (2019), decomposition results via the LMDI method are similar to the results of the decomposition via the Fisher ideal index. These two indexes can be used in the same analysis to provide further evidence or to complement the results. Leal et al. (2019) combine the LMDI method and the Fisher ideal index in order to perform a decomposition analysis by sectoral factors for the Australian sectors. The Fisher ideal index has most of the desirable properties on which the assessment of the IDA methods should be based, namely adaptability, strong theoretical foundation, easy use and interpretation, robustness to zero values, and consistency in aggregation (Ang, 2004). Furthermore, the Fisher ideal index computes decomposition without leaving any residual term. Therefore, the Fisher ideal index through its ability to compute an irreproachable decomposition of aggregate energy intensity into structural and efficiency components is considered attractive to observe the relevance of industry composition and efficiency effects (Zhang, 2013).

Throughout the years, the Fisher ideal index has been applied in a wide diversity of contexts. The index was firstly used in industrial energy demand studies by Liu and Ang (2003). In turn, Boyd and Roop (2004) performed the Fisher ideal index in the United States manufacturing sector to assess the sources of energy intensity change. Furthermore, this index is applied to assess the trends of economy-wide energy intensity (Metcalf, 2008). The Fisher ideal index can be applied to two strands. On the one hand, this index can be used to perform a decomposition analysis (Leal et al., 2019). On the other hand, beyond the decomposition, it can be included in regression techniques. Jimenez and Mercado (2014) applied the Fisher ideal index for 75 Latin American Countries to decompose the energy intensity into the relative contributions of energy efficiency and economic structure. Furthermore, the authors examined the main drivers of the energy indexes through panel data regression techniques. Leal and Marques (2020) and Özbuğday and Erbas (2015) computed the Fisher ideal index and included the energy efficiency index into the panel regression technique in order to analyze its effect on environmental degradation. Ullah et al. (2019) applied the Fisher ideal index to decompose the change in electricity intensity into efficiency and activity changes in Pakistan. Then the authors proceeded to analyze the drivers of the activity index, efficiency index, and intensity index, as well as Karimu et al. (2017). Deichmann et al. (2019) examined the determinants of the energy intensity index.

2.3 Sustainable development

Sustainable development is commonly measured and analyzed through Gross Domestic Product (GDP). Although GDP is the most used indicator, its creators (Kuznets, 1934) recognized it as insufficient to assess sustainable development. This inadequacy comes from the fact that GDP is powerless to measure welfare and sustainability (Aşici, 2013; Gaspar et al., 2017; Marques et al., 2016; Menegaki and Tugcu, 2018; Pais et al., 2019), as well as being inefficient for quantifying social welfare (Costanza et al., 2009; Li and Fang, 2014; Stockhammer et al., 1997). It is crucial to incorporate the three main dimensions of sustainability when measuring sustainable development, namely economic efficiency, social cohesion, and environmental responsibility (Aust et al., 2020; Dartey-Baah, 2014). This is one of the main limitations of the sustainable development measurements across the literature when measured through indicators such as the Human Development Indicator (Apinran et al., 2018; Sharma and Gani, 2004) and the Ecological Footprint (Balsalobre-Lorente et al., 2019; Doytch, 2020). Kwatra et al. (2020) provide a critical perspective of the evolution of sustainable development measurements.

Firstly computed by Daly and Cobb (1989), and later improved by Cobb and Cobb (1994), the ISEW arose as an alternative to measure sustainable development (Beça and Santos, 2010, 2014; Coscieme et al., 2020;

Gaspar et al., 2017; Marques et al., 2018; Menegaki et al., 2017; Menegaki and Tiwari, 2017; Menegaki and Tsagarakis, 2015). A study developed by Gaspar et al. (2017), which compares the sustainable development approach, measured by the ISEW, with the economic growth approach, measured by GDP, reveals that a sensitivity exists when the ISEW is used in place of GDP. Welfare is considered an ambiguous and multifaceted concept, and a composite indicator is essential to reflect it (Menegaki et al., 2017). The ISEW is an aggregate indicator for both current and future well-being, which includes environmental depletion through the costs of using available natural resources in its calculation. This makes it more trustworthy from an ecological point of view when compared with GDP. Therefore, according to Hák et al. (2016), the ISEW is suitable to analyze the accomplishment of the Sustainable Development Goals of the United Nations. The most crucial obstacles and opportunities of the ISEW calculation are presented by Bleys and Whitby (2015). The ISEW combines three spheres of sustainability, namely: environment, society, and economy (Menegaki et al., 2017). A summarized explanation of measuring sustainable development is provided by Gaspar et al. (2017), and Sánchez et al. (2020) provide the theories of sustainable economic well-being.

The ISEW is widely used in the energy economics literature, being analyzed in diverse contexts, mainly in the energy-growth nexus (Gaspar et al., 2017; Menegaki et al., 2017; Menegaki and Tiwari, 2017; Menegaki and Tugcu, 2016a,b, 2018), but also in the context of food consumption (Marques et al., 2018) and in the accomplishment of the United Nations Sustainable Development Goals (Coscieme et al., 2020). Besides that, the ISEW is applied for diverse groups of countries, namely sub-Saharan emerging economies (Menegaki and Tugcu, 2016a), 15 emerging economies (Menegaki and Tugcu, 2016b), American countries (Menegaki and Tiwari, 2017), Europe (Gaspar et al., 2017; Menegaki et al., 2017), Asian countries (Menegaki and Tugcu, 2018), Spain (O'Mahony et al., 2018), and Ecuador (Sánchez et al., 2020).

3. Empirical procedure

The purpose of this section is to present the features of the data and sample under analysis. It will also present and explain the methodology applied to fulfill the main objective of this research, which is to analyze the influence of globalization, its dimensions and measures, on sustainable development and on energy efficiency.

3.1 Data

This research is performed for the top 20 energy consumers per capita countries of 2017. These countries were selected using the BP Statistical Review of World Energy 2019 and the data availability of the variables included in this analysis. This group of 20 countries was divided into two groups. The division was made to capture the effect of different levels of globalization among the countries, giving rise to two groups, namely the high globalized countries (HGCs) and the low globalized countries (LGCs). Therefore, the countries were divided through the Overall Globalization Ranking obtained from the KOF Swiss Economic Institute. After collecting the score of each country in the Overall Globalization Ranking, the division was made through the calculation of the mean and median. Both mean and median calculation resulted in the same groups of countries, they are: (i) HGC consist of Austria, Belgium, The Czech Republic, Finland, France, Germany, the Netherlands, Norway, Sweden, and Switzerland (ii) LGC include Australia, Canada, Estonia, Ireland, Japan, Luxembourg, New Zealand, Slovakia, Slovenia, and the United States. In order to use the most available data, two time periods are under analysis. The computation of the ISEW is only available from 2000 to 2017 considering the availability of its components, so consequently, the model with the ISEW as the dependent variable was estimated for this time period. Nevertheless, considering available data from 1995 to 2017, for the remaining variables under analysis, the model with efficiency index as the dependent variable is estimated for this time period.

In order to capture economic, sustainable development, globalization, energy, and environmental effects, according with the literature, the variables under analysis are namely: (i) *GDP*, as a proxy of economic growth, collected from World Data Bank and measured in constant 2010 prices in US$; (ii) the *ISEW*, as a proxy of sustainable development, from own elaboration; (iii) the KOF Globalization Index Economic Dimension *de Facto* (*KOFECDF*); the KOF Globalization Index Economic Dimension *de Jure* (*KOFECDJ*); the KOF Globalization Index Social Dimension *de Facto* (*KOFSODF*); the KOF Globalization Index Social Dimension *de Jure* (*KOFSODJ*); the KOF Globalization Index Political Dimension *de Facto* (*KOFPODF*); and the KOF Globalization Index Political Dimension *de Jure* (*KOFPODJ*) obtained from the KOF Swiss Economic Institute (http:// globalisation.kof.ethz.ch/); (iv) oil consumption (*OIL*); coal consumption (*COAL*); gas consumption (*GAS*), and renewable energy consumption (*RES*), all collected from the BP Statistical Review of World Energy 2019. *OIL* is measured in Million tonnes (Mt), and *COAL, GAS,* and *RES* are measured in Millions of tonnes in oil equivalent (Mtoe); (v) CO_2 emissions, collected from the BP Statistical Review of World Energy 2019, is measured in Mt; (vi) Efficiency index (EF) (1990 $= 1$) is computed through the Fisher ideal index developed by Fisher (1921).

3.1.1 Index for Sustainable Economic Welfare

The ISEW is an aggregate indicator for present and future well-being based on economic, environmental, and social dimensions, and used as a proxy of sustainable development (Gaspar et al., 2017; Marques et al., 2018). The ISEW calculation is initiated with the personal consumption expenditure weighted for income inequality. This takes into account the advantages that rich countries benefit from economic growth compared to poorer ones. To this, positive magnitudes of welfare, namely public health and education expenditures, are added, and from this negative magnitudes of welfare, namely environmental costs (defensive costs), are subtracted. The ISEW computation used in the present study was adopted from Marques et al. (2018), Menegaki et al. (2017), Menegaki and Tsagarakis (2015), following the forma expression described in Eq. (11.1):

$$ISEW = C_w + G_{eh} + K_n + S - N - C_s \qquad (11.1)$$

where C_w denotes the weighted private consumption, G_{eh} denotes the non-defensive public expenditure, K_n represents the net capital growth, S consists of the unpaid work benefit, N measures the depletion of the natural environment, and C_s describes the cost of social issues.

Due to lack of data, as in Marques et al. (2018) and Menegaki et al. (2017), social costs were not included in the ISEW calculation. The parameters and calculation process of the ISEW are described in Table 11.1.

In the process of the calculation of the ISEW, extrapolation technique had to be used to calculate a few observations due to the lack of data. This technique was used to calculate the GINI coefficient for Japan for the years of 2016 and 2017.

TABLE 11.1 Index for Sustainable Economic Welfare components.

Variables	Sign	Calculation method	Source
Adjusted personal consumption with durables (C_w)	+	PC*(1-GINI)	PC: https://data.worldbank.org/indicator/NE.CON.PRVT.CD GINI: https://dataverse.harvard.edu/dataset.xhtml?persistentId=doi:10.7910/DVN/LM4OWF
Education expenditure (G_{eh})	+	Education expenditure*0,5	http://data.worldbank.org/indicator/NY.ADJ.AEDU.CD
Health expenditure (G_{eh})	+	Health expenditure*0,5	https://data.worldbank.org/indicator/SH.XPD.CHEX.PC.CD
Net capital growth (K_n)	±	FCA-CFC	FCA: http://data.worldbank.org/indicator/NE.GDI.TOTL.CD CFC: http://data.worldbank.org/indicator/NY.ADJ.DKAP.CD

Continued

TABLE 11.1 Index for Sustainable Economic Welfare components.—cont'd

Variables	Sign	Calculation method	Source
Mineral depletion (N)	–		http://data.worldbank.org/indicator/NY.ADJ.DMIN.CD
Energy depletion (N)	–		http://data.worldbank.org/indicator/NY.ADJ.DNGY.CD
Forest depletion (N)	–		http://data.worldbank.org/indicator/NY.ADJ.DFOR.CD
Damage from CO_2 emissions (N)	–		http://data.worldbank.org/indicator/NY.ADJ.DCO2.CD

PC denotes Final household consumption expenditure; FCA denotes fixed capital accumulation; CFC denotes consumption of fixed capital. The sources were last accessed in June 2020.

3.1.2 Efficiency index

To calculate the efficiency index, the following variables were collected: (a) Energy consumption by sector, collected from the International Energy Agency measured in kilotons of oil equivalent (ktoe); (b) Gross value added and GDP both collected from the United Nations Statistical Division and measured in constant 2015 prices in US$. The Fisher ideal index is computed as follows [The present study only analyses the efficiency index. Therefore, the computation presented is for the efficiency index. A complete computation of the three indexes can be found on (Leal and Marques, 2019b)].

Energy intensity is mathematically specified as a function of energy efficiency and economic activity (Metcalf, 2008). Therefore, aggregate energy intensity (e_t) is obtained through the total energy consumption (E_t) and the total output (GDP) (Y_t) in year t, that corresponds to the sum of energy intensity by sector (e_{it}) with the changes in the economy structure (s_{it}). Written as Eq. (11.2):

$$e_t = \frac{E_t}{Y_t} = \sum_i \left(\frac{E_{it}}{Y_{it}}\right)\left(\frac{Y_{it}}{Y_t}\right) = \sum_i e_{it}s_{it}, \qquad (11.2)$$

where E_{it} and Y_{it} denotes energy consumption and the measure of economic activity for sector i in year t, respectively.

The computation of the Fisher ideal index requires the construction of Laspeyres index (*L*) and Paasche index (*P*). Eq. (11.3) represents the calculation of the Laspeyres and Paasche indexes of efficiency (*EF*).

$$L_t^{EF} = \frac{\sum_i e_{it} s_{i0}}{\sum_i e_{i0} s_{i0}}; P_t^{EF} = \frac{\sum_i e_{it} s_{it}}{\sum_i e_{i0} s_{it}}, \tag{11.3}$$

Lastly, the efficiency index is obtained through the Fisher ideal index given by the Laspeyres and Paasche indexes, as written in Eq. (11.4).

$$F_t^{EF} = \left(L_t^{EF} * P_t^{EF}\right)^{\frac{1}{2}}. \tag{11.4}$$

The energy efficiency index is interpreted as follows. A positive variation of the index value denotes a decrease in the efficiency since it matches the increase of the energy used per unit of economic output. Therefore, a negative variation of the index value denotes an enhancement of the efficiency, since it matches the decrease in the energy used per unit of economic output. Thus, the efficiency index demonstrates the efficiency trajectory and the evolution through the years.

3.1.3 Preliminary tests

A preliminary analysis of the coefficients of the correlation matrix was performed to realize transformations on the variables to avoid correlation concerns. Therefore, the transformations performed, namely *GDP* and CO_2, were transformed into per capita, and *OIL*, *COAL*, and *GAS* were summed and gave rise to the *FF* (fossil fuels consumption) variable, that is *RES* were transformed into a percentage of primary energy consumption. Descriptive statistics of the variables under analysis for the high and LGCs and for both time periods are displayed in Tables A11.1 and A11.2.

To detect the characteristics of the panel data, some preliminary tests were carried out, namely the Pesaran (2004) cross-section dependence test (CD-test) (Tables A11.3 and A11.4). Cross-section dependence commonly occurs, and it is one of the challenges in panel data settings. Therefore, according with the literature, it is advised to give further attention to both the nature of the variables and the idiosyncrasies of the countries analyzed, mainly in macro panel (Eberhardt, 2011; Hoechle, 2007). The results from the CD-test strongly suggest the presence of cross-sectional dependence for all variables of both groups of countries, HGC and LGC, for the period of 1995–2017 (Table A11.3). For the period 2000–17, the presence of cross-sectional dependence is suggested for all variables of the HGC, while in the LGC, it is suggested for all variables except for *KOFECDF* and *KOFPODF* (Table A11.4).

Therefore, due to the presence of cross-sectional dependence, the second generation unit root test (CIPS) (Pesaran, 2007) was carried out. Whereas the presence of cross-sectional dependence could be a limitation for the first-generation unit root tests, the second generation unit root test has the advantage of being suitable in its presence. However, for the variable with an absence of cross-sectional dependence, the first-generation unit roots tests LLC (Levin et al., 2002), ADF-Fisher (Maddala and Wu, 1999), and ADF-Choi (Choi, 2001) were performed. The CIPS test is performed under the null hypothesis of nonstationarity. This test reveals that the variables under analysis are integrated in level I(0) and integrated of order one I(1), ensuring the absence of variables integrated of order two I(2) (Tables A11.5 and A11.6), and the first-generation unit root tests corroborate the results obtained.

The time span under analysis consists of more than two economic cycles, which give rise to the doubt of the occurrence of events in the series. Considering this, the unit root test with structural breaks is useful to corroborate the traditional unit root test and to reveal the break points in the series. These break points could be included as dummy variables in the estimations. Therefore, the Zivot and Andrews (1992) unit root test with structural breaks was carried out. To perverse space, the test is not displayed but is available from the authors on request.

Lastly, collinearity and multicollinearity among variables are computed to perform a comprehensive analysis of the data features. Hence, coefficients of correlation and the variance inflation factor statistics were assessed, revealing that neither collinearity nor multicollinearity are a concern. Considering the features of the data, it is concluded that the use of the ARDL model is suitable.

3.2 Methodology

Considering the time span of more than two economic cycles in which diverse events could have occur, the ARDL model developed by Pesaran et al. (1999); Pesaran and Smith (1995) has diverse characteristics that can be considered as a benefit. According with the literature, the ARDL model produces consistent and efficient parameter estimates (Marques et al., 2019a; Papageorgiou et al., 2016). Through the ARDL model, two important aspects are observed (Marques et al., 2019a; Pesaran and Shin, 1999), namely the dynamics of the short-run of the dependent variables adjustment when variations or shocks in explanatory variables occurs and the long-run equilibrium. Beyond this, the ARDL model allows the analysis of the dynamics of variables, disaggregating the impacts in both the short- and long-run. It also allows the inclusion of dummy variables to control the events occurred in the series. Besides that, the ARDL model has the advantage of being able to deal with variables with different integration orders, such variables I(0), I(1), and borderline, which is a feature of the data under analysis. This feature consents variables with long memory patterns to be conducted properly. Therefore, the ARDL model was used.

The general ARDL model is provided in Eq. (11.5).

$$\Delta\varphi_t = \beta_i + \vartheta_{i1}\Delta\theta_t + \sum_{p=1}^{k}\beta_{1i}\varphi_{t-p} + \sum_{p=1}^{k}\beta_{2i}\theta_{t-p} + \mu_{i,t}, \qquad (11.5)$$

where φ_t denotes the vector of dependent variables, β_i represents the intercept, ϑ_{i1} denotes the semi-elasticities, θ_t denotes the vector of independent variables, β_{1i} consists of the error correction mechanism (ECM), β_{2i} denotes the elasticities, and $\mu_{i,t}$ denotes the error term.

Following the general ARDL model structure Eq. (11.5), the following two equation represent the models estimated with the ISEW and Efficiency Index as dependent variables. Therefore, Eq. (11.6) for ISEW and Eq. (11.7) for Efficiency Index.

$$DLISEW_{it} = \alpha_0 + \sum_{j=0}^{n}\eta_{i1}DLCO2\_PC_{it} + \sum_{j=0}^{n}\eta_{i2}DLFF\_P_{it} + \sum_{j=0}^{n}\eta_{i3}DLRES\_P_{it} + \sum_{j=0}^{n}\eta_{i4}DLEF_{it}$$

$$+ \sum_{j=0}^{n}\eta_{i5}DLKOFECDF_{it} + \sum_{j=0}^{n}\eta_{i6}DLKOFECDJ_{it} + \sum_{j=0}^{n}\eta_{i7}DLKOFSODF_{it} + \sum_{j=0}^{n}\eta_{i8}DLKOFSODJ_{it}$$

$$+ \sum_{j=0}^{n}\eta_{i9}DLKOFPODF_{it} + \sum_{j=0}^{n}\eta_{i10}DLKOFPODJ_{it} + \omega_{i1}LISEW_{it-1} + \omega_{i2}LCO2\_PC_{it-1} + \omega_{i3}LFF\_P_{it-1}$$

$$+ \omega_{i4}LRES\_P_{it-1} + \omega_{i5}LEF\_P_{it-1} + \omega_{i6}LKOFECDF_{it-1} + \omega_{i7}LKOFECDJ_{it-1} + \omega_{i8}LKOFSODF_{it-1}$$

$$+ \omega_{i9}LKOFSODJ_{it-1} + \omega_{i10}LKOFPODF_{it-1} + \omega_{i11}LKOFPODJ_{it-1} + \varepsilon_{it},$$

$$(11.6)$$

$$DLEF_{it} = \alpha_0 + \sum_{j=0}^{n}\eta_{i1}DLGDP\_PC_{it} + \sum_{j=0}^{n}\eta_{i2}DLCO2\_PC_{it} + \sum_{j=0}^{n}\eta_{i3}DLFF\_P_{it} + \sum_{j=0}^{n}\eta_{i4}DLRES\_P_{it}$$

$$+ \sum_{j=0}^{n}\eta_{i5}DLKOFECDF_{it} + \sum_{j=0}^{n}\eta_{i6}DLKOFECDJ_{it} + \sum_{j=0}^{n}\eta_{i7}DLKOFSODF_{it} + \sum_{j=0}^{n}\eta_{i8}DLKOFSODJ_{it}$$

$$+ \sum_{j=0}^{n}\eta_{i9}DLKOFPODF_{it} + \sum_{j=0}^{n}\eta_{i10}DLKOFPODJ_{it} + \omega_{i1}LEF_{it-1} + \omega_{i2}LGDP\_PC_{it-1} + \omega_{i3}LCO2\_PC_{it-1}$$

$$+ \omega_{i4}LFF\_P_{it-1} + \omega_{i5}LRES\_P_{it-1} + \omega_{i6}LKOFECDF_{it-1} + \omega_{i7}LKOFECDJ_{it-1} + \omega_{i8}LKOFSODF_{it-1}$$

$$+ \omega_{i9}LKOFSODJ_{it-1} + \omega_{i10}LKOFPODF_{it-1} + \omega_{i11}LKOFPODJ_{it-1} + \varepsilon_{it},$$

$$(11.7)$$

In these equations, prefix D indicates first differences, α_0 exemplifies the intercept, n is the lag order, η_i represents the estimated parameters in the short-run, ω_i denotes the estimated parameters in the long-run, and ε_{it} symbolizes the error term.

A set of panel data specification tests were carried out in order to provide a suitable estimation. Hausman tests assess random effects against fixed effects, aiming to detect the most appropriate estimator to deal with the specific characteristics of the panel data. This test is performed under the null hypothesis that random effects are appropriate. The results are displayed in Table 11.2.

According with the results, the Hausman test supports the appropriateness of the fixed effect estimator for all models. Considering that, the phenomena of heteroscedasticity, autocorrelation, and contemporaneous correlation were

TABLE 11.2 Hausman test and F-test.

	HGC—ISEW	HGC—EF
F-test	5.68***	4.98***
Hausman test	41.61***	37.95***
	LGC—ISEW	**LGC—EF**
F-test	1.88*	6.43***
Hausman test	16.06*	46.41***

***, **, and * denotes significance level at 1%, 5%, and 10%, respectively.

assessed through the Modified Wald test, Wooldridge test, and the Pesaran test (Pesaran, 2004), Frees test (Frees, 1995), and Friedman test (Friedman, 1937), respectively, among cross sections were evaluated to decide on the most robust estimator (Table 11.3).

The presence of heteroscedasticity, autocorrelation, and contemporaneous correlation is revealed for almost all the models. The presence of all the features mentioned is suggested for the models of both groups of countries with ISEW as the dependent variable, and for the LGC with the efficiency index as the dependent variable. Whereas for HGC with the efficiency index as the dependent variable, the presence of autocorrelation and contemporaneous correlation is suggested, and there is an absence of heteroscedasticity.

TABLE 11.3 Specification tests.

	HGC—ISEW	HGC—EF
Pesaran test	11.638***	5.725***
Frees test	1.518***	0.398***
Friedman test	74.996***	55.418***
Modified Wald test	35.28***	6.98
Wooldridge test	48.232***	81.317***
	LGC—ISEW	**LGC—EF**
Pesaran test	10.806***	1.061
Frees test	0.729***	0.130*
Friedman test	60.361***	28.458***
Modified Wald test	79.17***	691.37***
Wooldridge test	29.005***	5.289**

***, **, and * denotes significance level at 1%, 5%, and 10%, respectively.

Therefore, the model specification tests indicate that the Driscoll-Kraay estimator (Driscoll and Kraay, 1998) is suitable for handling these data features. The presence of cross-section dependence provokes inconsistent estimates. However, the Driscoll-Kraay estimator computes consistent and robust estimated standard errors in the presence of cross-section dependence. Furthermore, this estimator assumes that the error structure is heteroskedastic, autocorrelated, and correlated (Driscoll and Kraay, 1998; Sarkodie and Strezov, 2019). This study employs a panel data regression with the Driscoll-Kraay estimator.

4. Results

The estimations and results of the relationship between globalization and energy efficiency, and globalization and sustainable development, through the ARDL model and Driscoll-Kraay estimator, are presented in Table 11.4 for both groups of countries, HGCs and LGCs.

TABLE 11.4 Estimations of Autoregressive Distributed Lag model.

	High globalized countries		Low globalized countries	
	ISEW	EF	ISEW	EF
DLGDP_pc		−0.9543***		−0.7635***
DLCO$_2$_pc		0.4272***	0.7364**	0.3808***
DLFF_p		−0.2145***	−0.6969**	−0.2103**
DLRES_p	−0.1193**		−0.1878**	−0.0574**
DLEF	−0.7930**			
DLKOFECDF			-0.9248***	
DLKOFECDJ			0.4725***	
DLKOFSODF	1.7996***	−0.3183***		
DLKOFSODJ			1.3013***	
DLKOFPODF		−0.2642**	0.1749*	0.0716*
DLKOFPODJ				
LISEW (−1)	*−0.3171****		*−0.2419****	
LEF (−1)	−0.4604**	*−0.3673****		*−0.4363****
LGDP_pc (−1)		−0.2774***		−0.1590***
LCO$_2$_pc (−1)		0.1231**		0.1368***
LFF_p (−1)	0.5455***	−0.1994**		

Continued

TABLE 11.4 Estimations of Autoregressive Distributed Lag model.—cont'd

	High globalized countries		Low globalized countries	
	ISEW	EF	ISEW	EF
LRES_p (−1)	−0.0671***		−0.0226*	−0.0281**
LKOFECDF (−1)		−0.0405*	−0.2443***	
LKOFECDJ (−1)	−0.9486***	−0.2265**	0.3687*	0.1265**
LKOFSODF (−1)		−0.3000**		−0.1424***
LKOFSODJ (−1)				
LKOFPODF (−1)				
LKOFPODJ (−1)	3.5561***	0.3127**		−0.2852***
C	−5.4001*	6.6127***	5.7408***	4.9282***
Dum_2001		0.0171***	−0.2947***	
Dum_2002			−0.1392***	
Dum_2014		−0.0374***		
Elasticities				
LGDP_pc (−1)		−0.7554***		−0.3644***
LEF (−1)	−1.4518***			
LCO₂_pc (−1)		0.3351***		0.3136***
LFF_p (−1)	1.72***	−0.5428***		
LRES_p (−1)	−0.2116***		−0.0934*	−0.0644***
LKOFECDF (−1)		−0.1102*	−1.01***	
LKOFECDJ (−1)	−2.991***	−0.6167**	1.524*	0.2900**
LKOFSODF (−1)		−0.8170**		−0.3263***
LKOFSODJ (−1)				
LKOFPODF (−1)				
LKOFPODJ (−1)	11.2128***	0.8513**		−0.6536***

***, **, and * denotes significance level at 1%, 5%, and 10%, respectively; prefix D denotes Differences; prefix L denotes logarithm; Dum denotes dummy; C denotes constant; Values in bold and italic are the ECM.

Taking into account the results of the Zivot and Andrews (1992) unit root test with structural breaks, the break points obtained through the test and in both the preceding and following years were tested in the models. Therefore, the dummies included in the models are supported by the unit root test with structural breaks, and it demonstrated adherence to actual events. For instance, the dummy in 2001 may control the implementation of the Millennium Development Goals. Furthermore, simultaneously, in this year, the OECD countries created their own International Development Goals. The semi-elasticities and the elasticities were computed. With semi-elasticities an increase of one percentual point (pp) of *LKOFSODF*, ceteris paribus, follows a 1.8 pp increase of ISEW on HGC, and an increase of 1 pp of *LKOFSODJ*, ceteris paribus, follows a 1.3 increase of ISEW on LGC. In turn, an increase of 1 pp of *LGDP*, ceteris paribus, follows a 0.95 pp and 0.76 pp decrease of the efficiency index of HGC and LGC, respectively. Regarding the elasticities, an increase of 1% of *LKOFECDJ* causes a decrease of 2.99% and an increase of 1.52% on ISEW of HGC and LGC, respectively. Furthermore, an increase of 1% of *LKOFPODJ* follows a 0.85% increase and a decrease of 0.65% on the efficiency index of HGC and LGC, respectively.

Referring back to Table 11.4, keeping in mind that the results of the efficiency index should be interpreted as: an increase on the index represents a decreasing in energy efficiency, while a decrease in the index is characterized by an improvement in energy efficiency. Therefore, the negative effect of *LGDP* on *LEF* means that GDP improves efficiency, reducing the index value. The same effect is provoked by fossil fuels and renewable energy consumption, while CO_2 emissions increase the value of the index, worsening the efficiency. With regard to globalization, in the HGC, globalization mostly has a negative effect on the efficiency index, which represents an improvement of the efficiency. In the LGC, *LKOFPODF* and *LKOFECDJ* decrease efficiency, while *KOFSODF* and *KOFPODJ* increase it. Concerning the ISEW, renewable energy consumption reduces the ISEW in both HGC and LGC. Simultaneously, globalization mostly has a positive effect on the ISEW of both groups of countries. Overall, the results suggest diverse drivers on sustainable development. The ECM of all models is highly significant, and with a moderate speed of adjustment to long-run equilibrium.

5. Discussion

Motivated by the will to understand the interaction between globalization, energy efficiency, and sustainable development, this chapter analyses two samples of countries according to their level of globalization. Therefore, this study introduces the novelty of analyzing globalization through its three dimensions and two new measures on energy efficiency and sustainable development. This analysis is performed for a set of 20 countries—the top 20 energy consumers per capita countries, divided into HGCs and LGCs. To the best of our knowledge, this is the first work analyzing the interactions of each dimension and measure of globalization on both efficiency index and ISEW. Overall, the estimations reveal the advantage of analyzing globalization using the new measures, through obtaining different effects in nature or magnitude of the same dimension of globalization, such as stated by Martens et al. (2015).

5.1 Efficiency index

The results of the estimations with the efficiency index as the dependent variable, shown in Table 11.4, reveal that GDP and fossil fuel consumption increase energy efficiency, while emissions decrease it, such as noted by (Marques et al., 2019b), in both groups of countries, HGC and LGC. Economic growth has an improvement effect on energy efficiency in both groups of countries. Considering the United Nations Sustainable Development Goals, mainly Goal 12—Responsible Consumption and Production, and Goal 13—Climate Action, and the Millennium Development Goals focused on the promotion of environmental sustainability, they push the economies toward energy efficiency, and the economies still search for techniques to produce more output with less cost and less energy. Therefore, with a growing GDP, the investment in efficient technology and Research and Development (R&D) is higher and consequently improves energy efficiency. In turn, CO_2 emissions decrease energy efficiency as energy efficiency improves environmental quality and induces decarbonization.

Regarding globalization, it mostly improves energy efficiency by reducing the index value in the HGC. Only political globalization de jure decreases energy efficiency, while economic globalization both de jure and de facto, and social and political globalization de facto improve energy efficiency. In turn, in the LGC, social globalization de facto and political globalization de jure improve energy efficiency reducing the efficiency index, while political globalization de facto and economic globaliz de jure decrease energy efficiency increasing the efficiency index. With a negative effect on the efficiency index in both groups of countries, HGC and LGC, in the long-run, the improving influence of social globalization de facto on energy efficiency could be explained by both patent applications and high technology exports. High technology exports represent products with high R&D intensity, which means, that the countries invest in R&D of efficient technology. Furthermore, the patent applications also represent innovative ideas, that could be related to efficient and green technology. In the HGC social globalization, de facto improves efficiency in both the short- and long-run, while in the LGC it is only in the long-run, meaning that the LGC started to invest in products with high R&D intensity more recently than the HGC.

Concerning economic globalization, both measures de jure and de facto induce energy efficiency in the HGC in the long-run. On the one hand, economic globalization de facto improves energy efficiency through Foreign Direct Investment (FDI), while on the other hand, economic globalization de jure improves it through trade taxes that consists of income from taxes on international trade. Both could be invested in efficient technology. In contrast, in the LGC in the long-run economic globalization, de jure reduces energy efficiency. This effect could be explained by investment restrictions that could stipulate investment in efficient technology and be directed to cheaper technology. The measures of political globalization reveal opposite effects on the groups of countries under analysis. In the short-run, political globalization de facto affects the efficiency index, improving energy efficiency in the HGC while decreases it in the LGC. In the long-run political globalization, de jure affects the efficiency index, decreasing the energy efficiency in the HGC, and improving it in the LGC. With an improving effect on energy efficiency, the effect of political globalization de jure in the LGC could be explained by strong energy efficiency and sustainability goals established by international treaties for these countries. In contrast, in the HGC, political globalization de jure decreases energy efficiency, which could be due to bilateral investment treaties. HGCs have a high number of bilateral investment treaties.

5.2 Sustainable Economic Welfare (ISEW)

The ISEW was used as a proxy to measure sustainable development. The results of the estimations with the ISEW as the dependent variable, shown in Table 11.4, reveal diverse drivers of sustainable development. In both groups of countries, and in both the short- and long-run, renewable energy consumption decreases sustainable development. This effect could be revealing the high implementation costs associated with renewables, such as noted by Menegaki et al. (2017). Furthermore, the strategy of green hydrogen development, which may consist of a high burden for the economies, could aggravate this effect. This finding is rather alarming, and consequently, policymakers should rethink the strategies of renewables penetration. In turn, the efficiency index reveals a negative effect on the ISEW in both the short- and long-run in the HGC, which means that a decrease of the efficiency index (which represents an improvement of energy efficiency) increases the ISEW. This effect denotes that energy efficiency, producing the same output with less energy, saving economic input, and promoting environmental quality, is beneficial for sustainable development in the HGC. In turn, in the LGC, the efficiency index does not seem to be affecting the ISEW, which could be revealing that the LGC should invest more in energy efficiency and efficient technology.

On the subject of globalization, mostly, it is beneficial for sustainable development in both HGC and LGC. Economic globalization is the only dimension with a harmful effect on sustainable development. In the HGC, economic globalization de jure decreases sustainable development, while in the LGC, it is economic globalization de facto that decreases it. In the HGC, the negative effect of economic globalization de jure could be explained by trade regulation that includes import and export costs. In the LGC, the negative effect of economic globalization de facto could be explained by an increase in economic activity through FDI, and trade in goods, which consequently provokes an increase in environmental damage and energy depletion, which leads to a reduction in the ISEW. In contrast, social and political globalization are drivers of the ISEW, conducting to sustainable development in both groups of countries under analysis. Regarding social globalization, in the HGC, social globalization de facto improves the ISEW, as well as improving energy efficiency. Improving energy efficiency results in a reduction in environmental damage or defensive costs, consequently increasing sustainable development. In the LGC, social globalization de jure induces sustainable development, which could reflect the expenditure on education. Incidentally, political globalization de jure in the HGC and political globalization de facto in the LGC improve sustainable development. This could reflect the diffusion of governmental policies, such as education and health. Overall, globalization is a driver for sustainable growth.

6. Conclusion

This chapter was undertaken to evaluate three current and relevant research issues: globalization, energy efficiency, and sustainable development, for which a comprehensive debate of the literature is provided. To do that, this chapter provides new empirical insights into the analysis of energy efficiency and sustainable development. It does this by assessing the role of the phenomenon of globalization through each dimension, economic, social, and political, and by using the measures, de jure and de facto, for each. To the best of our knowledge, this approach is innovative. This analysis is performed for the time span from 1995 to 2017 for energy efficiency, and from 2000 to 2017 for sustainable development. The top 20 energy consumers per capita countries were chosen for the study. This analysis provides insights to policymakers to develop suitable and assertive measures and policies to achieve sustainable development. This research reveals that most of the globalization dimensions and measures improve energy efficiency in the HGC and sustainable development in both HGC and LGC.

Concerning energy efficiency, measured by efficiency index, GDP assumes a driving role in improving energy efficiency in both groups of countries. Considering this, both HGC and LGC might continue to invest and direct part of GDP toward efficiency measures and efficient technology. Additionally, considering energy efficiency as a tool to mitigate environmental degradation, a reduction in CO_2 emissions leads to an increase in energy efficiency. Regarding globalization, in order to improve energy efficiency, measures should be developed to manage political globalization de jure in the HGC, and economic globalization de jure in the LGC. The LGC might reconsider investment restrictions in order to combat the detrimental effect of economic globalization de jure on energy efficiency. In turn, in the HGC, to combat the detrimental effect of political globalization de jure, bilateral investment treaties should ensure environmental preservation and promote energy efficiency measures.

Concerning sustainable development, measured through the ISEW, renewable energy consumption reveals an unexpected effect, reducing sustainable development. With this in mind, the strategy of renewables penetration should be reconsidered in both HGC and LGC. In regard to the globalization phenomenon, the results suggest that most of the dimensions and measures of globalization are beneficial for sustainable development, that is, overall globalization is a driver of sustainable development. However, economic globalization de jure in HGC and economic globalization de facto in LGC undermine sustainable development. Therefore, both HGC and LGC should focus their measures and policies on economic globalization in order to promote sustainable development. In the case of HGC, it is suggested that trade regulations could be improperly designed, and it should be redesigned. While in the case of LGC, it is suggested that FDI and trade in goods and services induce economic activity, which consequently could be increasing defensive and environmental damage costs.

Appendix

TABLE A11.1 Descriptive statistics from 1995 to 2017—efficiency index

	High globalized countries				
	Mean	Std. dev.	Min.	Max.	Obs.
EF	0.8113	0.1405	0.5118	1.1545	230
GDP_pc	48079.06	17974.28	13462.99	91565.73	230
CO2_pc	9.17E-06	2.98E-06	4.47E-06	1.46E-05	230
FF_p	66.3131	20.5673	29.9946	98.8285	230
RES_p	19.536	19.8065	0.2068	70.0054	230
KOFECDF	73.9462	9.9082	43.7157	92.0529	230
KOFECDJ	86.6182	4.4914	68.196	94.7231	230
KOFSODF	82.9711	5.7338	66.4892	95.1544	230
KOFSODJ	84.3414	5.3839	68.5726	92.6337	230
KOFPODF	91.8538	3.7365	84.0649	98.0264	230
KOFPODJ	95.5783	4.747	66.8212	100	230

	Low globalized countries				
	Mean	Std. Dev.	Min.	Max.	Obs.
EF	0.8111	0.2514	0.2765	2.1529	230
GDP_pc	41243.33	23966.81	7205.101	111968.4	230
CO2_pc	0.0000132	5.40e-06	5.50e-06	0.0000275	230
FF_p	83.3317	13.3325	58.6107	99.9788	230
RES_p	10.4968	11.1732	0.02119	39.5422	230
KOFECDF	65.0015	17.4401	27.8283	92.3781	230
KOFECDJ	80.5243	8.9836	46.6112	97.3884	230
KOFSODF	78.8434	8.5545	54.5939	92.1187	230
KOFSODJ	83.679	5.575	64.9952	93.2102	230
KOFPODF	9.593	12.9355	43.0076	94.8829	230
KOFPODJ	81.8009	11.1257	39.6222	95.5342	230

Max. denotes Maximum; Min. denotes Minimum; Std. Dev. denotes Standard deviation; Obs denotes Observations; EF denotes efficiency index; pc denotes per capita; FF denotes fossil fuels; KOFECDF denotes KOF Globalization Index Economic Dimension de Facto; KOFECDJ denotes KOF Globalization Index Economic Dimension de Jure; KOFSODF denotes KOF Globalization Index Social Dimension de Facto; KOFSODJ denotes KOF Globalization Index Social Dimension de Jure; KOFPODF denotes KOF Globalization Index Political Dimension de Facto; KOFPODJ denotes KOF Globalization Index Political Dimension de Jure.

TABLE A11.2 Descriptive statistics from 2000 to 2017—ISEW

High globalized countries

	Mean	Std. Dev.	Min.	Max.	Obs.
ISEW	4.31E+11	5.01E+11	2.97E+10	1.87E+12	180
CO_2_pc	9.01E-06	2.99E-06	4.47E-06	1.46E-05	180
FF_p	65.652	20.4767	29.9946	98.18945	180
RES_p	20.0835	19.8137	0.3688	70.0054	180
EF	0.76862	0.1223	0.5118	1.0906	180
KOFECDF	76.7725	7.6865	60.991	92.0529	180
KOFECDJ	86.2908	3.9418	74.4174	94.7231	180
KOFSODF	84.4701	5.072	71.8423	95.1544	180
KOFSODJ	86.0051	3.8582	73.1227	92.6337	180
KOFPODF	92.0462	3.7351	84.3439	98.0264	180
KOFPODJ	96.3892	3.3583	81.5323	100	180

Low globalized countries

	Mean	Std. Dev.	Min.	Max.	Obs.
ISEW	1.19E+12	2.50E+12	2.98E+09	1.10E+13	180
CO_2_pc	1.32E-05	5.51E-06	5.50E-06	2.75E-05	180
FF_p	83.0118	13.3154	58.6107	99.9155	180
RES_p	10.7502	11.0202	0.0845	38.6843	180
EF	0.7587	0.2040	0.2764	1.4859	180
KOFECDF	66.9945	16.5009	29.8409	92.3781	180
KOFECDJ	81.7494	6.9768	62.5938	97.3884	180
KOFSODF	80.8765	7.5105	59.5395	92.1187	180
KOFSODJ	85.5427	3.7712	73.9270	93.2102	180
KOFPODF	80.7622	11.7220	43.2533	93.4251	180
KOFPODJ	83.9222	9.4142	59.9408	95.5342	180

Max. denotes Maximum; Min. denotes Minimum; Std. Dev. denotes Standard deviation; Obs denotes Observations; pc denotes per capita; FF denotes fossil fuels; EF denotes efficiency index; KOFECDF denotes KOF Globalization Index Economic Dimension *de Facto*; KOFECDJ denotes KOF Globalization Index Economic Dimension *de Jure*; KOFSODF denotes KOF Globalization Index Social Dimension *de Facto*; KOFSODJ denotes KOF Globalization Index Social Dimension *de Jure*; KOFPODF denotes KOF Globalization Index Political Dimension *de Facto*; KOFPODJ denotes KOF Globalization Index Political Dimension *de Jure*.

TABLE A11.3 Cross-section dependence test from 1995 to 2017—efficiency index

	High globalized countries			Low globalized countries		
	CD-test	Corr	Abs (corr)	CD-test	Corr	Abs (corr)
LEF	30.06***	0.934	0.934	28.91***	0.899	0.899
LGDP_pc	30.84***	0.959	0.959	30.36***	0.944	0.944
LCO_2_pc	27.36***	0.850	0.850	11.80***	0.367	0.569
LFF_p	21.61***	0.672	0.672	12.88***	0.400	0.691
LRES_p	23.20***	0.721	0.721	21.81***	0.678	0.678
LKOFECDF	31.06***	0.966	0.966	10.46***	0.325	0.490
LKOFECDJ	19.44***	0.604	0.758	12.81***	0.398	0.453
LKOFSODF	30.80***	0.957	0.957	30.15***	0.937	0.937
LKOFSODJ	27.96***	0.869	0.869	28.64***	0.890	0.890
LKOFPODF	9.63***	0.299	0.370	3.68***	0.114	0.456
LKOFPODJ	26.86***	0.835	0.835	21.71***	0.675	0.805

*** denotes significance level at 1%; prefix L denotes Logarithm.

TABLE A11.4 Cross-section dependence test (CD-test) from 2000 to 2017—ISEW

	High globalized countries			Low globalized countries		
	CD-test	Corr	Abs (corr)	CD-test	Corr	Abs (corr)
LISEW	27.45***	0.965	0.965	22.70***	0.797	0.797
LCO$_2$_pc	25.69***	0.903	0.903	12.38***	0.435	0.674
LFF_p	18.20***	0.64	0.64	13.91***	0.489	0.793
LRES_p	21.31***	0.749	0.749	22.18***	0.779	0.779
LEF	25.96***	0.912	0.912	24.81***	0.872	0.872
LKOFECDF	24.41***	0.858	0.858	1.34	0.047	0.546
LKOFECDJ	17.82***	0.626	0.729	5.55***	0.195	0.339
LKOFSODF	26.99***	0.948	0.948	25.54***	0.897	0.897
LKOFSODJ	16.30***	0.573	0.573	22.13***	0.778	0.778
LKOFPODF	8.34***	0.293	0.413	0.77	0.027	0.46
LKOFPODJ	21.61***	0.759	0.759	18.78***	0.66	0.764

*** and ** denotes significance level at 1% and 5%, respectively; prefix L denotes Logarithm.

TABLE A11.5 Second generation unit root test (CIPS) from 1995 to 2017—efficiency index

	High globalized countries		Low globalized countries	
	Without trend	With trend	Without trend	With trend
LEF	−0.924	−0.161	1.014	2.405
DLEF	−9.501***	−8.528***	−8.579***	−8.165***
LGDP_pc	−0.246	1.165	0.272	2.710
DLGDP_pc	−4.833***	−3.035***	−4.576***	−3.585***
LCO$_2$_pc	−2.658***	−3.436***	0.284	−0.307
DLCO$_2$_pc	−9.920***	−8.343***	−7.288***	−6.194***
LFF_p	−1.806**	−2.773***	−1.598*	0.096
DLFF_p	−11.185***	−9.875***	−8.150***	−7.810***
LRES_p	−0.834	−2.969***	−1.242	−0.854
DLRES_p	−12.946***	−11.723***	−10.287***	−9.913***
LKOFECDF	−0.363	−0.257	−1.332*	−0.119
DLKOFECDF	−9.533***	−8.357***	−5.394***	−4.550***
LKOFECDJ	−1.048	−1.523*	−1.105	0.800
DLKOFECDJ	−11.306***	−9.866***	−9.616***	−9.227***
LKOFSODF	−2.400***	−1.721**	−3.479***	−1.702**
DLKOFSODF	−9.770***	−9.351***	−9.455***	−8.052***
LKOFSODJ	−1.538*	−1.510*	−3.063***	−1.271
DLKOFSODJ	−8.728***	−7.106***	−9.786***	−8.845***
LKOFPODF	−3.209***	−0.479	−4.036***	−4.475***
DLKOFPODF	−9.811***	−8.100***	−9.734***	−8.299***
LKOFPODJ	−2.765***	−2.898***	−2.325**	−2.040**
DLKOFPODJ	−10.007***	−9.388***	−8.518***	−7.729***

*, **, and *** denotes significance level at 10%, 5%, and 1%, respectively; prefix D denotes Differences; prefix L denotes Logarithm.

TABLE A11.6 Second Generation Unit Root test (CIPS) from 2000 to 2017—ISEW

	High globalized countries		Low globalized countries	
	Without trend	With trend	Without trend	With trend
LISEW	−1.454*	−0.288	0.555	1.309
DLISEW	−6.346***	−4.812***	−3.715***	−1.378*
LCO$_2$_pc	−3.076***	−1.595*	−0.868	−0.255
DLCO$_2$_pc	−7.31***	−5.551***	−6.243***	−4.275***
LFF_p	−1.636*	−1.673**	−0.766	0.414
DLFF_p	−8.399***	−6.547***	−6.273***	−5.162***
LRES_p	−1.097	−1.462*	−1.964**	−0.281
DLRES_p	−9.760***	−8.941***	−7.342***	−6.354***
LEF	−0.971	−0.664	−0.245	1.577
DLEF	−7.268***	−6.074***	−6.222***	−6.229***
LKOFECDF	−0.049	0.726	−0.628	−1.296*
DLKOFECDF	−6.877***	−5.775***	−7.175***	−6.629***
LKOFECDJ	−2.849***	−2.282**	−0.999	0.499
DLKOFECDJ	−8.958***	−7.922***	−7.264***	−7.477***
LKOFSODF	−1.84**	−1.496*	−3.036***	−1.723**
DLKOFSODF	−7.306***	−5.922***	−7.971***	−6.019***
LKOFSODJ	−3.627***	−1.963**	−3.472***	−2.624***
DLKOFSODJ	−7.850***	−7.669***	−8.245***	−6.662***
LKOFPODF	0.142	2.644	−0.975	−0.813
DLKOFPODF	−4.745***	−3.979***	−7.206***	−7.240***
LKOFPODJ	−3.675***	−2.554***	−1.496*	−2.637***
DLKOFPODJ	−7.947***	−6.243***	−6.882***	−5.412***

*, **, and *** denotes significance level at 10%, 5%, and 1%, respectively; prefix D denotes Differences; prefix L denotes Logarithm.

Acknowledgments

The authors would like to gratefully acknowledge the generous financial support of the NECE-UBI—Research Unit in Business Science and Economics, Portugal, Project no. UIDB/04630/2020, and a PhD fellowship (2020.06026.BD), both sponsored by the FCT - Portuguese Foundation for the Development of Science and Technology, Ministry of Science, Technology and Higher Education, Portugal.

References

Ahmed, Z., Cary, M., Phong, H., 2021. Accounting asymmetries in the long-run nexus between globalization and environmental sustainability in the United States: an aggregated and disaggregated investigation. Environ. Impact Assess. Rev. 86, 106511.

Aluko, O., Opoku, E., Ibrahim, M., 2021. Investigating the environmental effect of globalization: insights from selected industrialized countries. J. Environ. Manage. 281, 111892.

Ang, B.W., 2004. Decomposition analysis for policymaking in energy: which is the preferred method? Energy Policy 32, 1131—1139.

Ang, B.W., Liu, F.L., 2001. A new energy decomposition method: perfect in decomposition and consistent in aggregation. Energy 26, 537−548.

Apinran, M.O., Taşpınar, N., Gökmenoğlu, K.K., 2018. Impact of foreign direct investment on human development index in Nigeria. Bus. Econ. Res. J. 9, 1−13.

Aşici, A.A., 2013. Economic growth and its impact on environment: a panel data analysis. Ecol. Indic. 24, 324−333.

Aust, V., Morais, A.I., Pinto, I., 2020. How does foreign direct investment contribute to Sustainable Development Goals? Evidence from African countries. J. Clean. Prod. 245.

Baek, J., Cho, Y., Koo, W.W., 2009. The environmental consequences of globalization: a country-specific time-series analysis. Ecol. Econ. 68, 2255−2264.

Balsalobre-Lorente, D., Gokmenoglu, K.K., Taspinar, N., Cantos-Cantos, J.M., 2019. An approach to the pollution haven and pollution halo hypotheses in MINT countries. Environ. Sci. Pollut. Res. 26, 23010−23026.

Beça, P., Santos, R., 2014. A comparison between GDP and ISEW in decoupling analysis. Ecol. Indic. 46, 167−176.

Beça, P., Santos, R., 2010. Measuring sustainable welfare: a new approach to the ISEW. Ecol. Econ. 69, 810−819.

Beumer, C., Figge, L., Elliott, J., 2018. The sustainability of globalisation: including the "social robustness criterion". J. Clean. Prod. 179, 704−715.

Bilgili, F., Ulucak, R., Koçak, E., İlkay, S.Ç., 2020. Does globalization matter for environmental sustainability? Empirical investigation for Turkey by Markov regime switching models. Environ. Sci. Pollut. Res. 27, 1087−1100.

Bleys, B., Whitby, A., 2015. Barriers and opportunities for alternative measures of economic welfare. Ecol. Econ. 117, 162−172.

Bolwig, S., Ponte, S., du Toit, A., Riisgaard, L., Halberg, N., 2010. Integrating poverty and environmental concerns into value-chain analysis: a conceptual framework. Dev. Policy Rev. 28, 173−194.

Choi, I., 2001. Unit root tests for panel data. J. Int. Money Finance 20, 249−272.

Choi, I., 2001. Unit root tests for panel data. J. Int. Money Finance 20, 249−272.

Cobb, C.W., Cobb, J.B., 1994. The Green National Product: A Proposed Index of Sustain- Able Economic Welfare.

Copeland, B.R., Taylor, M.S., 2013. Trade and the environment: theory and evidence. In: Palgrave Handbook of International Trade. Prince, New Jersey, pp. 423−496.

Coscieme, L., Mortensen, L.F., Anderson, S., Ward, J., Donohue, I., Sutton, P.C., 2020. Going beyond gross domestic product as an indicator to bring coherence to the sustainable development. Goals. J. Clean. Prod. 248, 119232.

Costanza, R., Hart, M., Talberth, J., Posner, S., 2009. Beyond GDP: The Need for New Measures of Progress. Pardee Pap.

Daly, H.E., Cobb, J.B., 1989. For the Common Good: Redirecting the Economy toward Community, the Environment, and a Sustainable Future. Beacon Press.

Dargahi, H., Khameneh, K.B., 2019. Energy intensity determinants in an energy-exporting developing economy: case of Iran. Energy 168, 1031−1044.

Dartey-Baah, K., 2014. Effective leadership and sustainable development in Africa: is there "really" a link? J. Glob. Responsib.

Deichmann, U., Reuter, A., Vollmer, S., Zhang, F., 2019. The relationship between energy intensity and economic growth: new evidence from a multi-country multi-sectorial dataset. World Dev. 124, 104664.

Doytch, N., 2020. The impact of foreign direct investment on the ecological footprints of nations. Environ. Sustainability Indic. 124658.

Dreher, A., 2006. Does globalization affect growth? Evidence from a new index of globalization. Appl. Econ. 38, 1091−1110.

Dreher, A., Gaston, N., Martens, P., 2008. Measuring Globalisation. Gauging its Consequences Springer, New York.

Driscoll, J.C., Kraay, A.C., 1998. Consistent covariance matrix estimation with spatially dependent panel data. Rev. Econ. Stat. 80, 549−560.

Eberhardt, M., 2011. Panel time-series modeling: new tools for analyzing xt data. In: 2011 UK Stata Users Group Meeting.

Figge, L., Martens, P., 2014. Globalisation continues: the Maastricht globalisation index revisited and updated. Globalizations 11, 875−893.

Fisher, I., 1921. The best form of index number. Q. Publ. Am. Stat. Assoc. 17, 533−537.

Frees, E.W., 1995. Assessing cross-sectional correlation in panel data. J. Econom. 69, 393−414.

Friedman, M., 1937. The use of ranks to avoid the assumption of normality implicit in the analysis of variance. J. Am. Stat. Assoc. 32, 675−701.

Gaspar, J., dos, S., Marques, A.C., Fuinhas, J.A., 2017. The traditional energy-growth nexus: a comparison between sustainable development and economic growth approaches. Ecol. Indic. 75, 286−296.

Gigliotti, M., Schmidt-Traub, G., Bastianoni, S., 2019. The sustainable development goals. In: Fath, B. (Ed.), Encyclopedia of Ecology, second ed. Elsevier, Oxford, pp. 426−431.

Gozgor, G., Mahalik, M.K., Demir, E., Padhan, H., 2020. The impact of economic globalization on renewable energy in the OECD countries. Energy Policy 139, 111365.

Guan, J., Kirikkaleli, D., Bibi, A., Zhang, W., 2020. Natural resources rents nexus with financial development in the presence of globalization: is the "resource curse" exist or myth? Resour. Policy 66, 101641.

Gygli, S., Haelg, F., Potrafke, N., Sturm, J.E., 2019. The KOF globalisation index − revisited. Rev. Int. Organ. 14, 543−574.

Hák, T., Janoušková, S., Moldan, B., 2016. Sustainable Development Goals: a need for relevant indicators. Ecol. Indic. 60, 565−573.

Hoechle, D., 2007. Robust standard errors for panel regressions with cross-sectional dependence. Stata J 7, 281−312.

Intriligator, M.D., 2004. Globalization of the world economy: potential benefits and costs and a net assessment. J. Policy Model. 26, 485−498.

Jimenez, R., Mercado, J., 2014. Energy intensity: a decomposition and counterfactual exercise for Latin American countries. Energy Econ. 42, 161−171.

Karimu, A., Brännlund, R., Lundgren, T., Söderholm, P., 2017. Energy intensity and convergence in Swedish industry: a combined econometric and decomposition analysis. Energy Econ. 62, 347−356.

Kuznets, S., 1934. National income, 1929-1932. Natl. Bur. Econ. Res. 1−12.

Kwatra, S., Kumar, A., Sharma, P., 2020. A critical review of studies related to construction and computation of Sustainable Development Indices. Ecol. Indic. 112, 106061.

Leal, P.A., Marques, A.C., Fuinhas, J.A., 2019. Decoupling economic growth from GHG emissions: decomposition analysis by sectoral factors for Australia. Econ. Anal. Policy 62, 12−26.

Leal, P., Marques, A., 2019a. Are de jure and de facto globalization undermining the environment? Evidence from high and low globalized EU countries. J. Environ. Manage. 250, 109460.

Leal, P., Marques, A., 2019b. Rediscovering the EKC hypothesis on the high and low globalized OECD countries. In: Shahbaz, M., Balsalobre, D. (Eds.), Green Energy and Technology. Springer International Publishing, pp. 85—114.

Leal, P.H., Marques, A.C., 2020. Rediscovering the EKC hypothesis for the 20 highest CO2 emitters among OECD countries by level of globalization. Int. Econ.

Levin, A., Lin, C.F., Chu, C.S.J., 2002. Unit root tests in panel data: asymptotic and finite-sample properties. J. Econom. 108, 1—24.

Li, G., Fang, C., 2014. Global mapping and estimation of ecosystem services values and gross domestic product: a spatially explicit integration of national "green GDP" accounting. Ecol. Indic. 46, 293—314.

Liu, F.L., Ang, B.W., 2003. Eight methods for decomposing the aggregate energy-intensity of industry. Appl. Energy 76, 15—23.

Liu, M., Ren, X., Cheng, C., Wang, Z., 2020. The role of globalization in CO_2 emissions: a semi-parametric panel data analysis for G7. Sci. Total Environ. 718, 137379.

Maddala, G.S., Wu, S., 1999. A comparative study of unit root tests with panel data and a new simple test. Oxford Bull. Econ. Stat. Spec. Issue 61, 631—652.

Marques, A.C., Fuinhas, J.A., Gaspar, J.D.S., 2016. On the nexus of energy use - economic development: a panel approach. Energy Procedia 106, 225—234.

Marques, A.C., Fuinhas, J.A., Pais, D.F., 2018. Economic growth, sustainable development and food consumption: evidence across different income groups of countries. J. Clean. Prod. 196, 245—258.

Marques, A.C., Fuinhas, J.A., Pereira, D.S., 2019a. The dynamics of the short and long-run effects of public policies supporting renewable energy: a comparative study of installed capacity and electricity generation. Econ. Anal. Policy 63, 188—206.

Marques, A.C., Fuinhas, J.A., Tomás, C., 2019b. Energy efficiency and sustainable growth in industrial sectors in European Union countries: a nonlinear ARDL approach. J. Clean. Prod. 239.

Martens, P., Caselli, M., De Lombaerde, P., Figge, L., Scholte, J.A., 2015. New directions in globalization indices. Globalizations 12, 217—228.

Menegaki, A.N., Marques, A.C., Fuinhas, J.A., 2017. Redefining the energy-growth nexus with an index for sustainable economic welfare in Europe. Energy 141, 1254—1268.

Menegaki, A.N., Tiwari, A.K., 2017. The index of sustainable economic welfare in the energy-growth nexus for American countries. Ecol. Indic. 72, 494—509.

Menegaki, A.N., Tsagarakis, K.P., 2015. More indebted than we know? Informing fiscal policy with an index of sustainable welfare for Greece. Ecol. Indic. 57, 159—163.

Menegaki, A.N., Tugcu, C.T., 2018. Two versions of the index of sustainable economic welfare (ISEW) in the energy-growth nexus for selected Asian countries. Sustainnable Prod. Consum. 14, 21—35.

Menegaki, A.N., Tugcu, C.T., 2016a. Rethinking the energy-growth nexus: proposing an index of sustainable economic welfare for Sub-Saharan Africa. Energy Res. Soc. Sci. 17, 147—159.

Menegaki, A.N., Tugcu, C.T., 2016b. The sensitivity of growth, conservation, feedback & neutrality hypotheses to sustainability accounting. Energy Sustainable Dev. 34, 77—87.

Metcalf, G.E., 2008. An empirical analysis of energy intensity and its determinants at the state level. Energy J. 29, 1—26.

Mishkin, F.S., 2009. Globalization and financial development. J. Dev. Econ. 89, 164—169.

O'Mahony, T., Escardó-Serra, P., Dufour, J., 2018. Revisiting ISEW valuation approaches: the case of Spain including the costs of energy depletion and of climate change. Ecol. Econ. 144, 292—303.

Opoku, A., 2019. Biodiversity and the built environment: implications for the sustainable development goals (SDGs). Resour. Conserv. Recycl. 141, 1−7.

Özbuğday, F.C., Erbas, B.C., 2015. How effective are energy efficiency and renewable energy in curbing CO_2 emissions in the long run? A heterogeneous panel data analysis. Energy 82, 734−745.

Pais, D.F., Afonso, T.L., Marques, A.C., Fuinhas, J.A., 2019. Are economic growth and sustainable development converging? Evidence from the comparable genuine progress indicator for organisation for economic co-operation and development countries. Int. J. Energy Econ. Policy 9, 202−213.

Papageorgiou, T., Michaelides, P.G., Tsionas, E.G., 2016. Business cycle determinants and fiscal policy: a Panel ARDL approach for EMU. J. Econ. Asymmetries 13, 57−68.

Pesaran, M.H., 2007. A simple panel unit root test in the presence of cross-section dependence. J. Appl. Econom. 22, 265−312.

Pesaran, M.H., 2004. General Diagnostic Tests for Cross Section Dependence in Panels General Diagnostic Tests for Cross Section Dependence in Panels.

Pesaran, M.H., Pesaran, M.H., Shin, Y., Smith, R.P., 1999. Pooled mean group estimation of dynamic heterogeneous panels. J. Am. Stat. Assoc. 94, 621−634.

Pesaran, M.H., Shin, Y., 1999. Econometrics and Economic Theory in the 20th Century, Econometric Society Monographs. Cambridge University Press.

Pesaran, M.H., Smith, R., 1995. Estimating long-run relationships from dynamic heterogeneous panels. J. Econom.

Raab, M., Ruland, M., Schönberger, B., Blossfeld, H.P., Hofäcker, D., Buchholz, S., Schmelzer, P., 2008. GlobalIndex: a sociological approach to globalization measurement. Int. Sociol. 23.

Rahman, M.M., 2020. Environmental degradation: the role of electricity consumption, economic growth and globalisation. J. Environ. Manage. 253, 109742.

Rahman, M.M., 2017. Do population density, economic growth, energy use and exports adversely affect environmental quality in Asian populous countries? Renew. Sustain. Energy Rev. 77, 506−514.

Sánchez, M., Ochoa, M.W.S., Toledo, E., Ordóñez, J., 2020. The relevance of index of sustainable economic wellbeing. Case study of Ecuador. Environ. Sustainability Indic. 6, 100037.

Sarkodie, S.A., Strezov, V., 2019. Effect of foreign direct investments, economic development and energy consumption on greenhouse gas emissions in developing countries. Sci. Total Environ. 646, 862−871.

Sethi, P., Chakrabarti, D., Bhattacharjee, S., 2020. Globalization, financial development and economic growth: perils on the environmental sustainability of an emerging economy. J. Policy Model. 1−16.

Shahbaz, M., Mallick, H., Mahalik, M.K., Sadorsky, P., 2016. The role of globalization on the recent evolution of energy demand in India: implications for sustainable development. Energy Econ. 55, 52−68.

Shahbaz, M., Shahzad, S.J.H., Mahalik, M.K., 2018. Is globalization detrimental to CO2 emissions in Japan? New threshold analysis. Environ. Model. Assess. 23, 557−568.

Sharma, B., Gani, A., 2004. The effects of foreign direct investment on human development. Glob. Econ. J. 4, 1850025.

Stockhammer, E., Hochreiter, H., Obermayr, B., Steiner, K., 1997. The index of sustainable economic welfare (ISEW) as an alternative to GDP in measuring economic welfare. The results of the Austrian (revised) ISEW calculation 1955-1992. Ecol. Econ. 21, 19−34.

Suki, N.M., Sharif, A., Afshan, S., Suki, N.M., 2020. Revisiting the Environmental Kuznets Curve in Malaysia: the role of globalization in sustainable environment. J. Clean. Prod. 264, 121669.

Sun, C., Chen, L., Zhang, F., 2020. Exploring the trading embodied CO_2 effect and low-carbon globalization from the international division perspective. Environ. Impact Assess. Rev. 83, 106414.

Ullah, A., Neelum, Z., Jabeen, S., 2019. Factors behind electricity intensity and efficiency: an econometric analysis for Pakistan. Energy Strateg. Rev. 26, 100371.

Ulucak, R., Koçak, E., Erdoğan, S., Kassouri, Y., 2020. Investigating the non-linear effects of globalization on material consumption in the EU countries: evidence from PSTR estimation. Resour. Policy 67.

Vujakovic, P., 2010. How to measure globalization? A new globalization index (NGI). Atl. Econ. J. 38, 237.

Wang, L., Vo, X.V., Shahbaz, M., Ak, A., 2020. Globalization and carbon emissions: is there any role of agriculture value-added, financial development, and natural resource rent in the aftermath of COP21? J. Environ. Manage. 268, 110712.

Woo, B., Jun, H.J., 2020. Globalization and slums: how do economic, political, and social globalization affect slum prevalence? Habitat Int. 98, 102152.

Yameogo, C.E.W., Omojolaibi, J.A., Dauda, R.O.S., 2021. Economic globalisation , institutions and environmental quality in Sub-Saharan Africa. Res. Glob. 3, 100035.

You, W., Lv, Z., 2018. Spillover effects of economic globalization on CO_2 emissions: a spatial panel approach ☆. Energy Econ. 73, 248–257.

Zhang, F., 2013. The energy transition of the transition economies: an empirical analysis. Energy Econ. 40, 679–686.

Zivot, E., Andrews, D.W.K., 1992. Further evidence on the great crash, the oil price shock, and the unit root hypothesis. J. Bus. Econ. Stat. 10, 251–270.

Chapter 12

Renewable energy consumption, human capital index, and economic complexity in 16 Latin American countries: evidence using threshold regressions

Rafael Alvarado[1], Cristian Ortiz[2], Pablo Ponce[1], Elisa Toledo[3]

[1]*Carrera de Economía and Centro de Investigaciones Sociales y Económicas, Universidad Nacional de Loja, Loja, Ecuador;* [2]*Esai Business School, Universidad Espíritu Santo, Samborondon, Ecuador;* [3]*Departamento de Economía, Universidad Técnica Particular de Loja, Loja, Ecuador*

1. Introduction

Latin America still depends heavily on the natural resources rents and experiences an accelerated urbanization process (Alvarado et al., 2020). Recently, several countries in the region have embarked on a transition from a fossil fuel energy matrix to an energy matrix based on renewable energy. Statistics about the region's energy consumption suggest that the share of renewable energy in total consumption has increased in recent years (International Energy Organization, 2020). The reversal of the trend in polluting energy consumption constitutes a successful result that can be reinforced over time if political, social, and business commitment to the environment increases. This process results from a greater environmental awareness of companies and consumers, of the region's natural resources to generate renewable energy, and of political decisions aimed at achieving greater environmental sustainability. In parallel, Latin America is a region that makes serious efforts to improve industrial and labor specialization. Those responsible for designing the policy propose productive diversification as a necessary mechanism for sustainable growth. Policies that promote industrial specialization and the workforce's qualification contribute to the search for a more sustainable development model.

Energy-Growth Nexus in an Era of Globalization. https://doi.org/10.1016/B978-0-12-824440-1.00001-1

287

There is a broad consensus in the academic and political debate that dependence on traditional resources and energies limits the achievement of economic development with inclusion and environmental sustainability (Alvarado and Iglesias, 2017; Hosseini, 2020; Ezbakhe and Perez-Foguet, 2020; Clausen and Rudolph, 2020).

The benefits of increased renewable energy consumption have divers dimensions. Several empirical investigations show that increases in demand for this type of energy lead to a decrease in greenhouse gas emissions (Balsalobre-Lorente et al., 2018; Belaïd and Zrelli, 2019; Alvarado et al., 2019; Deng et al., 2020). Strategies to promote renewable energy consumption require the integration of a broad set of legal mechanisms, environmental awareness, and a firm business commitment. In this process, the role of human capital is relevant to promote society's commitment to the environment and accelerate the search for a new development model. Environmental sustainability, based on the consumption of energy from renewable sources, has other social and economic benefits for society. For example, increasing renewable energy facilitates the access to rural households' energy, increases productive efficiency, and generates new sources of employment (Ram et al., 2020). Likewise, the high costs that environmental pollution causes in health and agriculture drive the change in the energy matrix with less pollution (Ocal and Aslan, 2013; Astariz and Iglesias, 2015). In this context, the choice in consumption of the type of energy would be determined by minimizing costs that the consumer must pay. However, in Latin America, renewable energy consumption has a positive effect on economic growth (Alvarado et al., 2019). The natural resources of the region allow hydroelectric, wind, and solar energy to be produced at low production costs. Technological advancement and knowledge have improved significantly in recent years. This fact enables the costs of production, distribution, and commercialization of renewable energy to decrease as research and development create new technologies applied to energy.

In this context, the objective of this research is to examine the effect of the human capital index and the economic complexity index on the consumption of renewable energy. Our first hypothesis states that as knowledge increases, environmental awareness will increase, and understanding the importance of environmental sustainability will lead to increased renewable energy consumption. The second hypothesis states that the diversification of export capacity, measured by the index of economic complexity, will lead to an increase in energy consumption from nonpolluting sources. To capture the effect of other variables on renewable energy consumption, we added a set of control variables: globalization, public spending, trade, economic growth, urbanization rate, and the shadow economy. The structural characteristics of Latin America determine the inclusion of these covariates. Various investigations have shown that the intensity of some relationships between variables depends on the level of development and the productive structure. Specifically, the strength of the relationship between energy and economic variables is heterogeneous across countries. To verify the two research hypotheses, we estimated a fixed-effects panel data threshold model formalized by Hansen (1999, 2000). The application of threshold regressions has some advantages over other models. First, the effect of the human capital index and the economic complexity index on renewable energy consumption differ depending on the real output and the globalization index. Even though Latin American countries depend heavily on the exploitation of natural resources, there are differences in the orientation toward the international market. Consequently, threshold regressions capture the impact of independent variables on the dependent above and below the thresholds. The econometric estimates results show that the choice of the methodological strategy is adequate because we found that the impact between the variables differs according to the thresholds used. In the empirical literature, some research has used the instrumental framework of panel threshold regressions to estimate the factors that influence renewable energy consumption (Yang et al., 2019; Wang and Wang, 2020; Shahbaz et al., 2020; Uzar, 2020).

Our research contributes to the previous literature on the determinants of renewable energy consumption in two directions. First, we estimate the effect of the human capital index and the economic complexity index on renewable energy consumption through a nonlinear model, which has been little explored in the recent empirical literature. Second, to our knowledge, empirical research evaluating the link between the two independent variables and renewable energy is still limited. We found that the human capital index and the economic complexity index have a heterogeneous impact on renewable energy consumption below and above the thresholds. Our estimates support a set of environmental policy implications for Latin American countries, where there are large amounts of water resources that facilitate the construction of power plants, and geography offers opportunities for wind power generation. Governments, businesses, and academics face a unique opportunity to promote environmental friendly energy consumption. Environmental awareness has increased significantly, and new technologies can offer new possibilities to reduce environmental pollution (Sinha et al., 2020). Taking advantage of the new markets generated by the latest sustainable energy technologies can contribute to sustainable economic growth, create employment, achieve energy security, and promote environmental sustainability in the region (Viviescas et al., 2019).

The chapter has five additional sections. The second section contains a brief review of the previous theoretical and empirical literature. The third section describes the statistical sources. The fourth presents the econometric strategy. The fifth section presents the results, and we conducted a discussion of the findings with the results of the previous literature. Finally, the last section shows the conclusions and the implications of environmental policy.

2. Literature review

The Kyoto Protocol signed in 1997 by more than 100 countries, opened the way to a global commitment to promote renewable energy production and consumption of energy. The Paris Agreement extended the commitment of several countries to reduce greenhouse gas emissions and achieve sustainable development. Sustainable Development Goal 7 (SDG7) promotes accessible and nonpolluting energy, which has directed various energy policies from sustainable sources. The development of this type of energy is the key to face climate change and achieve a low carbon economy (Chen, 2018). The empirical literature has focused its interest on analyzing the link between variables that measure economic performance, through economic growth, and renewable energy consumption (Apergis and Payne, 2010b; Alper and Oguz, 2016; Tuna and Tuna, 2019; Chen et al., 2020; Rahman and Velayutham, 2020). The existing literature has allowed us to identify four hypotheses based on the causal relationship between renewable energy consumption and economic growth. First, the feedback hypothesis implies a bidirectional causal relationship between renewable energy consumption and economic growth (Alvarado et al., 2019; Apergis and Payne, 2010a,b; Menegaski, 2011). In this case, conservative energy policies can harm national income. Second, the growth hypothesis proposes a one-way causality that goes from energy consumption to economic growth (Tiwari, 2011; Fang, 2011; Chien and Hu, 2007). This fact means that energy plays a fundamental role in the production process as a complement to the traditional factors of capital and labor, and affects economic growth. Third, the conservation hypothesis suggests a one-way cause of economic growth to renewable energy consumption (Sadorsky, 2009). Fourth, the neutrality hypothesis indicates that there is no causal link between renewable energy consumption and output. However, the real output is not the only determinant of renewable energy consumption. In recent years, environmental awareness has increased, particularly in new businesses and the young population. The need to adopt policies aimed at promoting clean energy use has greater support from society. The available knowledge and technologies allow energy to be produced at a cost more accessible to households.

In recent years, the literature has focused on searching for new variables that allow analyzing other aspects that influence renewable energy consumption (Sener et al., 2018; Pursiheimo et al., 2019; Przychodzen and Przychodzen, 2020). In the search for new determinants, this research proposes the human capital index and the economic complexity index. One of the factors influencing environmental awareness to promote responsible consumption of sustainable energy is knowledge (Yao et al., 2019; Ponce et al., 2019; Sarkodie et al., 2020; Akintande et al., 2020). Furthermore, the accumulation of human capital in various areas facilitates the production, distribution, and consumption of energy. In parallel, several policies aim to promote productive diversification, particularly in the export capacity of the countries.

The economic complexity index could affect energy consumption for various reasons. One of the arguments is that diversification and complexity are a clear indication that the goods produced and exported by a country include a wide range of options. In this sense, productive diversification would lead to reducing dependence on fossil fuels that are highly polluting. Therefore, the relationship between the economic complexity index and the consumption of renewable energy should be positive. In the empirical literature, the possible impact of economic complexity on renewable energy consumption has been systematically omitted. An exception is a recent research by Mealy and Teytelboym (2020). They link the productive capacity of countries with the green economy and demonstrate that complexity can affect green industrial policy. The societies cannot achieve sustainable development without productive specialization and by improving the endowments of the workforce. The specialization of the workforce and the diversification produced accelerates the production, distribution, and consumption of renewable energy in a highly globalized economy. In the process of economic, social, and environmental development, energy efficiency policies are vital to the transition toward a green economy (Ringel et al., 2016; Matraeva et al., 2019). In this sense, incorporating the environmental dimension as a pillar of development requires more considerable momentum from the academy to achieve the sustainable development proposed in the Sustainable Development Goals.

Additionally, the accelerated process of globalization influences countries' performance in economic, political, and social aspects (Gozgor et al., 2020). These changes allow economies to decrease the demand for energy from fossil fuels and incorporate clean technologies in consumption activities (Baek et al., 2009). Globalization improves access to technology, increases the productivity of energy sources, and lowers the cost of producing and marketing it (Gozgor et al., 2020). Furthermore, the literature on globalization often shows that this phenomenon has significant effects on increased trade and promotes financial and capital flows between countries (Koengkan et al., 2019). For example, Lean and Smyth (2010), Ozturk and Acaravci (2013), and Shahbaz et al. (2013) indicate that trade openness reduces fossil energy consumption by facilitating the use of imported technology and stimulating environmental quality within countries. In contrast, Aissa et al. (2014) and Narayan and Smyth (2009) do not find a significant relationship between exports as an indicator of economic globalization and polluting energy consumption. However, there are currently more accurate measures of globalization, such as the KOF Swiss Economic Institute Globalization Index, which captures the economic, social, and political dimensions of globalization. Shahbaz et al. (2016) examine the effects of the KOF globalization index on energy consumption in India and find that the level of globalization reduces energy consumption through energy efficiency channels. These results are consistent with the findings for a sample of 25 developed countries, where the authors demonstrate that globalization promotes the clean energy use (Shahbaz et al., 2018a). Similar results are reported in Ireland and the Netherlands during the period 1970—2015 (Shahbaz et al., 2018b).

Consumer patterns around the world lead to a link between household consumption and energy consumption. Consumption-oriented behavior is a challenge for environmental sustainability, particularly in developed countries. Empirical research has shown that a source of energy consumption is household consumption, which generates a sustained increase in polluting emissions (Zhang et al., 2017; Ding et al., 2017). Another source of energy consumption is public spending, which can be used as an instrument of environmental policy directing spending on energy consumption from renewable sources. Lin and Zhu (2019) find that fiscal spending on research and development and spending on education contribute to achieving green economic growth. The sustained increase in household consumption is associated with urbanization processes, where households consume more energy when the population's quality of life improves (Alvarado et al., 2020; Wang et al., 2020). Furthermore, cities' consumption patterns are strengthened if the countries' industrial capacity allows them to generate goods and services at low cost to consumers. The manufacturing industry consumes a significant part of the total energy in countries with high industrialization (Reddy and Ray, 2010; Sheinbaum-Pardo et al., 2012; Zhao et al., 2014). Manufacturing consolidation increases energy consumption, whether renewable or nonrenewable. Hence, public policy and consumer choice should be guided by environmentally friendly energy.

Another possible determinant of renewable energy consumption is the underground economy. This fact occurs because, in developing countries, there are a significant amount of informal and hidden economic activities that consume energy from solar panels and other sources. The size of the shadow economy generates a particular interest among researchers and policymakers (González-Fernández and González-Velasco, 2014; Orviská et al., 2006), mainly because it escapes environmental controls because they are economic activities not registered by the government. Despite this importance, the relationship between the underground economy and energy consumption, the link between the two variables has not been extensively developed by the empirical literature (Basbay et al., 2016). It is essential to consider that the informal economy represents a significant part of the official product, particularly in developing countries (Hassan and Schneider, 2016). Studying the causal link between energy consumption and economic growth without considering unregistered income can lead to biased results (Benkraiem et al., 2019). The underground economy's size affects the effectiveness of economic policies aimed at reducing environmental pollution and promoting the sustainability of development.

3. Statistical sources

This research uses variables from different databases published by international institutions. The databases used collect aggregated information from 16 Latin American countries during 1990–2017. We used a balanced data panel, and the availability of data limited the choice of the sample of countries and the period analyzed. We consider renewable energy as a percentage of total energy consumption because to achieve environmental sustainability is necessary that the relative weight of nonpolluting energy increases over time. The two independent variables are the human capital index and the economic complexity index. The first variable evaluates the knowledge on average that the workforce reaches. The hypothesis proposed is that as the human capital index increases, renewable energy consumption also increases. The increase in environmental awareness offered by the accumulation of human capital support this hypothesis. The second variable captures the diversity and ubiquity of exports as a measure of diversification of the economy. The argument supporting the possible positive relationship between renewable energy consumption and the index of economic complexity is that the diversification of export capacity indicates the diversification of the entire economy. Exports reflect the productive industrial of the economy. The diversification facilitates the entry of new companies to the market and stimulates the activities with the highest performance. A higher index of economic complexity implies that the economic sectors have more productive chains, and the interaction between industries and sectors is high. Therefore, diversification encourages companies to take advantage of new technologies to reduce costs.

The covariates included in the research aim to capture the effect of other additional factors analyzed in the previous empirical study (Meleddu and Pulina, 2018; Martinho, 2018; Nicolli and Vona, 2019; Benkraiem et al., 2019; Murshed, 2020). For example, several Latin American countries have implemented multiple policies aimed to increasing the benefits of globalization. The integration, capital flow, and people agreements between the countries of the region and the external agreements with other countries reflect this trend. The impact of trade flows on nonrenewable energy consumption should be similar to the effect caused by globalization. However, the effect of trade is more limited to imports and exports. Furthermore, authors such as Gozgor et al. (2020) have noted that globalization has a positive impact on renewable energy consumption. In practice, the government can influence energy producers and distributors. The state is one of the largest consumers of energy in any economy. Including public spending as a percentage of gross domestic product (GDP) in the model captures the influence of government final consumption spending and the relative weight of the state's size on renewable energy consumption. Furthermore, it is well known that Latin America is highly dependent on natural resources and that the demand for these raw materials has increased in recent years (Alvarado and Iglesias, 2017; Smart, 2020; García et al., 2020). In parallel, several recent empirical investigations use data from official statistical institutions and confirm that the underground economy has a high share of total product, both in the informal sector and in clandestine activities (Salinas et al., 2018; Salinas et al., 2019; Medina and Schneider, 2019). Table 12.1 extends the description of the variables and defines the symbology, unit of measurement, and data source.

TABLE 12.1 Description of variables and data sources.

Variable	Symbol	Definition	Measure	Data source
Dependent variable				
Renewable energy consumption	RE_{it}	Renewable energy consumption is the share of renewable energy in total final energy consumption.	% of total final energy consumption	World Bank.
Independent variable				
Human capital index	HCI_{it}	It is a measure based on years of schooling and returns to education.	Index	Pen World Table 9.1
Economic complexity index	ECI_{it}	ECI is a measure of the intensity of knowledge of an economy, considering the intensity of knowledge of its exports.	Index	Observatory of economic complexity.
Covariants				
Globalization index	GI_{it}	The index measures the economic, social and political dimensions of globalization.	Index.	KOF Swiss Economic Institute.

Continued

TABLE 12.1 Description of variables and data sources.—cont'd

Variable	Symbol	Definition	Measure	Data source
Public spending	PS_{it}	Includes all government current expenditures for purchases of goods and services. It also includes most expenditures on national defense and security.	% of GDP.	World Bank.
Trade	TR_{it}	Trade is the sum of exports and imports of goods and services measured as a share of gross domestic product.	% of GDP.	World Bank.
Economic growth	EG_{it}	The set of goods and services that the economy produces.	Logarithm.	World Bank.
Urban population	UR_{it}	This variable is the percentage of the population concerning the total that lives in the urban area.	% total.	World Bank.
Shadow economy	SE_{it}	It includes all economic activities that are hidden from official authorities for monetary, regulatory and institutional reasons.	% of GDP.	International Monetary Fund.

Table 12.2 presents the descriptive statistics for each variable and the partial correlation matrix between pairs of variables. A relevant fact from Table 12.2 is that the standard deviation can be used to analyze the variability between countries (between) and within them (within). We found that there is more considerable variability between countries than within them, except for the variable economic globalization index, where there is more significant heterogeneity within countries. This result indicates that globalization has varied significantly during the period analyzed. In the correlation matrix, the asterisk indicates the statistical significance at 5%. The correlation coefficient

TABLE 12.2 Descriptive statistics and correlation matrix of variables.

	RE	HCI	ECI	GLI	SH	PS	UR	TR	GDP
Mean	34.40	2.35	-0.22	58.16	36.51	13.20	2.10	55.15	25.07
Std. Dev. (Overall)	18.51	0.35	0.51	9.09	12.49	3.61	0.87	20.40	1.55
Std. Dev. (Between)	18.22	0.30	0.49	5.27	11.39	3.32	0.69	17.60	1.57
Std. Dev. (Within)	5.55	0.19	0.18	7.52	5.84	1.62	0.55	11.18	0.29
Min	7.60	1.50	-1.34	36.79	7.45	5.68	-1.51	13.75	22.26
Max	69.15	3.10	1.36	78.40	71.34	30.16	4.45	133.07	28.51
N	448	448	448	448	448	448	448	448	448
N	16	16	16	16	16	16	16	16	16
T	28	28	28	28	28	28	28	28	28
RE	1								
	—								
HCI	-0.56*	1							
	(0.001)	—							
ECI	-0.14	-0.02	1						
	(0.040)	(0.100)	—						

GLI	−0.29* (0.002)	0.68* (0.00)	0.15 (0.123)	1 —					
SH	0.32* (0.001)	−0.47* (0.001)	−0.26* (0.009)	−0.45* (0.004)	1 —				
PS	0.001 (0.030)	0.11 (0.321)	0.23* (0.005)	0.04 (0.046)	−0.07 (0.036)	1 —			
UR	0.23* (0.000)	−0.51* (0.001)	−0.16 (0.106)	−0.58* (0.001)	0.25* (0.008)	−0.035 (0.077)	1 —		
TR	0.26* (0.000)	−0.08 (0.34)	−0.37* (0.002)	0.10 (0.022)	−0.09 (0.044)	−0.21* (0.002)	0.27* (0.000)	1 —	
GDP	−0.49* (0.000)	0.36* (0.003)	0.59* (0.002)	0.23* (0.002)	−0.40* (0.004)	0.12 (0.102)	−0.37* (0.001)	−0.56* (0.001)	1 —

*indica la significancia estadística al 5%.

TABLE 12.3 Multicolinearity statistics.

Variable	VIF	SQRT VIF	Tolerance	Squared
HCI	1.85	1.36	0.5403	0.4597
ECI	3.47	1.86	0.2879	0.7121
GLI	2.12	1.46	0.4708	0.5292
SH	2.88	1.70	0.3473	0.6527
PS	1.75	1.32	0.5716	0.4284
URB	1.16	1.08	0.8593	0.1407
TR	1.92	1.39	0.5200	0.048
LnGDP	2.15	1.47	0.4650	0.5350

Note: the null and alternative hypotheses are H_0: $\alpha_1 = \alpha_2$ y H_1: $\alpha_1 \neq \alpha_2$.

shows a basic look at the direction of the relationship between the variables. However, in this research, we examine the link between variables using a nonlinear mechanism.

The independent and covariate variables do not show multicollinearity, as shown by the variance inflation factor (VIF) test. The absence of collinearity between the explanatory variables is a necessary condition for obtaining unbiased and consistent estimators. The VIF value is the inverse of the partial correlation coefficient between pairs of the return variables. Therefore, if the partial correlation coefficient tends to one, the VIF tends to infinity. In the empirical literature, when the VIF values exceed 5, it is considered that there are collinearity problems (Sinha et al., 2020). In practice, the estimators' bias occurs mainly because, in the presence of partial correlations close to one, the variance and covariance are significant, causing the estimators to lose all precision and consistency. Table 12.3 shows that there is no problem with a high partial correlation in the variables of the model. Besides, we report the statistics associated with the VIF.

4. Econometric strategy

In this investigation, we use threshold regressions to examine the effect of the human capital index and the economic complexity index on renewable energy consumption. The choice of this methodology has two foundations. First, the threshold model proposed by Chan (1993) and formalized by Hansen (1999) evaluates the nonlinear effect of the return variables on the dependent variable.

If we introduce a simple function related to the independent variables to the consumption of renewable energy, the initial model is as follows:

$$RE_{it} = f(HCI_{it}, ECI_{it}, Z_{it}) \qquad (12.1)$$

In addition to the variables defined in Table 12.1, Z_{it} is a covariant matrix that includes the globalization index, the shadow economy, public spending, the urbanization rate, trade, and output. The second argument that justifies the choice of the econometric strategy is that the previous empirical evidence suggests that the effect of the determinants of renewable energy consumption differs according to certain thresholds (Yang et al., 2019; Chen et al., 2020; Wang and Wang, 2020). Consequently, the estimators obtained by nonlinear methods are more consistent than the parameters obtained by linear methods. In this sense, we propose a threshold regression model with panel data. Explicitly, we set two thresholds in the econometric estimates: the real output and the globalization index. There are at least two criteria for choosing the two thresholds. First, the size of the economy plays a vital role in export capacity or product diversification. The size of the market facilitates the creation and consolidation of new companies. Furthermore, following the environmental Kuznets curve, in the long term, increases in the output would lead to a decrease in environmental pollution (Grossman and Krueger, 1995). Second, the benefits of globalization (access to technology at a lower cost, financial, capital, and goods flows) in the consumption of renewable energy should impact if the opening is high and the countries have fully integrated into the global economic dynamics. Eq. (12.2) formulates the threshold regression model with panel data as follows:

$$RE_{it} = \psi_{0i} + \psi_1 HCI_{it}I(q_{it} \leq \gamma) + \psi_2 ECI_{it}I(\gamma < q_{it}) + \sum \psi_m Z_{it} + \varepsilon_{it} \qquad (12.2)$$

In Eq. (12.2), i represents the country ($i = 1, 2, ..., 16$) and t denotes the time ($t = 1, 2, ..., 28$). The variable q_{it} represents the threshold variable, while γ is the specific threshold. Furthermore, ε_{it} is the stochastic error term that captures the set of unobservable characteristics invariant in time, characteristic of panel data. The term ε_{it} is independently and identically distributed with zero mean and the variance is determined by $\sigma^2(\varepsilon_{it} \sim i.i.d(0, \sigma^2))$. The parameters ψ_i are the estimators to be obtained employing the nonlinear regressions and measure the impact of the regressors on renewable energy consumption. Also, $I()$ represents a threshold indicating function. In practice, the first step is to evaluate the existence of a nonlinear effect in the model through a process that allows us to determine if the threshold values are statistically significant. The null hypothesis states that $\psi_1 = \psi_2$ and the alternative hypothesis proposes that $\psi_1 \neq \psi_2$. When there is not enough evidence to accept the null hypothesis, it implies that the sample is divided into two groups; otherwise, linear estimation is the best option to obtain the estimators. In practice, there may be two thresholds in the model. Hansen (1999) points out that the best way to form the confidence intervals for γ is to establish a nonrejection region using the likelihood ratio (LR) statistic to test the null hypothesis. When LR is too large, and the P-value exceeds the confidence interval, the null hypothesis is rejected. The LR is given as $LR = -2\ln(1 - \sqrt{1 - \alpha})$, and with a significance level of 5%, the $LR = 7.35$, whose value is used to test the reliability of the threshold value. Likewise, to make our threshold values more robust and avoid errors in verifying or rejecting the null hypothesis, we followed a bootstrap procedure with 300 repetitions. The independent variables and the threshold variables are assumed with fixed values within the repeated samples using bootstraps. This bootstrap procedure produces first-order asymptotic distribution, which means that the P-values constructed from the bootstrap are asymptotically valid.

5. Results and discussion

Table 12.4 reports the test of the threshold effect of the GDP and globalization index variables. We found a statistically significant threshold for both variables since their probability is less than 0.05. This result conditions subsequent estimates because the basic results suggest that the effect of the human capital index and the economic complexity index have a nonlinear impact on renewable energy consumption in the sample of countries analyzed. We report the critical values at 1%, 5%, and 10%, however, we make the inferences with a significance level of 5% traditional. The existence of a threshold effect indicates that the impact of the independent variables on renewable energy consumption differs below and above the thresholds. A possible explanation for the significance of thresholds in the GDP and globalization index variables is that, in the initial stages of development, environmental sustainability does not necessarily matter in energy production. However, when economies exceed an output threshold, there are more resources to encourage increased

TABLE 12.4 Threshold effect test (bootstrap = 300).

Threshold variable	Threshold effect	F	P- value	Critical value of F 1%	5%	10%
Per capita output	Single	163.85**	0.000	82.685	70.212	63.704
	Double	42.74	0.283	58.882	66.716	86.182
Globalization index	Single	73.26**	0.0407	102.022	65.228	56.813
	Double	12.53	0.1767	84.08	39.352	33.6707

**indicates the statistical significance at 1%.

environmental awareness, and the economy has more tools to take advantage of the new technologies available. Likewise, the significance of the globalization index threshold suggests that for the effect of human capital and diversification to have a more significant impact on renewable energy, globalization must be high. Recent research has shown that thresholds can condition renewable energy determinants for some variables (Chen et al., 2020; Alvarado et al., 2020). Using the bootstrap method with 300 repetitions, we determined the existence of thresholds using the F statistic values proposed by Hansen (1999, 2000), which are compared with the critical values. We found evidence in favor of the existence of a threshold at a significance level of 5%. Statistics show that the second threshold is not statistically significant for both variables that set the threshold.

Fig. 12.1 illustrates the threshold of the product's logarithm in the upper panel and the threshold of globalization in the lower panel. Following the proposed methodology, the threshold value found represents the value of the output and the globalization index when the LR statistic is equal to zero. When the LR statistic is intercepted with the dotted line, it denotes the threshold's existence at 95% confidence. The logarithm of GDP threshold occurs around 23.52, and the globalization index threshold occurs about 70.47 points. In practice, the threshold effects cannot be omitted from the estimates, and their inclusion generates nonlinear estimators that are more consistent and unbiased than traditional linear estimators.

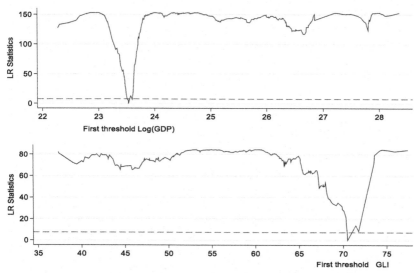

FIGURE 12.1 Likelihood ratio statistic of one threshold.

TABLE 12.5 Threshold value estimation.

Threshold variable	Threshold	Threshold estimation value	Interval Lower	Upper
GDP	Th-1	23.529	23.505	23.557
GI	Th-1	70.474	70.405	70.488

Table 12.5 shows that there is a single threshold that is statistically significant. This fact implies that the null hypothesis of the linear relationship at a level of 1% significance is rejected. The results suggest that, in the 16 countries of Latin America, energy consumption has a nonlinear relationship with its determining factors. Both estimated thresholds are within the confidence interval. Consequently, the estimates were made with a single threshold.

The nonlinear effect of the human capital index and economic complexity on the consumption of renewable energy formalized in Eq. (12.2) is reported in Table 12.6. The use of each variable threshold divides the model into two submodels and calculates estimators with higher precision (Wang and Wang, 2020). The results obtained show that the human capital index and the economic complexity index have a threshold impact on the consumption of renewable energy in the 16 countries of Latin America during the period analyzed. The results of Table 12.6 can be grouped into two parts. First, we find strong empirical evidence showing that human capital and economic complexity have a statistically significant and positive effect. We found that $\psi_2 > \psi_1$, indicating that the diversification of export capacity has a more significant impact on renewable energy consumption with respect to knowledge. Likewise, the results show that above the threshold, the effect of the output on renewable energy consumption is higher than below the threshold. This fact implies that as the output increases, the energy consumption from nonpolluting sources increases at higher rates. Second, when the threshold is the globalization index, we find human capital and economic complexity have a positive impact on renewable energy. Furthermore, the globalization index has a negative effect below the threshold and a positive effect above the threshold whose results are statistically significant.

In both regressions, the shadow economy has a negative relationship with renewable energy consumption, indicating that undeclared activities, whether informal or clandestine, reduce the use of environmentally friendly energy. A similar result occurs with public spending. Increases in government final consumption spending lead to lower consumption of renewable energy. This result is opposite to what is expected; however, the established patterns of energy consumption may not allow an immediate transition from polluting

TABLE 12.6 Coefficients estimates of threshold regression.

Threshold = real output		Threshold = globalization index	
ln GDP < 23.529	2.317***	GLI < 70.474	−0.316***
	(−6.54)		(−10.73)
ln GDP ≥ 23.529	2.859***	GLI ≥ 70.474	0.224***
	(−6.94)		(−9.11)
ECI	3.154**	ECI	4.135***
	(2.93)		(3.46)
HCI	1.002**	HCI	1.857*
	(2.04)		(1.99)
SE	−0.291***	SE	−0.0993*
	(−7.20)		(−2.37)
PS	−0.395**	PS	−0.475***
	(−3.24)		(−3.67)
UR	0.304	UR	0.472
	(−0.56)		(0.81)
TR	−0.0857***	TR	−0.0766***
	(−4.67)		(−3.79)
Constant	299.4***	Constant	73.44***
	(9.01)		(12.99)
Observations	448		448
Adjusted R^2	0.508		0.527

t statistics in parentheses, and * $P < .05$, ** $P < .01$, *** $P < .001$

energy to clean energy. The urbanization rate has a positive but not significant impact on the consumption of renewable energy. This fact suggests that the consumption patterns of urban households are already established and may take several years to change. However, in the urban area, it is possible to promote environmental awareness campaigns in which the human capital is higher than the rural area. Therefore, promoting renewable energy consumption in cities is perhaps one more offer and strengthening the commitment to the environment. Our results suggest that trade does not benefit renewable energy consumption, given that for each percentage point that trade increases, clean energy consumption decreases by around 8%. This result is contradictory to the results obtained from previous research (Lean and Smyth, 2010; Ozturk and Acaravci, 2013). Latin America has recently increased international cooperation agreements on trade, investment, and capital and financial flows. Several of these agreements are aimed at strengthening the exploitation of natural resources. In particular, there are broad agreements for the exploitation and export of energy resources, which may explain these results. In practice, our results suggest that when output increases, there are more resources to import or adopt new technologies, which is reinforced by research and development within economies. Innovation in more environmentally friendly products and processes would lead to reduced environmental degradation. Hence, when the actual output exceeds the threshold, it has a more significant impact on renewable energy consumption. Eqs. (12.3) and (12.4) summarize the results obtained in the two estimates.

$$RE_{it} = 3.154 \ ECI_{it} + 1.002 \ HCI_{it} - 0.291 \ SE_{it} - 0.395 \ PS_{it}$$
$$+ 0.304 \ UR_{it} - 0.0857 \ TR_{it} + 2.317 \ \ln GDP_{it} I(\ln GDP_{it} < 23.529)$$
$$+ 2.859 \ \ln GDP_{it} I(\ln GDP_{it} \geq 23.529) + \varepsilon_{it}$$

$$(12.3)$$

$$RE_{it} = 4.135 \ ECI_{it} + 1.857 \ HCI_{it} - 0.0993 \ SE_{it} - 0.475 \ PS_{it} + 0.472 \ UR_{it}$$
$$- 0.0766 \ TR_{it} - 0.316 \ GLI_{it} I(GLI_{it} < 70.474)$$
$$+ 0.224 \ GLI_{it} I(GLI_{it} \geq 70.474) + \varepsilon_{it}$$

$$(12.4)$$

Table 12.7 classifies the countries in the two samples according to the thresholds of the output and the globalization index. We report the result for 1990, 2000, and 2017. Some countries remain at the same threshold, while other countries have variations over time. The existence of changes in the countries in the threshold implies that the dynamics of the consumption of renewable energy has changed over the 28 years analyzed.

The 16 Latin American countries included in this research have a different behavior concerning the analyzed variables. The classification in Table 12.6 offers an overview of the countries according to the thresholds. When the threshold is output, most countries are above the threshold, where the impact on renewable energy is highest. While when the threshold is the globalization index, most of the countries are below the threshold. These results suggest that, if the output of the countries of the region increases, the consumption of renewable energy will increase, and if the countries analyzed are more inserted in the processes of globalization, the renewable energy consumption will increase.

TABLE 12.7 Distribution of Latin American countries according to the threshold.

Year	GDP < 23.529	GDP ≥ 23.529
1990	Bolivia, Costa Rica, El Salvador, Nicaragua, Paraguay.	Argentina, Brazil Chile, Colombia, Dominican Republic, Ecuador, Guatemala, Mexico, Peru, Uruguay, Venezuela.
2000	Bolivia, El Salvador, Nicaragua.	Costa Rica, Paraguay, Argentina, Brazil Chile, Colombia, Dominican Republic, Ecuador, Guatemala, Mexico, Peru, Uruguay, Venezuela.
2017	Nicaragua.	Costa Rica, Paraguay, Argentina, Brazil Chile, Colombia, Dominican Republic, Ecuador, Guatemala, Mexico, Peru, Uruguay, Venezuela, El Salvador, Bolivia.
	GLI < 70.474	**GLI ≥ 70.474**
1990	Argentina, Bolivia, Brazil, Chile, Colombia, Costa Rica, Dominican Republic, Ecuador, El Salvador, Guatemala, Mexico, Nicaragua, Paraguay, Peru, Uruguay, Venezuela.	
2000	Argentina, Bolivia, Brazil, Chile, Colombia, Costa Rica, Dominican Republic, Ecuador, El Salvador, Guatemala, Mexico, Nicaragua, Paraguay, Peru, Uruguay, Venezuela.	
2017	Argentina, Bolivia, Brazil, Colombia, Dominican Republic, Ecuador, El Salvador, Guatemala, Nicaragua, Paraguay, Peru, Venezuela.	Chile, Uruguay, Costa Rica, Mexico

6. Conclusions and policy implications

In this research, we examine the effect of the economic complexity index and the human capital index on the consumption of renewable energy for a sample of 16 Latin American countries. We estimate this relationship in a highly dependent context on natural resources and a change in energy consumption patterns due to increased environmental awareness and greater availability of technology at affordable costs. Besides, the region has a set of natural characteristics that allow it to seek the sustainability and energy security that renewable energy sources offer. The benefits of changing the energy matrix are not limited to, reducing polluting emissions and achieving SDG7. The opportunities opened up by the renewable energy market can enhance long-term economic growth, improve access to renewable energy in rural areas, and promote energy efficiency and energy security. We find that the effect of human capital and diversification on renewable energy is nonlinear. Two thresholds were used in the estimates: the real output and the globalization index. The adoption of this econometric strategy allows us to offer a clear picture of how to increase the consumption of renewable energy, and the consequent decrease in polluting gas emissions. We find that the effect of the human capital index and the economic complexity index is positive below and above the threshold, but the force of the impact is more significant when globalization and output exceed the threshold. These results offer essential tools to policymakers, academics, and the business sector committed to sustainable development. The consolidation of renewable energy consumption requires maintaining the current commitment and adopting new technologies to reduce the costs of producing clean energy. In practice, the reduction of tariffs and tax rates on renewable energy products could promote the entry of new technologies to the markets of the countries analyzed. In terms of environmental sustainability, economic growth and globalization are two mechanisms to promote renewable energy consumption in Latin American countries.

Acknowledgments

The authors express their gratitude to the *Club de Investigación de Economía* (CIE). Loja Ecuador.

References

Aïssa, M., Jebli, M., Youssef, S., 2014. Output, renewable energy consumption and trade in Africa. Energy Policy 66, 11–18.

Akintande, O.J., Olubusoye, O.E., Adenikinju, A.F., Olanrewaju, B.T., 2020. Modeling the determinants of renewable energy consumption: evidence from the five most populous nations in Africa. Energy 206, 117992.

Alper, A., Oguz, O., 2016. The role of renewable energy consumption in economic growth: evidence from asymmetric causality. Renew. Sustain. Energy Rev. 60, 953–959.

Alvarado, R., Iglesias, S., 2017. Sector externo, restricciones y crecimiento económico en Ecuador. Problemas del Desarrollo 48 (191), 83–106.

Alvarado, R., Ortiz, C., Bravo, D., Chamba, J., 2020. Urban concentration, non-renewable energy consumption, and output: do levels of economic development matter? Environ. Sci. Pollut. Control Ser. 27 (3), 2760–2772.

Alvarado, R., Ponce, P., Alvarado, R., Ponce, K., Huachizaca, V., Toledo, E., 2019. Sustainable and non-sustainable energy and output in Latin America: a cointegration and causality approach with panel data. Energy Strat. Rev. 26, 100369.

Apergis, N., Payne, J., 2010a. Renewable energy consumption and economic growth: evidence from a panel of OECD countries. Energy Policy 38 (1), 656–660.

Apergis, N., Payne, J., 2010b. Renewable energy consumption and growth in Eurasia. Energy Econ. 32 (6), 1392–1397.

Astariz, S., Iglesias, G., 2015. The economics of wave energy: a review. Renew. Sustain. Energy Rev. 45, 397–408.

Baek, J., Cho, Y., Koo, W., 2009. The environmental consequences of globalization: a country-specific time-series analysis. Ecol. Econ. 68 (8–9), 2255–2264.

Balsalobre-Lorente, D., Shahbaz, M., Roubaud, D., Farhani, S., 2018. How economic growth, renewable electricity and natural resources contribute to CO_2 emissions? Energy Policy 113, 356–367.

Basbay, M., Elgin, C., Torul, O., 2016. Energy consumption and the size of the informal economy. Econ. Open Access Open Assess. E-J. 10 (2016–14), 1–28.

Belaïd, F., Zrelli, M.H., 2019. Renewable and non-renewable electricity consumption, environmental degradation and economic development: evidence from Mediterranean countries. Energy Policy 133, 110929.

Benkraiem, R., Lahiani, A., Miloudi, A., Shahbaz, M., 2019. The asymmetric role of shadow economy in the energy-growth nexus in Bolivia. Energy Policy 125, 405–417.

Chan, K.S., 1993. Consistency and limiting distribution of the least squares estimator of a continuous threshold autoregressive model. Ann. Stat. 21, 520–533.

Chen, C., Pinar, M., Stengos, T., 2020. Renewable energy consumption and economic growth nexus: evidence from a threshold model. Energy Policy 139, 111295.

Chen, Y., 2018. Factors influencing renewable energy consumption in China: an empirical analysis based on provincial panel data. J. Clean. Prod. 174, 605–615.

Chien, T., Hu, J., 2007. Renewable energy and macroeconomic efficiency of OECD and non-OECD economies. Energy Policy 35 (7), 3606–3615.

Clausen, L.T., Rudolph, D., 2020. Renewable energy for sustainable rural development: synergies and mismatches. Energy Policy 138, 111289.

Deng, Q., Alvarado, R., Toledo, E., Caraguay, L., 2020. Greenhouse gas emissions, non-renewable energy consumption, and output in South America: the role of the productive structure. Environ. Sci. Pollut. Control Ser. 1–15.

Ding, Q., Cai, W., Wang, C., Sanwal, M., 2017. The relationships between household consumption activities and energy consumption in China—an input-output analysis from the lifestyle perspective. Appl. Energy 207, 520–532.

Ezbakhe, F., Perez-Foguet, A., 2020. Decision analysis for sustainable development: the case of renewable energy planning under uncertainty. Eur. J. Oper. Res. (in press).

Fang, Y., 2011. Economic welfare impacts from renewable energy consumption: the China experience. Renew. Sustain. Energy Rev. 15 (9), 5120–5128.

García, J.C., de Barrios, M.L., Diniz, M.J., 2020. Exports to China and economic growth in Latin America, unequal effects within the region. Int. Econ. (in press).

González-Fernández, M., González-Velasco, C., 2014. Shadow economy, corruption and public debt in Spain. J. Pol. Model. 36 (6), 1101–1117.

Gozgor, G., Mahalik, M.K., Demir, E., Padhan, H., 2020. The impact of economic globalization on renewable energy in the OECD countries. Energy Policy 139, 111365.

Grossman, G.M., Krueger, A.B., 1995. Economic growth and the environment. Q. J. Econ. 110 (2), 353–377.

Hansen, B.E., 1999. Threshold effects in non-dynamic panels: estimation, testing, and inference. J. Econom. 93 (2), 345–368.

Hansen, B.E., 2000. Sample splitting and threshold estimation. Econometrica 68 (3), 575–603.

Hassan, M., Schneider, F., 2016. Size and development of the shadow economies of 157 worldwide countries: updated and new measures from 1999 to 2013. J. Global Econ. 4 (3), 1–15.

Hosseini, S.E., 2020. An outlook on the global development of renewable and sustainable energy at the time of covid-19. Energy Res. Soc. Sci. 101633.

International Energy Organization, 2020. Data and Statistics. Available at: https://www.iea.org/data-and-statistics.

Koengkan, M., Poveda, Y., Fuinhas, J., 2019. Globalisation as a motor of renewable energy development in Latin America countries. GeoJournal 1–12.

Lean, H., Smyth, R., 2010. Multivariate Granger causality between electricity generation, exports, prices and GDP in Malaysia. Energy 35 (9), 3640–3648.

Lin, B., Zhu, J., 2019. Fiscal spending and green economic growth: evidence from China. Energy Econ. 83, 264–271.

Martinho, V., 2018. Interrelationships between renewable energy and agricultural economics: an overview. Energy Strat. Rev. 22, 396–409.

Matraeva, L., Solodukha, P., Erokhin, S., Babenko, M., 2019. Improvement of Russian energy efficiency strategy within the framework of "green economy" concept (based on the analysis of experience of foreign countries). Energy Policy 125, 478–486.

Mealy, P., Teytelboym, A., 2020. Economic complexity and the green economy. Res. Policy 103948.

Medina, L., Schneider, F., 2019. Shedding Light on the Shadow Economy: A Global Database and the Interaction with the Official One (2019). CESifo Working Paper No. 7981. Available at SSRN: https://ssrn.com/abstract=3502028.

Meleddu, M., Pulina, M., 2018. Public spending on renewable energy in Italian regions. Renew. Energy 115, 1086–1098.

Menegaki, A., 2011. Growth and renewable energy in Europe: a random effect model with evidence for neutrality hypothesis. Energy Econ. 33 (2), 257–263.

Murshed, M., 2020. Are trade liberalization policies aligned with renewable energy transition in low and middle income countries? An instrumental variable approach. Renew. Energy 151, 1110–1123.

Narayan, P., Smyth, R., 2009. Multivariate Granger causality between electricity consumption, exports and GDP: evidence from a panel of Middle Eastern countries. Energy Policy 37 (1), 229–236.

Nicolli, F., Vona, F., 2019. Energy market liberalization and renewable energy policies in OECD countries. Energy Policy 128, 853–867.

Ocal, O., Aslan, A., 2013. Renewable energy consumption–economic growth nexus in Turkey. Renew. Sustain. Energy Rev. 28, 494–499.

Orviská, M., Čaplánová, A., Medved, J., Hudson, J., 2006. A cross-section approach to measuring the shadow economy. J. Pol. Model. 28 (7), 713–724.

Ozturk, I., Acaravci, A., 2013. The long-run and causal analysis of energy, growth, openness and financial development on carbon emissions in Turkey. Energy Econ. 36, 262–267.

Ponce, P., Alvarado, R., Ponce, K., Alvarado, R., Granda, D., Yaguana, K., 2019. Green returns of labor income and human capital: empirical evidence of the environmental behavior of households in developing countries. Ecol. Econ. 160, 105–113.

Przychodzen, W., Przychodzen, J., 2020. Determinants of renewable energy production in transition economies: a panel data approach. Energy 191, 116583.

Pursiheimo, E., Holttinen, H., Koljonen, T., 2019. Inter-sectoral effects of high renewable energy share in global energy system. Renew. Energy 136, 1119–1129.

Rahman, M., Velayutham, E., 2020. Renewable and non-renewable energy consumption-economic growth nexus: new evidence from South Asia. Renew. Energy 147, 399–408.

Ram, M., Aghahosseini, A., Breyer, C., 2020. Job creation during the global energy transition towards 100% renewable power system by 2050. Technol. Forecast. Soc. Change 151, 119682.

Reddy, B.S., Ray, B.K., 2010. Decomposition of energy consumption and energy intensity in Indian manufacturing industries. Energy Sustain. Dev. 14 (1), 35–47.

Ringel, M., Schlomann, B., Krail, M., Rohde, C., 2016. Towards a green economy in Germany? The role of energy efficiency policies. Appl. Energy 179, 1293–1303.

Sadorsky, P., 2009. Renewable energy consumption and income in emerging economies. Energy Policy 37 (10), 4021–4028.

Salinas, A., Muffatto, M., Alvarado, R., 2018. Informal institutions and informal entrepreneurial activity: new panel data evidence from Latin American countries. Acad. Enterpren. J. 24 (4), 1–17.

Salinas, A., Ortiz, C., Muffatto, M., 2019. Business regulation, rule of law and formal entrepreneurship: evidence from developing countries. J. Entrepreneur. Public Policy 8 (in press).

Sarkodie, S.A., Adams, S., Owusu, P.A., Leirvik, T., Ozturk, I., 2020. Mitigating degradation and emissions in China: the role of environmental sustainability, human capital and renewable energy. Sci. Total Environ. 137530.

Sener, S., Sharp, J., Anctil, A., 2018. Factors impacting diverging paths of renewable energy: a review. Renew. Sustain. Energy Rev. 81, 2335–2342.

Shahbaz, M., Lahiani, A., Abosedra, S., Hammoudeh, S., 2018b. The role of globalization in energy consumption: a quantile cointegrating regression approach. Energy Econ. 71, 161–170.

Shahbaz, M., Lean, H., Farooq, A., 2013. Natural gas consumption and economic growth in Pakistan. Renew. Sustain. Energy Rev. 18, 87–94.

Shahbaz, M., Mallick, H., Mahalik, M., Sadorsky, P., 2016. The role of globalization on the recent evolution of energy demand in India: implications for sustainable development. Energy Econ. 55, 52–68.

Shahbaz, M., Raghutla, C., Chittedi, K.R., Jiao, Z., Vo, X.V., 2020. The effect of renewable energy consumption on economic growth: evidence from the renewable energy country attractive index. Energy 207, 118162.

Shahbaz, M., Shahzad, S., Mahalik, M., Sadorsky, P., 2018a. How strong is the causal relationship between globalization and energy consumption in developed economies? A country-specific time-series and panel analysis. Appl. Econ. 50 (13), 1479–1494.

Sheinbaum-Pardo, C., Mora-Pérez, S., Robles-Morales, G., 2012. Decomposition of energy consumption and CO_2 emissions in Mexican manufacturing industries: trends between 1990 and 2008. Energy Sustain. Dev. 16 (1), 57–67.

Sinha, A., Sengupta, T., Alvarado, R., 2020. Interplay between technological innovation and environmental quality: formulating the SDG policies for next 11 economies. J. Clean. Prod. 242, 118549.

Smart, S., 2020. The political economy of Latin American conflicts over mining extractivism. Extr. Ind. Soc. 7 (2), 767–779.

Tiwari, A., 2011. A structural VAR analysis of renewable energy consumption, real GDP and CO_2 emissions: evidence from India. Econ. Bull. 31 (2), 1793–1806.

Tuna, G., Tuna, V., 2019. The asymmetric causal relationship between renewable and NON-RENEWABLE energy consumption and economic growth in the ASEAN-5 countries. Resour. Policy 62, 114–124.

Uzar, U., 2020. Is income inequality a driver for renewable energy consumption? J. Clean. Prod. 255, 120287.

Viviescas, C., Lima, L., Diuana, F.A., Vasquez, E., Ludovique, C., Silva, G.N., et al., 2019. Contribution of Variable Renewable Energy to increase energy security in Latin America: complementarity and climate change impacts on wind and solar resources. Renew. Sustain. Energy Rev. 113, 109232.

Wang, Q., Lin, J., Zhou, K., Fan, J., Kwan, M.P., 2020. Does urbanization lead to less residential energy consumption? A comparative study of 136 countries. Energy 202, 117765.

Yang, X., He, L., Xia, Y., Chen, Y., 2019. Effect of government subsidies on renewable energy investments: the threshold effect. Energy Policy 132, 156–166.

Yao, Y., Ivanovski, K., Inekwe, J., Smyth, R., 2019. Human capital and energy consumption: evidence from OECD countries. Energy Econ. 84, 104534.

Zhang, Y.J., Bian, X.J., Tan, W., Song, J., 2017. The indirect energy consumption and CO_2 emission caused by household consumption in China: an analysis based on the input–output method. J. Clean. Prod. 163, 69–83.

Zhao, Y., Ke, J., Ni, C.C., McNeil, M., Khanna, N.Z., Zhou, N., et al., 2014. A comparative study of energy consumption and efficiency of Japanese and Chinese manufacturing industry. Energy Policy 70, 45–56.

Chapter 13

Quest for energy efficiency: the role of human capital and firm internationalization

Muhammad Shujaat Mubarik[1], Navaz Naghavi[2]

[1]College of Business Management (CBM), Institute of Business Management (IoBM), Karachi, Pakistan; [2]School of Accounting & Finance, Faculty of Business & Law, Taylor's University, Lakeside Campus, Subang Jaya, Malaysia

1. Introduction

Due to the catastrophic effects of fossil fuel energy on the environment, there is a global campaign for abandoning it by substituting it with environmentally friendly (green) energy. World leaders agreed in Paris accord to take sustainable measures for reducing the global temperature to the level of the preindustrialization era (IPCC, 2014). The global seriousness to control the rising temperatures and get rid of the disastrous effects of fossil fuels is forcing firms to adopt green energy sources. The issue has greater ramifications for the firms involved in the internationalization as firms involved in the internationalization process have to ensure global environmental compliances. Such firms also tend to face the global legitimating actors, such as multilateral agencies or international nongovernmental organizations (Marano and Kostova, 2016; Marano and Tashman, 2012). Similarly, firms operating in international markets also encounter external institutional pressures from the host country governments, society, regulators, and markets. Nevertheless, such pressures and challenges can greatly vary from country to country (Meyer et al., 2011; Mubarik et al., 2016).

A number of researchers (Meyer et al., 2011; Rasiah and Mubarik, 2020) argue that firms involved in the internationalization process tend to use green energy. These studies consider internationalization as a positive influencer of green energy consumption. Despite their convincing contextual explanation as to how the internationalization process can lead a firm to adopt green energy, the empirical evidence on this dyad is absent from the literature. Specifically, in the presence of the fact that international pressures and challenges greatly vary from country to country and can even have a conflicting prescription in

Energy-Growth Nexus in an Era of Globalization. https://doi.org/10.1016/B978-0-12-824440-1.00018-7

311

some cases. In order to align with both global and local practices, firms are relooking at their energy strategies and trying to take all the possible measures to reduce nongreen energy consumption. On the other hand, another stream of authors claims that the internationalization process does not make any difference in the energy usage of a firm.

In some cases, the firms involved in internationalization consume higher nongreen energy as compared to the other firms. Their argument is simple, "for meeting the demand of international markets, a firm has to produce more." The need for further production is met with the immediately available energy, i.e., "nongreen energy" (Mubarik and Naghavi, 2020). These diverged opinions of the scholars have created a state of ambivalence. It calls for the study to examine the impact of internalization on a firm's energy consumption. Understanding the nature of this impact can significantly help the stakeholders to devise policies for enhancing green energy consumption.

Further, the second debatable issue in the context of energy consumption is firm's human capital development. Some scholars argue (Lan and Munro, 2013; Mubarik and Naghavi, 2020) consider HC as a key resource of a firm that significantly affects a firm's energy consumption patterns. It is considered as a precursor to reducing nongreen energy consumption and pushes for using green energy. Shahbaz et al. (2018a,b) argue that a firm with relatively stronger human capital levels tends to adopt production technologies that consume less energy and have green energy compatibility. Similarly, firms with higher levels of human capital have a higher tendency to adopt green energy sources (Shahbaz et al., 2019).

Mubarik and (2020) notes, "surprisingly, a great body of literature focuses upon the association of human capital and energy consumption at a macro level; however, this relationship has been lesser discussed at the firm level. Particularly, the question as to how human capital development is linked with the consumption of renewable energy is yet to be addressed." Their findings reveal an inverted U-shaped association between HC and fossil fuel energy consumption and a U-shaped association between HC and green energy consumption.

Nevertheless, the question of how human capital and internationalization affect a firm's energy consumption is yet unanswered. This serves as the impetus of this chapter. Hence, this study aims to investigate the impact of internationalization and human capital on a firm's energy consumption (green and nongreen). In doing so, this study contributes to the literature on internationalization and human capital.

The rest of the study has been divided into four sections. The subsequent section reviews the literature on human capital, energy consumption, and technological innovation. Section 3, methodology, undertakes the research methodology applied to test the proposed framework. The last section concludes the chapter by providing theoretical, managerial, and policy implications.

2. Literature review

2.1 Human capital and energy consumption

The seminal work of Grossman and Krueger (1991) accelerated the interest of researchers and policymakers by offering the environmental Kuznets curve (EKC), which is now considered as one of the significant theoretical bedrock in studying the environment-growth dyad. The prime submission of the EKC is that economic growth and environmental degradation have an inverted U-shaped relationship. According to Grossman and Krueger (1991), the environment quality gets deteriorated in the early phases of growth. Whereas after a certain time, it gets reversed, and the quality of the environment starts improving with an increase in economic growth (Onafowora and Owoye, 2014). It infers that environmental improvement is considerably linked with the long-term economic growth of a country. We draw the same inferences at the firm level and link it with human capital. We argue that in the short run, an increase in the firm's production increases its nongreen energy consumption, whereas it tends to slow down in the medium run and reduces in the long run provided that the firm improves its human capital. In the absence of higher levels of human capital, a firm may not be adopting green energy consumption even in the long run (Balsalobre-Lorente and Shahbaz, 2016). Hence, HC plays a decisive role in decreasing nongreen energy consumption and substituting it with green energy. A major stream of literature supports this notion. For example, Manderson and Kneller (2012) assert that improving HC decreases nongreen energy consumption. Likewise, Blackman and Kildegaard (2010) demonstrate HC as a major determinant of green energy consumption in Mexican firms. They claim that educated employees could play an instrumental role in adopting energy efficiency technologies and practices. Relatively less educated employees may not be fully aware of the detrimental effects of nongreen energy consumption, leading to an insensitive behavior toward green energy consumption. In short number of studies (e.g., Manderson and Kneller, 2012; Mubarik and Naghavi, 2020; Mubarik et al., 2016) show that firms' energy consumption patterns are heavily influenced by its human capital. Nonetheless, studies (e.g., Shahbaz et al., 2019) argue that improving HC lead to the consumption of nongreen energy. Specifically, Mubarik and Naghavi (2020) note, "...increasing levels of human capital requires to undertake more tasks requiring more energy to consume. Since the immediate, accessible and cheaper source of energy is nongreen, it is used to meet the increase in energy demand." These ambivalent results of the studies require reinvestigation of the HC energy consumption dyad. We argue that the association between HC and nongreen energy (NGE) consumption is inverted U-shape. We also claim that HC and green energy (GE) consumption is U-shaped.

2.2 Firm internationalization and energy consumption

Although the internationalization of a firm is gauged with a number of factors, the degree to which a firm exports (export intensity) is one of the commonly used proxies to measure the earlier phases of internationalization. Traditionally exports are directly linked to energy consumption. Scholars consider export as one of the major sources of nongreen energy consumption. However, recently the scholars have challenged this notion, considering exports as one of the factor that can push a firm to adopt green energy. Scholars (e.g., Bansal, 2005; Suarez-Perales et al., 2017; Hartmann and Vachon, 2018) claim that firms involved in internationalization have a higher tendency to use green energy as compared to the noninternationalizing firms. As noted Gómez-Bolaños et al. (2019), "International firms are exposed to the institutional pressures of all the countries in which they are present together with global norms and global legitimating actors. Hence, firms need to deploy efforts aimed at attaining legitimacy and maintaining their competitive positions Firms may use alternative practices to attain legitimacy in an international context, such as reinforcing their environmental disclosure." Aragón-Correa et al. (2016) mention that internationalizing firms have higher tendency to disclose the environmental performance, including energy consumption and their sources. The claim, "multinational firms go beyond local environmental standards by transferring advanced environmental technology to their subsidiaries, thus coping with the regulatory demands of the strictest countries in which they operate." Keeping in view intercountry heterogeneities in levels of developments, regulations, and tolerance for corruption, the impact of internationalization on the firm energy consumption is not straightforward. It is normally expected that a firm may face tougher situation concerning environment if it is entering in a country which has lesser tolerance for environmental degraded activities and have higher compliant of rules and regulations. On the other hand, if a firm is exporting to a country with higher tolerance for environmental regulations, specifically in regard to energy consumption and its sources, that firm may not face any kind of pressures to use the green energy for producing production. It implies that impact of exportability on green energy consumption depends upon the country to which a firm is exporting. It is also argued that due to foreignness, firm may face higher scrutiny in terms of their environmental compliance in the country with low tolerance for environmental regulations (Fig. 13.1).

3. Methodology

3.1 Model

Following two equations have been formed to examine the impacts of HC and internationalization on firm's energy consumption (green and nongreen).

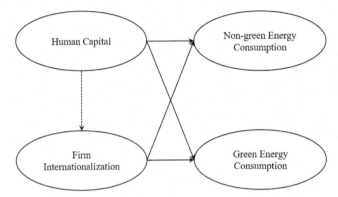

FIGURE 13.1 HC-internationalization-energy triad. *Source: Author.*

$$NGC_{it} = \acute{\alpha} + \beta_1 HC_{it} + \beta_2 HC_{it\,2}^2 + \beta_3 HC * FINT_{it} + \beta_4 Size_{it} + \beta_5 Ind_{it} + \vartheta_{it}.......(1) \quad (13.1)$$

$$GC_{it} = \acute{\alpha} + \beta_1 HC_{it} + \beta_2 HC_{it\,2}^2 + \beta_3 HC * FINT_{it} + \beta_4 Size_{it} + \beta_5 Ind_{it} + \vartheta_{it}........(2) \quad (13.2)$$

whereas:

NGC is firm's nongreen energy consumption. It has been measured as the share of nonrenewable energy in the total final energy consumption (% of total energy consumption).

GC represents firm's green energy consumption, which has been measured as the share of renewable energy in the total final energy consumption (% of total energy consumption).

HCD represent human capital development measured as the average of employees' related experience, education, number of training received in a year.

HC represents human capital of a firm. It has been operationalized as the average of employees' related experience, education, number of training received in a year.

FINT is firm's internationalization, measured as the average of firms' exports, and access to international market.

Size is firm size. The firms having employees more than 250 have been categorized as large firms, between 100 and 249 as medium and less than 100 as small.

Ind is type of industry.

3.2 Data

We collected the firm-level data from 417 manufacturing sector firms for the last four years, 2016 to 2020. The data of these firms were collected each year by Mohammad Ali Jinnah University, Pakistan's Business Research Center (Tables 13.1 and 13.2).

TABLE 13.1 Measurement of variables.

Variables	Measurement	Past researches
Renewable energy consumption	Share of renewable energy in the total final energy consumption (% of total energy consumption)	Hanif et al. (2019)
Nonrenewable energy consumption	Share of nonrenewable energy in the total final energy consumption (% of total energy consumption)	Hanif et al. (2019)
Human capital	Average of employees related experience, education, number of training received in a year	Mubarik (2015), Mubarik et al. (2018)
Firm internationalization	Average of firms exports, and access to the international market	

Reproduced from Mubarik, M.S., Naghavi, N., 2020. Human capital, green energy, and technological innovations: firm-level analysis. In: Econometrics of Green Energy Handbook (pp. 151–164). Springer, Cham.

TABLE 13.2 Breakup of sample.

Industry	Number	%
Textile	102	24
Leather	84	20
Food	95	23
Small sscale eengineering	65	16
Sports	71	17
Total	**417**	

Reproduced from Mubarik, M.S., Naghavi, N., 2020. Human capital, green energy, and technological innovations: firm-level analysis. In: Econometrics of Green Energy Handbook (pp. 151–164). Springer, Cham.

3.3 Method

For estimating the modeled relationships, we employed Panel Feasible Generalized Square (FGLS). This approach is applied to combine the variances in the cross sections and considered highly robust against heteroscedasticity compared to other panel random effect or panel fixed-effect models (Davidson and MacKinnon, 1993; Hassan et al., 2019). Similarly, due to the changes in standard errors of cross sections, autocorrelation could also be tackled in FGLS. Below is the mathematical representation of the model:

$$\beta_{FGLS} = \left(x'\vartheta^{-1}x\right)^{-1} x'\vartheta^{-1}\omega$$

$$Var(\beta_{FGLS}) = \left(x'\vartheta^{-1}x\right)^{-1}$$

$$\vartheta = \sum_{n*n} \otimes M_{K_iyK_i}$$

$$\sum_{i,k} = \beta_i\beta_k/L$$

Here ϑ can be adjusted to include heteroscedasticity and autocorrelation while computing the coefficients and their standard errors.

4. Findings and discussion

The results of both models have been exhibited in Tables 13.3 and 13.4. The green energy model results show that the coefficient of HC is negative and significant in the overall sample ($\beta = -1.012$, $P < .05$) and disaggregated analysis. Further, the coefficients of HC square are also significant but positive in the total ($\beta = 0.241$, $P < .05$) as well as in the disaggregated sample. It confirms the existence of a U-shaped relationship between HC and green energy consumption. Findings illustrate that consumption of green energy reduces at the earlier stages of HC development, whereas after passing certain stages of HC development, it starts increasing. The HC nongreen energy association results show that HC has a positive and significant coefficient, whereas the HC square has a negatively significant coefficient. These results confirm the presence of a U-shaped relationship between HC and nongreen energy consumption. The point from which HC energy consumption tends to change, the turning point, has been depicted in Table 13.4. The coefficients of HC in the total sample and industry—wise analysis show that the turning point for green energy consumption comes earlier compared to the NGE turning point.

We generate the variable HC×FINT to examine the intervening role of firm internationalization in the association between HC and energy consumption. The results, exhibited in Table 13.4, reveal an essential facet. The coefficients of HC×FINT for green energy consumption are not only significant and positive but also are higher in magnitude as compare to the direct effect of HC on green energy. The same situation prevails in nongreen energy, where the effects of HC×FINT in reducing nongreen energy are much higher than the HC. These findings imply that HC development, when combined with internationalization, becomes a lethal weapon to combat nongreen energy consumption and to promote green energy consumption.

TABLE 13.3 Renewable energy.

| | Total sample | | By industry | | | | | | | |
| | | | Textile | | Leather | | Sports | | Small eng | |
Variable(s)	GE	NGE	GE	NGE	GE	NGE	GE	GE	NGE	GE
HC	-1.012 *(0.000)*	1.432 *(0.000)*	-0.540 *(0.001)*	1.014 *(0.031)*	-0.390 *(0.007)*	0.930 *(0.012)*	-0.510 *(0.001)*	0.730 *(0.014)*	-0.58 *(0.000)*	0.860 *(0.000)*
HC×FINT	0.183 *(0.000)*	-0.091 *(0.000)*	0.120 *(0.000)*	-0.072 *(0.000)*	0.11 *(0.027)*	-0.16 *(0.000)*	0.234 *(0.006)*	-0.181 *(0.000)*	0.091 *(0.021)*	-0.082 *(0.000)*
HC square	0.241 *(0.001)*	-0.011 *(0.015)*	0.210 *(0.008)*	-0.041 *(0.000)*	0.109 *(0.029)*	-0.160 *(0.009)*	0.050 *(0.012)*	-0.051 *(0.043)*	0.122 *(0.000)*	-0.151 *(0.000)*
Size	0.081 *(0.000)*	0.111 *(0.001)*	0.150 *(0.000)*	0.082 *(0.000)*	0.051 *(-0.001)*	0.072 *(0.000)*	0.101 *(0.003)*	0.170 *(0.000)*	0.092 *(0.000)*	0.192 *(0.000)*
Adj R square	0.620	0.681	0.690	0.730	0.681	0.730	0.620	0.670	0.730	0.710
Wald test	73.51 *(0.004)*	87.00 *(0.005)*	79.87 *(0.032)*	95.23 *(0.009)*	68.52 *(0.019)*	86.59 *(0.021)*	76.00 *(0.018)*	101.25 *(0.005)*	87.25 *(0.009)*	97.82 *(0.011)*

Note: Values in Italics are *P*-values.
GE, Green energy; *NGE*, nongreen energy.

TABLE 13.4 Threshold point of HC.

	GE		NGE	
	HC	HC^2	HC	HC^2
Total sample	−1.4	0.24	1.87	−0.09
Textile	−0.68	0.22	1.35	−0.11
Leather	−0.6	0.14	0.95	−0.16
Sports	−0.64	0.13	0.76	−0.07
Small engineering	−0.93	0.08	1.08	−0.13

Our results echo the findings of other studies. The main findings of our study are that developing human capital leads to an increase in the energy demand, and nongreen energy could be the immediately available and cheaper option. In such a case, despite having awareness about the negative impacts of nongreen energy, the firm tends to use it. Nevertheless, in the medium and long run, employees' awareness to improve human capital leads firms to explore and adopt green energy venues. HC in this case not only puts pressure to adopt green energy but also plays an instrumental role in the adoption of technologies, which are green energy compatible and in promotion of energy conservation (Mubarik and Naghavi, 2020).

Results on the impact of the HC internationalization dyad on energy consumption are important. It shows that internationalization plays a profound role in pushing firms to adopt green energy. The international markets serve as the push factor to adopt green energy consumption, whereas HC serves as the internal pull factor. In this context, both internationalization and HC development supplement each other and pave the way for the firm to adopt green energy sources. Some scholars challenge the impact of internationalization on green energy consumption (Reza et al., 2021). The claim that the extent to which internationalization pushes a firm for green energy adoption depends upon the host country regulations and tolerance for nongreen energy. By looking into our data, we could analyze that most firms are exporting to developed countries or countries with lesser tolerance for nongreen energy. Therefore, these firms are being continuously pushed by their host markets to "go green."

After reaching human capital, tends to substitute nonrenewable energy consumption with renewable sources (Doğan and Değer, 2018). Our results show a significant interacting role of technological innovation ($\beta = 0.07$, $P < .05$) in the relationship between human capital and energy consumption. It implies that the increase in technological innovation leads human capital to consume more green energy. Whereas, results in Table 13.4 show that an increase in technological innovation ($\beta = -0.13$, $P < .05$) discourages the usage of nonrenewable energy. In condensed form, innovation in interaction with human capital increases the use of renewable energy and makes nonrenewable energy less attractive. These results concur with the study of Reza et al. (2021). Further as expected an increase in the size of the firms leads to an increase in the consumption of both renewable and nonrenewable energy.

5. Conclusion and implications

The chapter aimed to examine the two major associations. The first was to examine as to how human capital influences the consumption of green and nongreen energy. Second, what is the role that firm internationalization plays in the association between HC and energy consumption. Data of the study were collected from 417 manufacturing sector firms of Pakistan. The FMOLS was employees to estimate the modeled relationship. The results confirm the quadratic relationship of HC with energy consumption (U-shaped relationship with green energy and inverted U-shaped relationship with nongreen energy). Our findings highlight that internationalization plays a very strong role in pushing firms to adopt green energy sources for production. These findings led us to conclude that if a firm is involved in internationalization, the improvement in its human capital will significantly impact its energy consumption patterns compared to the noninternationalizing firm. Further, we conclude that firms with a comparatively stronger human capital are close to the turning points from where the increase in HC can further lead to accelerating in green energy consumption.

The study has some profound implications. The first implication of the study is the development of human capital for promoting green energy. Although most firms tend to involve their HC in developmental activities through training, the lesser could be seen about green energy-focused HC development. Firms can adopt twofold strategies in this regard. First, human capital development should have some linkage with greenness, especially green energy. For example, if a firm is imparting training related to the technologies, it must also keep a part in that training focusing on green technologies. Secondly, the firm should include awareness about greenness as one of the criteria factors of HR recruitment.

References

Aragón-Correa, J.A., Marcus, A., Hurtado-Torres, N., 2016. The natural environmental strategies of international firms: old controversies and new evidence on performance and disclosure. Acad. Manag. Perspect. 30 (1), 24–39.

Balsalobre-Lorente, D., Shahbaz, M., 2016. Energy consumption and trade openness in the correction of GHG levels in Spain. Bull. Energ. Econ. 4 (4), 310–322.

Bansal, P., 2005. Evolving sustainably: a longitudinal study of corporate sustainable development. Strat. Manag. J. 26 (3), 197–218.

Blackman, A., Kildegaard, A., 2010. Clean technological change in developing-country industrial clusters: Mexican leather tanning. Environ. Econ. Pol. Stud. 12 (3), 115—132.

Davidson, R., MacKinnon, G.J., 1993. Estimation and Inference in Econometrics. Oxford University Press, New York.

Doğan, B., Değer, O., 2018. The role of economic growth and energy consumption on CO_2 emissions in E7 countries. Theor. & Appl. Econ. 25 (2(615)), 231—246.

Gómez-Bolaños, E., Hurtado-Torres, N.E., Delgado-Márquez, B.L., 2020. Disentangling the influence of internationalization on sustainability development: evidence from the energy sector. Bus. Strat. Environ. 29 (1), 229—239.

Grossman, G.M., Krueger, A.B., 1991. Environmental Impacts of a North American Free Trade Agreement (No. W3914). National Bureau of Economic Research.

Hanif, I., Aziz, B., Chaudhry, I.S., 2019. Carbon emissions across the spectrum of renewable and nonrenewable energy use in developing economies of Asia. Renew. Energy 143, 586—595.

Hartmann, J., Vachon, S., 2018. Linking environmental management to environmental performance: the interactive role of industry context. Bus. Strat. Environ. 27 (3), 359—374.

Hassan, M.S., Bukhari, S., Arshed, N., 2019. Competitiveness, governance and globalization: what matters for poverty alleviation? Environ. Dev. Sustain. 1—27.

Intergovernmental Panel on Climate Change (IPCC), 2014. Climate Change 2014 - Synthesis Report. https://www.ipcc.ch/pdf/assessment-report/ar5/syr/SYR_AR5_FINAL_full_ wcover. pdf (Accessed 23 April 2018).

Lan, J., Munro, A., 2013. Environmental compliance and human capital: evidence from Chinese industrial firms. Resour. Energy Econ. 35 (4), 534—557.

Manderson, E., Kneller, R., 2012. Environmental regulations, outward FDI and heterogeneous firms: are countries used as pollution havens? Environ. Resour. Econ. 51 (3), 317—352.

Marano, V., Kostova, T., 2016. Unpacking the institutional complexity in adoption of CSR practices in multinational enterprises. J. Manag. Stud. 53 (1), 28—54.

Marano, V., Tashman, P., 2012. MNE/NGO partnerships and the legitimacy of the firm. Int. Bus. Rev. 21 (6), 1122—1130.

Meyer, K.E., Mudambi, R., Narula, R., 2011. Multinational enterprises and local contexts: the opportunities and challenges of multiple embeddedness. J. Manag. Stud. 48 (2), 235—252.

Mubarik, M.S., 2015. Human capital and performance of small & medium manufacturing enterprises: a study of Pakistan. Doctoral dissertation. University of Malaya. http://studentsrepo. um.edu.my/6573/.

Mubarik, M.S., Govindaraju, C., Devadason, E.S., 2016. Human capital development for SMEs in Pakistan: is the "one-size-fits-all" policy adequate? Int. J. Soc. Econ. 43 (8), 804—822.

Mubarik, M.S., Chandran, V.G.R., Devadason, E.S., 2018. Measuring human capital in small and medium manufacturing enterprises: what matters? Soc. Indicat. Res. 137 (2), 605—623.

Mubarik, M.S., Naghavi, N., 2020. Human capital, green energy, and technological innovations: firm-level analysis. In: Econometrics of Green Energy Handbook. Springer, Cham, pp. 151—164.

Onafowora, O.A., Owoye, O., 2014. Bounds testing approach to analysis of the environment Kuznets curve hypothesis. Energy Econ. 44, 47—62.

Rasiah, R., Mubarik, M.S., 2020. Energy consumption and greening: strategic directions for Pakistan. The Lahore J. Econ. 25 (1), 59—88.

Reza, S., Mubarik, M.S., Naghavi, N., Nawaz, R.R., 2021. Internationalisation challenges of SMEs: role of intellectual capital. Intellect. Cap. 18 (3), 252—277.

Shahbaz, M., Shahzad, S.J.H., Mahalik, M.K., Sadorsky, P., 2018a. How strong is the causal relationship between globalization and energy consumption in developed economies? A country-specific time-series and panel analysis. Appl. Econ. 50 (13), 1479–1494.

Shahbaz, M., Mahalik, M.K., Shahzad, S.J.H., Hammoudeh, S., 2019. Does the environmental Kuznets curve exist between globalization and energy consumption? Global evidence from the cross-correlation method. Int. J. Finance Econ. 24 (1), 540–557.

Shahbaz, M., Lahiani, A., Abosedra, S., Hammoudeh, S., 2018b. The role of globalization in energy consumption: a quantile cointegrating regression approach. Energy Econ. 71, 161–170.

Suarez-Perales, I., Garces-Ayerbe, C., Rivera-Torres, P., Suarez-Galvez, C., 2017. Is strategic proactivity a driver of an environmental strategy? Effects of innovation and internationalization leadership. Sustainability 9 (10), 1870.

Chapter 14

Green growth and energy transition: An assessment of selected emerging economies

Suborna Barua

Department of International Business, University of Dhaka, Dhaka, Bangladesh

1. Introduction

The traditional focus on gross domestic product (GDP) growth as the measure for economic development has driven the world to face substantial environmental quality deterioration. Economies often follow a "growth first" approach. They prioritize achieving economic objectives such as the growth of GDPs and significantly undermine the environment's protection (Barua and Chiesa, 2019; Chiesa and Barua, 2019). In the advent of deteriorating quality and quantity of environmental resources, the world has already begun to feel the urgency to shift from the traditional view of economic development to sustainable development. Green growth, a concept closely related to sustainable development, is generally termed as environment-friendly economic growth, where countries at the national level are to aim and promote economic growth while not sacrificing the environment rather protecting it. Despite its significance, global understanding, and pursuit of the green growth concept remains substantially limited. While all economies need to prioritize green growth, it is more important for emerging or high-growth economies, particularly in the context of sustainable development goals (SDGs). A "faster growth" focus in emerging economies often comes with minimal or no environmental protection (Barua, 2021). As one of the key impediments of green growth, these economies heavily rely on fossil fuels, and nonrenewable energy consumptions, which impact the environment in several ways, for example, pollution, depletion of natural resources, and in the long run contributing to climate change causes global damage on economic activities and lives (Nathaniel et al., 2020; Nathaniel et al., 2021; Barua and Valenzuela, 2018). It is true that energy consumption is a key driver of economic growth of countries alongside other factors (Rahman et al., 2019; Rana and Barua, 2015;

Energy-Growth Nexus in an Era of Globalization. https://doi.org/10.1016/B978-0-12-824440-1.00003-5

Barua and Nath, 2021). But an unsustainable pattern of energy consumption would only threaten the future of the environment and the earth. All considered, the traditional approach of economic growth being pursued by economies globally, particularly the emerging ones, appears to contradict the philosophy of the SDGs set to be achieved by 2030. To achieve sustainable development, sustainable energy transition is essential. It requires gradual replacement of nonrenewable fossil fuel–based energy (e.g., oil, coal, and natural gas) production and consumption with renewable energy (e.g., solar, wind, and hydro).

This chapter reviews the current state of green growth and energy transition in the context of emerging economies. To do so, the chapter considers 10 key emerging economies of the world that are most cited in different available rankings and evaluations. The chapter analyzes and evaluates the trends and patterns of green growth and energy-related indicators that contribute to green growth in the context of the selected economies. As such, the chapter explores how the selected emerging economies perform in terms of energy transition that helps drive green growth. The availability of consistent and continuous time series data on green growth and related energy variables across countries is fairly limited. To support the overall discussion, this chapter analyses the best possible and most reliable publicly available data from the World Bank, the Organization for Economic Cooperation and Development (OECD), and the Global Green Growth Institute (GGGI), covering the period from 1990 to 2018. The chapter is organized in six sections: Section 2 reviews the definition and measurements of green growth and highlights the energy-related measures that are considered as an integral component of green growth; Section 3 provides a brief overview of the progress of global green growth and the related energy indicators; Section 4 elaborates on the trends and patterns on green growth and the related energy indicators in the context of the selected economies; Section 5 highlights the key challenges lying ahead for energy transition in emerging economies; followed by a conclusion in Section 6.

2. The meaning and measures of green growth and energy consideration thereof

The Economic and Social Commission for Asia and the Pacific (UNESCAP, 2020a,b) defines "green growth" as a process of achieving economic growth in order to reduce poverty and achieve without worsening environmental resource constraints and climate crisis (Barua et al., 2020). It could be viewed as complementary to achieving sustainable development, as environmental sustainability is given prime consideration while achieving economic growth and development. According to OECD (2011), green growth refers to the process of achieving economic growth and sustainability, ensuring that natural assets continue to provide the resources and environmental services on which our well-being relies. To achieve green growth, investment and innovation need to be catalyzed that drive sustained growth and create new economic opportunities. The GGGI defines green growth as a development approach that seeks to deliver economic growth that is both environmentally sustainable and socially inclusive. It seeks economic growth opportunities that are low-carbon and climate-resilient, prevents or remediate pollution, maintains healthy and productive ecosystems, and create green jobs, reduce poverty, and enhance social inclusion (GGGI, 2017). Green growth aims to turn resource constraints and the climate crisis into economic growth opportunities through investing in economic growth and well-being while using fewer resources and generating fewer emissions in the important domains of food production, transport and mobility, construction and housing, heavy industry, energy, and water (UNESCAP, 2013). The World Bank describes green growth as the pursuit of economic growth that is environmentally sustainable and creates a synergy among three aspects of sustainability—social, economic, and environment (The World Bank, 2012). In similarity to green growth, the United Nations Environment Program (UNEP) uses a slightly different term "green economy." It defines it as an economy that supports the growth of income and human well-being, while minimizing environmental risk and ecological scarcities (UNEP, 2011).

Several efforts are underway to develop green economic or growth performance measures applicable at the global, regional, and country levels. At the global level, the most discussed and well-recognized three green growth measures are those developed by the GGGI, the OECD, and the UNESCAP. Although the three measures have a fair degree of similarity, they appear to be noticeably heterogeneous in terms of the indicators and components considered and the methodology used. Tables 14.1−14.3 present the broad components, subcomponents under each broad component, and the summary of indicators under each subcomponent used for the three measures. The GGGI publishes an overall Green Growth Index (GGI) score and performance level rating based on its component-wise scores assigned based on indicator-wise data collected. The OECD (2014) and the UNESCAP (2013) record and track performance by indicator but do not construct any component or subcomponent-wise index or score. The tables clearly show that all three measures mainly focus on environmental and natural resources sustainability. However, there are some fundamental differences. For example, the GGGI considers social inclusion, while the other two do not; green economic opportunities are measured differently by the GGGI and the OECD; similarly, the OECD indicators have a more detailed break into subcomponents, and a wide range of indicator converge for each subcomponent, which is not the case for the other two; the UNESCAP indicators capture equitable distribution and access, structural transformation, and human rights issues which the other two do not cover. Overall, the tables clearly illustrate that although the aim is more or less the same across the three measures—measuring green economic growth and progresses, they have a significant degree of heterogeneity in methodology.

TABLE 14.1 Green growth measurement indicators recorded by Organization for Economic Cooperation and Development.

Broad component	Subcomponent	Coverage of indicators
Environmental and resource productivity	CO_2 productivity	Productivity, intensity, total emissions, emissions from air transports
	Energy productivity	Productivity, intensity, total energy supply, renewable energy supply, renewable electricity, energy consumption in agriculture, services, industry, transport, and other sectors
	Nonenergy material productivity	Productivity, biomass, nonmetallic minerals, and metals
Natural asset base	Freshwater resources	Permanent and seasonal surface water, conversion of permanent water to nonwater and seasonal water surface and vice versa
	Land resources	Lateral, seminatural, bare, and cropland, artificial surfaces, water, total built-up area availability, and changes
	Forest resources	Forest availability, resource stock, and under sustainable management certification
	Atmosphere and climate	Annual surface temperature changes
Environmental dimension of quality of life	Exposure to environmental risks	Population exposure to PM2.5, more than 10 and 30 $\mu g/m^3$, mortality and welfare cost of premature mortalities due to exposure to ambient 2.5, ozone, lead, and residential radon
	Access to drinking water and sewage treatment	Population with access to improved drinking water sources
Economic opportunities and policy responses	Technology and innovation: Patents	Development of environment-related technologies and their relative advantage
	Environmental taxes and transfers	Environmental tax revenues, feed-in tariff for renewables, and supports to and prices of fossil fuels and renewables

Continued

TABLE 14.1 Green growth measurement indicators recorded by Organization for Economic Cooperation and Development.—cont'd

Broad component	Subcomponent	Coverage of indicators
Socioeconomic context	Economic context	Real GDP and value added in industry, agricultural, and services
	Social context	Population, population age and gender distribution, fertility rate, life expectancy, migration, and population density

Source: OECD, 2017. Green Growth Indicators 2017. OECD Green Growth Studies. OECD Publishing, Paris. https://doi.org/10.1787/9789264268586-en.

TABLE 14.2 Green growth indicators used by Global Green Growth Institute to calculate Green Growth Index.

Broad category	Coverage of indicators
Efficient and sustainable resource use	Primary energy supply, renewables consumption, water efficiency, freshwater availability changes, organic soil carbon content, organic agriculture, domestic material consumption, and material footprint
Natural capital protection	Air pollution through PM2.5, total CO_2 emissions, agricultural non-CO_2 emissions, mortality from unsafe water, waste generation, forest, protected, biodiversity area availability, soil biodiversity, red list index, and tourism and recreation level activity.
Green economic opportunities	Adjusted net savings, exports of environmental goods, green employment, and environmental technology patents
Social inclusion	Population with access to safe water and sanitation and its urban-rural distribution, electricity and clean fuels, cellular and fixed internet subscriptions, women in parliament, and female account holders in financial institutions, equal gender pay, income inequality, youth not in education or employment, population receiving a pension, healthcare access and quality, and urban slum population.

Source: Acosta, L., Maharjan, P., Peyriere, H., Galotto, L., Mamiit, R., Ho, C., et al., 2019. Green Growth Index: Concepts, Methods and Applications. Global Green Growth Institute. Retrieved from http://greengrowthindex.gggi.org/wp-content/uploads/2019/12/Green-Growth-Index-Technical-Report_20191213.pdf.

TABLE 14.3 Green growth measurement indicators recorded by UNESCAP.

Headline indicators	Coverage of indicators
Equitable distribution and access	Distribution of access to water resources and quality, food security, energy, environmental footprints, clean air, ecosystem services, and distribution of burden of degradation, and land.
	Institutional and policy support for inclusion and participation in environmental decision making and justice, distribution of tax incomes, land rights, environmental reporting, and human rights.
Structural transformation	Production, value-added contributions, and trade of environmental goods and services, sustainable public procurement, green jobs, renewable energy use, green buildings, greenhouse gas (GHG) emissions, and material footprint.
	Institutional and policy support for social, green technology, education, training, research, and other innovation
Eco-efficiency	Resources efficiency, GHG emissions intensity, material consumption, waste generation, household water and energy use, waste recycling rates, renewable energy shares, and energy used in transports.
	Institutional and policy supports (e.g., subsidies) for efficiency/productivity improvement, including energy pricing and taxation, carbon pricing, savings, eco information disclosure, and corporate social responsibility
Investment in natural capital	Natural capital stocks and natural resources flow such as mineral energy, land, soil, water, biological, timber, and aquatic resources
	Policy and institutional support, legal consideration, and taxation on externalities of ecosystem service
Planetary limits	Renewables, nonrenewables, and emission limits and targets at regional, subregional, national, and or subnational levels
	Institutional and policy support for science-policy interface and stakeholder involvement in setting limits and targets, monitoring and feedback mechanisms

Source: UNESCAP, 2013. Green Growth Indicators: A Practical Approach for the Asia and the Pacific. UNESCAP. Retrieved from https://www.unescap.org/sites/default/files/publications/GGI_2014.pdf.

Furthermore, the OECD (2020) among the three provides relatively more consistent and continuous time series data on its indicators across a large number of countries, which is much helpful in evaluating trends and patterns of green growth. In addition, Table 14.4 shows that among the three, the OECD has a much wider range of energy-related indicators. Given this chapter's aim, the OECD (2020) database offers the most publicly available information about green growth and energy transition globally.

3. Global green growth and energy transition at the world level

The most updated green growth measure available is the GGI developed and published by the GGGI. The overall index (GGI) is scored and rated, combining scores assigned across three major constituents as explained in the earlier section—Efficient and sustainable resource use, Natural capital protection, Green economic opportunities, and Social inclusion. The index evaluates a total of 207 countries across each constituent, while the overall index score is available for 115 countries due to data limitations. Fig. 14.1 shows that according to the 2019 ranking, only 20% of the 115 countries ranked have a high performance, while about 35% have low to a very low level of green growth performance. Furthermore, most countries yet to have a moderate level of performance. Table 14.5 shows the top five counties for each of the five regions considered in the GGI. In general, Europe and Asia appear to lead the league as countries in the two regions have a significantly higher overall score than others. On the other hand, Africa is the poorest performer, according to the index scores. However, looking at the "level of performance" rating assigned by the GGGI, the countries across all regions except Europe perform only moderately. This means that even if tops the regional ranking does not mean it is doing great; rather the level of performance provides the information on its true performance. As such, the moderate level of performance of the toppers across regions except Europe would mean the other countries in those regions perform even lower. Overall, in a global context, Europe is the leader in green growth performance.

It is worth-reiterating that while all available measures of green growth discussed earlier include energy-related indicators, the OECD records and tracks the broadest range of energy-related indicators over time and across a large number of countries. The consideration of clean energy in green growth is paramount. Higher green growth-performing countries are generally likely to do better in terms of sustainable or green energy transition. A transition toward clean and sustainable energy means gradually increasing production, supply, and renewable energy consumption, replacing the nonrenewables or fossil fuels (Leach, 1992; Schellnhuber, 2001). This means a rebalancing of the energy mix, where renewable energy would take the whole or the majority share. However, that may not be the case when the relationship is examined

TABLE 14.4 Key energy-related indicators in the available green growth measures.

Organization for Economic Cooperation and Development Green Growth Indicators (OECD, 2017)	Green Growth Index Index 2019 (GGGI, 2020)	UNESCAP (2013)
−Energy productivity, gross domestic product (GDP) per unit of total primary energy supply (TPES) −Energy intensity, TPES per capita −Total primary energy supply, index 2000 = 100 −Total primary energy supply −Renewable energy supply, % TPES −Renewable electricity, % total electricity generation −Energy consumption in agriculture, % total energy consumption −Energy consumption in services, % total energy consumption −Energy consumption in industry, % total energy consumption −Energy consumption in transport, % total energy consumption −Energy consumption in other sectors, % total energy consumption −Energy-related tax revenue, % total environmental tax revenue −Energy-related tax revenue, % total environmental tax revenue	−Emissions priced above EUR 30 per tonne of CO_2, % total emissions −Petrol end-user price, USD per liter diesel end-user price, USD per liter residential electricity price, USD per kWh −Industry electricity price, USD per kWh mean feed-in tariff for solar photovoltaics electricity generation −Mean feed-in tariff for wind electricity generation −Fossil fuel consumer support, % total tax revenue −Fossil fuel consumer support, % energy-related tax revenue −Fossil fuel consumer	−Ratio of total primary energy supply to GDP (MJ per $2011 PPP GDP) −Share of renewables to total final energy consumption (percent)	−Access to energy −Energy footprints to the environment −Renewable energy use −Nonrenewable energy use −Mineral energy stock

Continued

TABLE 14.4 Key energy-related indicators in the available green growth measures.—cont'd

Organization for Economic Cooperation and Development Green Growth Indicators (OECD, 2017)	Green Growth Index Index 2019 (GGGI, 2020)	UNESCAP (2013)
−Emissions priced above EUR 60 per tonne of CO_2, % total emissions	support, % total fossil fuel support −Fossil fuel producer support, % total fossil fuel support −Fossil fuel general services support, % total fossil fuel support −Petroleum support, % total fossil fuel support −Coal support, % total fossil fuel support −Gas support, % total fossil fuel support −Electricity support, % total fossil fuel support	

Source: Author developed based on OECD, 2017. Green Growth Indicators 2017. OECD Green Growth Studies. OECD Publishing, Paris. https://doi.org/10.1787/9789264268586-en, Acosta, L., Maharjan, P., Peyriere, H., Galotto, L., Mamiit, R., Ho, C., et al., 2019. Green Growth Index: Concepts, Methods and Applications. Global Green Growth Institute Retrieved from http://greengrowthindex.gggi.org/wp-content/uploads/2019/12/Green-Growth-Index-Technical-Report_20191213.pdf. at GGGI, and UNESCAP, 2013. Green Growth Indicators: A Practical Approach for the Asia and the Pacific. UNESCAP. Retrieved from https://www.unescap.org/sites/default/files/publications/GGI_2014.pdf.

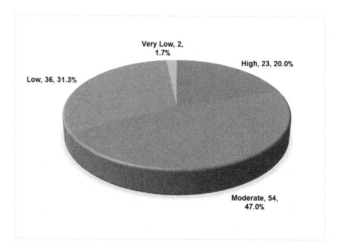

FIGURE 14.1   Distribution of countries by level of performance at GGGI. *Source: Author developed based on ranking by Acosta, L., Maharjan, P., Peyriere, H., Galotto, L., Mamiit, R., Ho, C., et al., 2019. Green Growth Index: Concepts, Methods and Applications. Global Green Growth Institute Retrieved from http://greengrowthindex.gggi.org/wp-content/uploads/2019/12/Green-Growth-Index-Technical-Report_20191213.pdf. at GGGI.*

between a countries actual economic growth and the status of energy transition. At the world level, economic growth is generally presumed to be gray rather than green. Fig. 14.2A and B show the relationship between the growth of the world economy and the share of renewable energy in total energy consumption at the world aggregate level. The figures clearly show that the share of renewable energy consumption is negatively associated with global economic growth. As the world economy grows faster, renewable consumption relative to nonrenewables declines. In other words, the figures reflect the increasing level of dependency of global economic growth on the increased use of fossil fuel or nonrenewable energy sources, which directly contradicts the concepts and aims of green growth and sustainable energy transition.

While global growth depends heavily and increasingly on nonrenewable fossil fuels, global energy efficiency patterns also do not offer much hope. Fig. 14.3A shows that alongside increases energy productivity since 2011, energy intensity also shows an increasing trend. An increased energy intensity over time indicates an increasingly larger amount of energy consumption in order to produce one economic unit of products (e.g., GDP). The figure also suggests a higher level of volatility in the energy intensity pattern, which perhaps reflects the global economy's inability to ensure consistent improvement in energy efficiency.

TABLE 14.5 Top five countries with the highest Green Growth Index score by region.

Region	Country	Score	Level of performance	Regional rank
Europe	Denmark	75.32	High	1
	Sweden	75.09	High	2
	Austria	72.32	High	3
	Finland	71.69	High	4
	Czech Republic	71.29	High	5
Asia	Singapore	58.43	Moderate	1
	Malaysia	55.88	Moderate	2
	Philippines	55.54	Moderate	3
	Georgia	55.45	Moderate	4
	China	55.41	Moderate	5
The Americas	Dominican Republic	55.10	Moderate	1
	United States	54.22	Moderate	2
	Canada	54.04	Moderate	3
	El Salvador	53.94	Moderate	4
	Mexico	52.71	Moderate	5
Oceania	New Zealand	52.17	Moderate	1
	Australia	47.89	Moderate	2
	Fiji	45.48	Moderate	3
Africa	Botswana	45.88	Moderate	1
	Tanzani	44.32	Moderate	2
	Mauritius	42.63	Moderate	3
	Morocco	42.61	Moderate	4
	Ghana	42.42	Moderate	5

Source: Green Growth Index (2019).

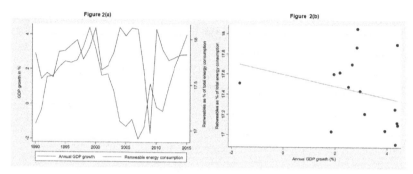

FIGURE 14.2   World economic growth and renewable energy consumption. *Source: The World Bank, 2020. World Development Indicators. Retrieved 22 June 2020 from http://datatopics. worldbank.org/world-development-indicators/.*

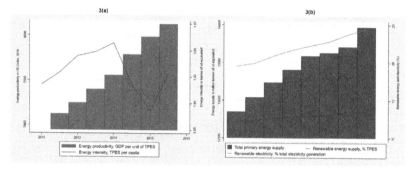

FIGURE 14.3   Energy supply, efficiency, and the share of renewables at the world level. *Source: Author developed based on OECD, 2020. Green Growth Indicators. OECD. Retrieved 30 June 2020 from https://stats.oecd.org/Index.aspx?DataSetCode=GREEN_GROWTH data.*

However, the increased dependency of global growth on nonrenewables and increased level of energy inefficiency should not be interpreted as the silence of the world economy in transitioning toward sustainable energy sources. Over the last decade, countries across the world and supranational agencies have introduced and adopted policies and regulations and directed a significant amount of investments to accelerate renewables production in the energy mix (BNEF, 2020). According to BNEF (2020), $363.3 billion are invested toward fostering renewables production across all regions globally only in 2019, where Europe and China take the lead. As an outcome of such investments over a period, an increased level of production of renewable energy is evident. As a reflection of growing investment in renewable energy generation, Fig. 14.3B shows that the share of renewable energy production and supply, such as renewable electricity, has consistently increased since 2010 with the increases in total primary energy supply (TPES). Nevertheless, these trends offer some hope about the global efforts and priority in fostering energy transition and aligning the traditional energy mix toward sustainable sources.

Achieving a sustainable energy transition requires aligning key economic activities with renewable energy. The OECD (2020) key economic activity-wise energy consumption data from 2010 to 2017 shows that Industry and Transport are the major consumers of energy sharing close to 30% each, while Services and Agriculture sector accounts for only about 8% and 2%, respectively. The share of each economic activity was at a consistently similar level over the 8 years. The data highlight that sustainable energy transition should pay the most attention to industrial activities and transports and communication activities. In general, a sustainable energy transition in the relevant sectors can be achieved once gradual improvements are made by introducing innovative energy and environment-related technologies supported by long-term oriented policy and regulatory frameworks. As such, a twofold approach is needed: (1) introducing new and innovative technologies that replace the existing less-efficient nonrenewable energy-related technologies and (2) bringing innovation in renewable technologies that make them cheaper and more efficient than before and relative to nonrenewable technologies.

Continuous technological advancement in any productive sector can improve energy efficiency and productivity and reduce the cost and adverse impacts of energy production, such as air pollution. Fossil fuel consumption in almost all economic activities is considered the primary source and driver of air pollution globally. They emit a massive amount of pollutants like carbon dioxide (CO_2) every day around the world. Fig. 14.4A shows that both total production and demand based carbon (CO_2) emissions are consistently on the rise since 2010, as global economic activities and growth expands. While global carbon intensity declined from 2014 to 2016 from their rising trends in 2013, it shows signs of a rebound after 2016. Fig. 14.4A overall shows the danger of overreliance on fossil fuels or nonrenewable energy.

On the other hand, Fig. 14.4B shows that the development of environment-related technologies consistently drops from 2010 to 2016 in both per capita and as % of all technologies measures. Technological innovation in the energy sector is key to make renewable energy more affordable and durable. The figure indicates that while the global economy relies heavily on fossil fuels and carries environmental harms such as pollution, environment-related technology development takes keeps declining. The trend directly conflicts with the concepts and the dreams of green growth supported by a sustainable energy transition.

4. The patterns of green growth and related energy development in emerging economies

As energy is the fuel for economic activities and growth, energy consumption from sustainable sources is necessary to achieve green growth. Meeting the

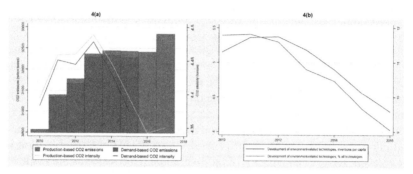

FIGURE 14.4 CO_2 emissions indicators and technological development at the world level. *Source: Author developed based on OECD, 2020. Green Growth Indicators. OECD. Retrieved 30 June 2020 from https://stats.oecd.org/Index.aspx?DataSetCode=GREEN_GROWTH data.*

rising energy demand and achieving universal access while ensuring an energy sector development that is environmentally friendly, socially sound, and economically feasible presents a significant challenge (UNESCAP, 2020a,b). This challenge is acute in emerging economies—economies that currently grow or are likely to grow at a faster rate in the near future (Manibog, 2017). As such, emerging economies are those countries that are considered to be the future engine and drivers of global economic growth (IMF, 2019). The challenge of shifting the energy mix to sustainable sources while fueling faster economic growth is a daunting task in these economies. This is because the demand for energy keeps rising at a faster rate to support higher growth in these economies, while investment in renewable energy and energy-related technology stands significantly low due to the economies' secondary priority to it and lack of macrofinancial capacity (Cubeddu et al., 2014; International Finance Corporation, 2016). Brining a paradigm shift in the energy mix by increasing the share of sustainable or renewable energy requires a considerable investment, which often the emerging economies either cannot afford or do not intend to mobilize since the investments could alternatively be utilized for accelerating economic growth (Brunnschweiler, 2010).

There is no global consensus about which economies are "absolutely" the emerging economies, i.e., there is no exclusive single list of countries that are considered the ultimate one by all. Emerging economies are also naively termed as emerging markets sometimes. Many supranational agencies, multinational financial institutions, and international rating agencies publish their own list of emerging economies, where countries are included based on the economic fundamentals and growth potential in the next few decades relative to other economies. A thorough review of several such lists published by the IMF (2015), FTSE (2014), MSCI (2015), HSBC (2018), JP Morgan (2016), S&P Dow Jones (2018), and Columbia University (2015) finds that the lists of emerging economies offer some degree of commonality in terms of

TABLE 14.6 List of selected emerging economies.

Country	Region	Economic level	Green Growth Index score	Green Growth performance level	Green Growth Index regional rank
Poland	Europe	UMIC	62.00	High	23
Philippines	Asia	LMIC	55.54	Moderate	3
China	Asia	UMIC	55.41	Moderate	5
Brazil	South America	UMIC	49.82	Moderate	8
Morocco	Africa	LMIC	42.61	Moderate	4
India	Asia	LMIC	40.81	Moderate	9
Indonesia	Asia	LMIC	40.81	Moderate	16
Turkey	Asia	UMIC	39.22	Low	18
Vietnam	Asia	LMIC	39.05	Low	19
Bangladesh	Asia	LMIC	[a]	[a]	[a]

LMIC, lower middle-income country; UMIC, upper middle-income country
[a] overall score for Bangladesh is not available at GGGI (Acosta et al., 2019); the table is sorted by Green Growth Index score.
Source: Author developed.

which countries are included, while the number of countries listed remains ranges from 16 to 27. To have an understanding about the status of energy transition that contributes to green growth in emerging economies, 10 countries are selected in this chapter as shown in Table 14.6 that are common across all the lists mentioned above of emerging economies. The inclusion of these countries in all the lists considered indicates a consensus and agreement of different agencies and institutions about their economic potential.

Furthermore, the countries selected represent a diverse regional background such as Asia, South America, Europe, and Africa, although the majority comes from Asia. This is consistent with many studies suggesting that Asia is likely to lead the world economy in the next few decades (IMF, 2017; Romei, 2020). Out of the list in Table 14.6, six countries are classified as lower middle-income countries (LMICs) and four are upper middle-income countries (UMICs). Table 14.6 also presents the GGI score and rating for each of the economies. Looking at the GGI indicators, Poland stands as the high green growth-performing country of all, while six countries fall in a moderate level of performance and Turkey and Vietnam perform poorly. Only one emerging economy is rated as "High." The GGI score and performance levels indicate that most emerging economies still have a long way to achieve real green growth.

This chapter focuses on how the selected economies perform in terms of the energy transition-related indicators included in the GGIs available globally. In other words, analyzing the energy-related constituent of different GGIs would provide information about how the selected economies perform in terms of energy transition that contribute to their green growth. Table 14.7 shows the overall energy transition situation of the selected economies in 2020. The Energy Transition Index (ETI) published by the World Economic Forum (2020) suggests a significantly poorer transition performance of the economies, as eight out of the 10 economies fall above a 50, with overall ETI scores rounding about less than 60% for all countries. As another indicator of the economies' poor performance, the transition readiness score in the index for the countries stands either below or just over 50%.

A poor energy transition performance would mean a lower utilization of renewables in the energy portfolio. Fig. 14.5 shows the trends of economic growth and renewable energy consumption from 2000 to 2015. Over the period, while economic growth has been relatively higher in China, India, and Bangladesh, growth trends appear to be volatile, particularly for Turkey, Morocco, the Philippines, and Poland. Regardless of the growth pattern, the

TABLE 14.7 Energy transition readiness, selected economies.

Country	System Performance (%)	Transition Readiness (%)	2020 ETI score (%)	2020 ETI rank[a]
Malaysia	64	55	59.45	38
Brazil	69	46	57.90	47
Morocco	61	51	56.50	51
Philippines	62	49	55.30	57
Vietnam	57	50	53.50	65
Poland	57	48	52.90	69
Indonesia	61	44	52.40	70
India	54	49	51.50	74
China	50	52	50.90	78
Bangladesh	54	43	48.40	87

[a]*a higher ETI rank indicates poorer performance and vice versa.*
Source: Energy Transition Index (ETI) 2020 World Economic Forum, 2020. Fostering Effective Energy Transition 2020 Edition. Insight Report. Retrieved 14 June 2020 from http://www3.weforum.org/docs/WEF_Fostering_Effective_Energy_Transition_2020_Edition.pdf.

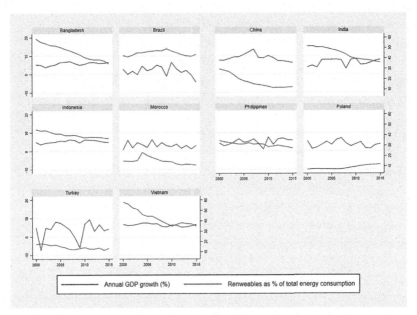

FIGURE 14.5 Economic growth and renewable energy consumption in selected emerging economies. *Source: Author developed based on The World Bank, 2020. World Development Indicators. Retrieved 22 June 2020 from http://datatopics.worldbank.org/world-development-indicators/.*

share of renewable energy in total energy consumption has declined consistently across the economies except for Poland. Countries like Bangladesh, Vietnam, India, China, and the Philippines have seen a steeper decline in renewable energy consumption over the period. However, the contribution of renewables in energy consumption remains historically low in Poland, Turkey, and Morocco (less than 15%). Overall, the patterns across the economies indicate that despite being tagged as an emerging economy, historical economic growth patterns remain noticeably volatile, and renewable energy consumption continues to decline in these economies. In fact, the share of renewables consumption stands at a significantly low level in most of the economies. This trend is inconsistent with the concepts of green growth and sustainable energy transition.

4.1 Energy supply, productivity, and efficiency in the selected economies

Fig. 14.6 shows energy supply trends in the selected countries from 1990 to 2018. In line with the observations made earlier, Fig. 14.6 shows a declining renewable energy supply trends in the selected economies. TPES increases

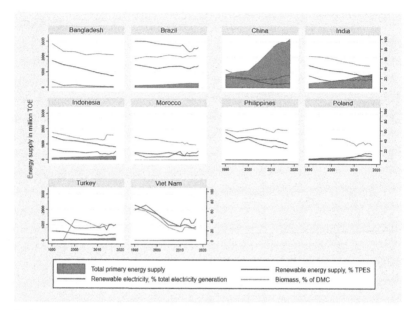

FIGURE 14.6   Energy supply in selected emerging economies. *Source: Author developed based on OECD, 2020. Green Growth Indicators. OECD. Retrieved 30 June 2020 from https://stats.oecd. org/Index.aspx?DataSetCode=GREEN_GROWTH data.*

over time, with the largest increase happened in China and India, followed by Brazil and Indonesia. However, the supply of renewable energy, particularly renewable electricity, as % of TPES declines over the period across economies except for Poland. A faster rate of decline is evident for renewable energy share, particularly in India, Bangladesh, the Philippines, and Vietnam. This is entirely in line with the declining trends of renewable energy consumption discussed in the previous section. Consistent with the renewable energy supply patterns, the share of biomass consumption—organic materials often used for power generation—out of total domestic material consumption also follows a declining trend.

Fig. 14.7 shows that both energy productivity and intensity continues to increase in the selected economies. While increased energy productivity is good news for the countries except for Brazil, increased energy intensity is not. An increasing energy intensity level over the period indicates an increasingly larger amount of energy is consumed to produce one unit of economic output commonly measured as GDP. In other words, increased energy intensity means a higher cost of converting energy into economic output, which reflects the rising level of energy inefficiency in the selected emerging economies. All considered, while the economies grow or aim to grow faster, their energy efficiency remains historically poor and continues to worsen over time. The pattern is inconsistent with the concepts of sustainable energy-driven green growth.

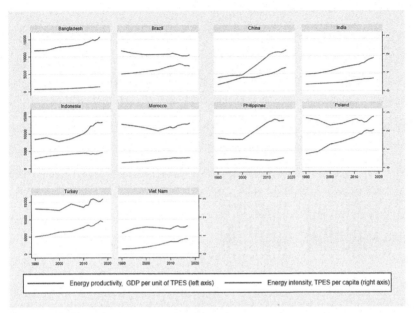

FIGURE 14.7   Energy productivity and intensity in selected emerging economies. *Source: Author developed based on OECD, 2020. Green Growth Indicators. OECD. Retrieved 30 June 2020 from https://stats.oecd.org/Index.aspx?DataSetCode=GREEN_GROWTH data.*

4.2 Energy consumption, prices, and taxes

Fig. 14.8 shows that Transport and Industrial sectors are the largest consumer of energy in the economies, and the consumption share, particularly of the Transports sector continues to increase over time. Across the economies, the consumption share of Industry in most economies remain at a similar level or decline slightly, while in all economies Transport sector's share expands noticeably. The patterns indicate the growing significance of transport and communication infrastructure to support faster economic growth in the emerging economies.

Countries across the world often apply different policy and pricing mechanisms to influence the levels of production, supply, and consumption of both renewable and nonrenewable energies (Polzin et al., 2015; Zamfir et al., 2016). Fig. 14.9 shows the energy taxation and pricing trends in the selected economies. For many of the countries, data on end-user price and taxation are not available. Based on wherever data available, it is clearly evident that end-user prices of nonrenewable energy or fossil fuels follow a declining trend in recent years in all economies. In Indonesia, Turkey, and Poland, petrol and diesel prices increased up to around 2010 and show a declining trend after that. Generally, a key driver of such price declining trends is a reduction in energy-related taxes. Fig. 14.9 shows that for countries where data are available,

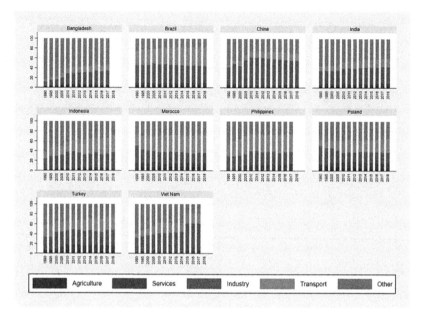

FIGURE 14.8   Energy consumption by economic sectors in the selected emerging economies. *Source: Author developed based on OECD, 2020. Green Growth Indicators. OECD. Retrieved 30 June 2020 from https://stats.oecd.org/Index.aspx?DataSetCode=GREEN_GROWTH data.*

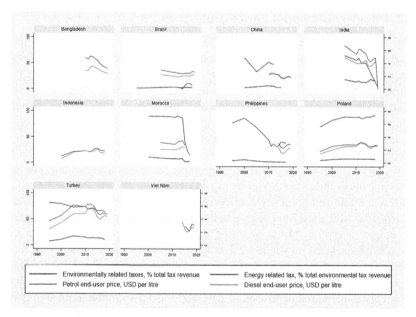

FIGURE 14.9   Taxes and prices of energy in selected emerging economies. *Source: Author developed based on OECD, 2020. Green Growth Indicators. OECD. Retrieved 30 June 2020 from https://stats.oecd.org/Index.aspx?DataSetCode=GREEN_GROWTH data.*

energy and environment-related tax revenue nosedives over time for all countries except Poland. Overall, the figure indicates that fossil fuels such as petrol and diesel have become cheaper over time with lower levels of taxes and prices. While this trend would help accelerate economic growth through enabling industries and economic activities to consume more fossil fuel—based energy at a lower price, it would nonetheless be detrimental to the environment and sustainable development.

Government policies often influence taxes and prices, intending to alter energy production and consumption patterns (Anwar, 2011). Often, they include energy support given directly or indirectly to consumers and producers. The OECD (2020) data show that governments in most of the economies considered in this chapter mostly provide consumer price supports directly or indirectly that stands as high as 98% as of 2019. Consumer supports are generally aimed at enabling consumers to purchase energy cheaper and boost economic activities (Keyuraphan et al., 2012). For example, in many emerging economies, most of the population are poor or low-income households, for whom the standard prices of energy (such as electricity) would be highly expensive. In such instances, governments in countries such as Bangladesh and India allow consumers to pay government-subsidized prices for energy such as electricity (IISD, 2012; Garg et al., 2020).

4.3 Fossil fuel supports in the selected economies

Based on the available data for six countries at OECD (2020), Fig. 14.10 shows the patterns and distribution of supports to nonrenewable fossil fuel—based energy. A relatively higher level of support was given historically in China, Indonesia, and India until around 2013, which begins to decline consistently in the following years. China's case is particularly noticeable since total fossil fuel support reached as high as 2000% of tax revenue, thanks to massive government subsidies and policy supports. Even at the latest, the country's support stands at over 500% in 2019. In the other three countries, fossil fuel supports remain relatively lower, just around 5% of total tax revenue.

The OECD (2020) data show that petroleum receives the majority of the supports across all countries (about 20% in Poland to 98% in India and China) as of 2019, except the fact that Poland particularly shows an increasing level of support for Coal (roughly 75%) from 2005 to 2019. While Turkey used to support coal earlier until 2013, it appears to shift majority supports gradually. On the other hand, Indonesia shows increased support for electricity since 2010, while India gives sudden enormous support for natural gas (over 350%) in 2019. Overall, it can be deduced that the emerging economies keep increasingly supporting fossil fuels over time.

In contrast to the supports for nonrenewables, how do the countries perform supporting renewable energy? Available data do not offer much hope. Available data for four of the selected economies presented in Fig. 14.11 show

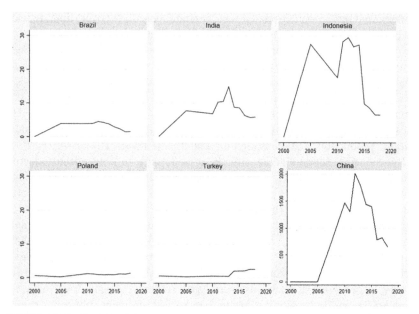

FIGURE 14.10 Total fossil fuel support in the selected economies, as % of total tax revenue. *Source: Author developed based on OECD, 2020. Green Growth Indicators. OECD. Retrieved 30 June 2020 from https://stats.oecd.org/Index.aspx?DataSetCode=GREEN_GROWTH data.*

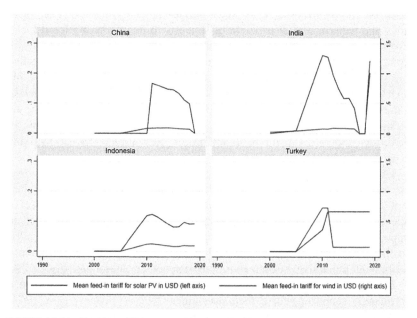

FIGURE 14.11 Feed-in tariff for renewables in selected emerging economies. *Source: Author developed based on OECD, 2020. Green Growth Indicators. OECD. Retrieved 30 June 2020 from https://stats.oecd.org/Index.aspx?DataSetCode=GREEN_GROWTH data.*

that the average feed-in tariff in USD for solar photovoltaics and wind consistently declines for China and Indonesia. At the same time, it remains flat for Turkey in recent years. There is an increasing level of feed-in tariff supports offered for renewables in India only. The trends overall suggest that compared to nonrenewables, renewable energy supports are broadly not encouraging in the selected economies; instead, they tend to decline over time. The pattern is essentially counteractive to sustainable energy transition and green growth efforts.

4.4 Reliance on fossil fuel drives up pollution

Increased reliance on and supports fossil fuel consumption contributes to increasing pollution and damages the environmental quality. Fig. 14.12 shows an increased pattern of CO_2 emissions in the selected economies over the period from 1990 to 2019. Carbon (CO_2) emissions have substantially grown in China and India, mainly driven by massive industrial and infrastructure development driven fossil fuel—based energy consumption. Energy-related emissions intensity shows an increasing trend in all the economies, demonstrating the rising degree of energy inefficiency in these economies. However, in some of the economies, productivity shows an uptrend, which indicates an increased level of output generated per unit of energy-related CO_2 emitted.

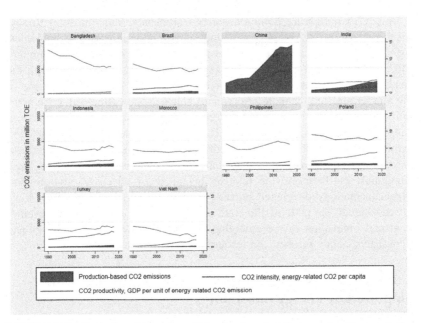

FIGURE 14.12 Production driven air pollution in the selected economies. *Source: Author developed based on OECD, 2020. Green Growth Indicators. OECD. Retrieved 30 June 2020 from https://stats.oecd.org/Index.aspx?DataSetCode=GREEN_GROWTH data.*

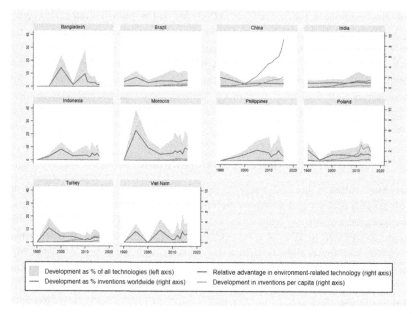

FIGURE 14.13 Development of environment-related technology. *Source: Author developed based on OECD, 2020. Green Growth Indicators. OECD. Retrieved 30 June 2020 from https://stats.oecd.org/Index.aspx?DataSetCode=GREEN_GROWTH data.*

A key to mitigating pollution, energy efficiency, and transition to sustainable energy are developing and adopting new and innovative technologies. Fig. 14.13 suggests a broadly increasing trend in developing environment-related technology compared to the 1990s across all measures, although the trends are somewhat declining in recent years. In general, the share of environment-related technologies remains below 20% in almost all economies with a relatively highly volatile pattern in most economies—particularly in Bangladesh, Vietnam, and Morocco. These patterns indicate a relatively lower priority and inconsistency (possibly driven by inconsistent policy efforts) of environment and energy-related technology development over time in these economies. Furthermore, the relative advantage of whatever environment-related technology developed remains poor in all economies except China. All considered, the patterns illustrate a low degree of technological development and innovation in the emerging economies, mostly due to lower and inconsistent policy priority and resources commitment.

5. Challenges and way forward to ensure green growth supported by energy transition

It is important that economies achieving higher growth align their national energy policy with the SDG7 targets (UNESCAP, 2020b). Emerging economies, including those considered in this chapter, have a wide range of challenges that prevent them from progressing toward sustainable energy transition and green growth. The challenges stem from the fact that often the two objectives of achieving higher economic growth and transitioning toward sustainable energy contradicts or at least perceived to be so. Some of the key challenges are highlighted in this section.

First, many emerging economies have a significant lack of financial and economic capacity that prevents investments in energy technologies (McMullen-Laird, 2016; Barua, 2019). Most emerging economies, particularly the LMICs, have a substantial lack of domestic financial resources, which often make them heavily reliant on foreign direct investments, overseas development assistance, and lending by supranational institutions such as the World Bank and the International Monetary Fund. *Second*, many emerging economies are economically fragile, mainly due to poor institutional, governance and market infrastructure (BIS, 1997). A fragile economic system disallows timely mobilization of adequate resources to the right causes like sustainable energy transition. *Third*, poor economic systems in emerging economies are often coupled with inefficient and underdeveloped financial ecosystems (Nilsen and Rovelli, 2001; Ahrend and Goujard, 2012). An underdeveloped and inefficient financial system does not or cannot mobilize resources to the right causes due to market failure, the lack of product and service innovation, and market depth. For example, in many emerging countries like Bangladesh, banks remain the primary sources of long-term financing, followed by equity markets. There is no market for corporate bonds, making it very difficult for many firms to raise large capital at a reasonable cost for large-sized private energy-sector projects (Rahman and Barua, 2016; Barua, 2020; (Chiesa et al., 2021)). *Fourth*, because many emerging economies' economic capacity is limited, achieving growth is taken as the first priority, and all or most of whatever resources available are direct toward achieving it. This process undermines sustainable energy's significance and its strong association with achieving environmental objectives and the SDGs. *Fifth*, many emerging economies have an underdeveloped regulatory and policy framework that systematically prevents the countries from considering sustainable energy transition and green growth (Capozza and Samson, 2019). Government ministries, offices, and policy-makers often are not aware or do not emphasize enough embedding green growth and sustainable energy considerations into the national development policies and plans. For example, the national energy policies currently being followed by the emerging economies are not adequately aligned with the SDG7 targets, which intends to ensure access to affordable, reliable, sustainable, and modern energy for all (UNESCAP, 2018). Furthermore, in many emerging economies, existing regulations do not penalize enough the fossil fuel–based polluters, particularly because doing so would discourage private investments from domestic and international sources (Oliva et al., 2019). Many industries in these economies continue consuming energy with a high level of inefficacy for years, but available regulations often have limited provisions to make them pay for the inefficiency and pollution they generate. *Sixth*, the levels of awareness and education about the significance of sustainable energy and green growth to protect the future are often low among the public, particularly the consumers and producers of energy in emerging economies (Malik and Abdullah, 2019). *Seventh*, renewable energy is considered expensive at the consumer level due to a high cost of generation that demotivates end-users and private energy producers to replace renewables with nonrenewables (Blazquez et al., 2018). *Eighth*, many emerging economies lack the ability to develop or innovate new technologies and seek technology transfer in most cases from foreign sources (Quitzow et al., 2019). The technological inability is mainly attributed to the economies' lower investments in energy and environmental technology-related research and innovation, mainly due to lower policy priority assigned to it and the lack of human and financial resources.

All considered, the challenges outlined are often interrelated and interact in a complicated way, and cannot be solved overnight. Mitigating the challenges requires a comprehensive set of continuously updated policy and regulatory efforts with a long-term orientation. A detailed energy transition roadmap coupled with a policy framework that considers environmental protection a top priority may help emerging economies to progress toward sustainable energy transition or zero-carbon economy and green growth in the long run.

6. Conclusion

Green growth in emerging economies is still a long walk to go. In order to progress and achieve green growth, the transition toward sustainable energy plays a key role. However, energy transition in emerging economies remains a daunting task. In contrast to encouraging energy transition, the world's most-cited emerging economies show rather a persistently increasing reliance on fossil fuels and a declining trend of renewable energy supply. Furthermore, the economies embrace a rising energy inefficiency level, with a greater level of support rendered to the fossil fuels. On the other hand, the share of and government supports for renewable energy consumption keeps declining in these economies. The selected economies seem to walk a path opposite to energy sustainability and green growth, mainly because of their supreme priority on achieving faster economic growth over achieving energy sustainability. In general, several challenges prevent emerging economies, including the ones considered in this chapter, from progressing toward sustainable energy-driven green growth; such as a lack of financial and economic resources and capacity, market imperfection and failure, fragility in the economic system, a lack of policy priority, research and innovation, insufficient regulatory and policy frameworks, and a lack of awareness and education among stakeholders. For most emerging economies globally, the primary focus remains on achieving higher economic growth targets every year regardless of its sustainability implications. This approach is purely detrimental to the environment and the efforts of worldwide SDG implementations. Since emerging economies are expected to see the largest growth in the next few decades and lead the world economy, it is essential that they do it sustainably to protect the future of the people and the planet. It must now be recognized by all that green growth supported by sustainable energy in emerging economies is no more a choice rather a necessity.

References

Acosta, L., Maharjan, P., Peyriere, H., Galotto, L., Mamiit, R., Ho, C., et al., 2019. Green Growth Index: Concepts, Methods and Applications. Global Green Growth Institute. Retrieved from. http://greengrowthindex.gggi.org/wp-content/uploads/2019/12/Green-Growth-Index-Technical-Report_20191213.pdf.

Ahrend, R., Goujard, A., 2012. International Capital Mobility and Financial Fragility-Part 3: How Do Structural Policies Affect Financial Crisis Risk? Evidence from Past Crises Across OECD and Emerging Economies (June 12, 2012).

Anwar, Y., 2011. Income tax incentives on renewable energy industry: case of geothermal industry in USA and Indonesia. Afr. J. Bus. Manage. 5 (31) https://doi.org/10.5897/ajbm11.421.

Barua, S., 2019. Financing sustainable development goals: a review of challenges and mitigation strategies. Bus. Strat. Dev. 1−17. https://doi.org/10.1002/bsd2.94 (Early view).

Barua, S., 2020. The Principles of Green Banking: Managing Environmental Risk and Sustainability. De Gruyter, Boston (MA).

Barua, S., 2021. Human Capital, Economic Growth, and Sustainable Development Goals: An Evaluation of Emerging Economies. In: Shahbaz, M., Mubarik, M.S., Mahmood, T. (Eds.), The Dynamics of Intellectual Capital in Current Era. Springer, Singapore. https://doi.org/10.1007/978-981-16-1692-1_6.

Barua, S., Chiesa, M., 2019. Sustainable financing practices through Green Bonds: what affects the funding size? Bus. Strat. Environ. 28 (6), 1131−1147.

Barua, S., Colombage, S., Valenzuela, E., 2020. Climate Change Impact on Foreign Direct Investment Inflows: A Dynamic Assessment at the Global, Regional and Economic Level. SSRN Electronic Journal. https://doi.org/10.2139/ssrn.3674777.

Barua, S., Nath, S.D., 2021. The impact of COVID-19 on air pollution: Evidence from global data. Journal of Cleaner Production, 126755. https://doi.org/10.1016/j.jclepro.2021.126755.

Barua, S., Valenzuela, E., 2018. Climate Change Impacts on Global Agricultural Trade Patterns: Evidence from the Past 50 Years. Proceedings of the 6th International Conference on Sustainable Development 2018, United Nations Sustainable Development Solutions Network and Columbia University, USA. In preparation.

BIS, M., 1997. Financial Stability in Emerging Market Economies. Bank for International Settlements, Basle.

Blazquez, J., Fuentes-Bracamontes, R., Bollino, C.A., Nezamuddin, N., 2018. The renewable energy policy Paradox. Renew. Sustainablr Energy Rev. 82, 1−5.

BNEF (2020), 2019. Clean Energy Investment Trends. Bloomberg New Energy Finance.

Brunnschweiler, C., 2010. Finance for renewable energy: an empirical analysis of developing and transition economies. Environ. Dev. Econ. 15 (3), 241−274. https://doi.org/10.1017/s1355770x1000001x.

Capozza, I., Samson, R., 2019. Towards Green Growth in Emerging Market Economies: Evidence from Environmental Performance Reviews.

Chiesa, M., Barua, S., 2019. The surge of impact borrowing: the magnitude and determinants of green bond supply and its heterogeneity across markets. J. Sustainable Finance Invest. 9 (2), 138−161. https://doi.org/10.1080/20430795.2018.1550993.

Chiesa, M.A., McEwen, B., Barua, S., 2021. Does a Company's Environmental Performance Influence Its Price of Debt Capital? Evidence from the Bond Market. The Journal of Impact and ESG Investing, jesg.2021.1.015. https://doi.org/10.3905/jesg.2021.1.015.

Columbia University, 2015. Emerging Market Global Players (EMGP). Accessed 2 February 2015 at. https://ccsi.columbia.edu/publications/emgp/.

Cubeddu, L., Culiuc, A., Fayad, G., Gao, Y., Kochhar, K., Kyobe, A., et al., 2014. Emerging Markets in Transition: Growth Prospects and Challenges. International Monetary Fund. Retrieved from. https://www.imf.org/external/pubs/ft/sdn/2014/sdn1406.pdf.

FTSE, 2014. FTSE Annual Country Classification Review. FTSE Group. Accessed 4 February 2015 at. https://research.ftserussell.com/products/downloads/FTSE-Country-Classification-Update_latest.pdf.

Garg, V., Viswanathan, B., Narayanaswamy, D., Beaton, C., Ganesan, K., Sharma, S., Bridle, R., 2020. Mapping India's Energy Subsidies 2020: Fossil Fuels, Renewables, and Electric Vehicles. International Institute for Sustainable Development. Retrieved from. https://www.iisd.org/sites/default/files/publications/india-energy-transition-2020.pdf.

GGGI, 2017. Accelerating the Transition to a New Model of Growth: GGGI Refreshed Strategic Plan 2015 − 2020. Retrieved from. https://www.gggi.org/.

GGGI, 2020. Green Growth Index 2019. Retrieved from. https://www.gggi.org/.

HSBC, 2018. The World in 2030. HSBC. Retrieved from. https://enterprise.press/wp-content/uploads/2018/10/HSBC-The-World-in-2030-Report.pdf.

IISD, 2012. A Citizens' Guide to Energy Subsidies in Bangladesh. International Institute for Sustainable Development. Retrieved from. https://www.iisd.org/gsi/sites/default/files/ffs_bangladesh_czguide.pdf.

IMF, 2017. Regional Economic Outlook. Asia and Pacific : Preparing for Choppy Seas. International Monetary Fund, Washington, DC. Retrieved from. https://www.imf.org/en/Publications/REO/APAC/Issues/2017/04/28/areo0517.

IMF, 2019. World Economic Outlook. International Monetary Fund, Washington. Retrieved from. https://www.imf.org/en/Publications/WEO/Issues/2019/10/01/world-economic-outlook-october-2019.

IMF, 2015. World Economic Outlook. International Monetary Fund. Accessed 7 December 2020 at. http://www.imf.org/external/pubs/ft/weo/2015/02/pdf/text.pdf.

International Finance Corporation, 2016. How Banks Can Seize Opportunities in Climate and Green Investment. EM Compass. Retrieved from. http://documents1.worldbank.org/curated/en/334501486539681923/pdf/112688-BRI-EMCompass-Note-27-Banks-and-Climate-Finance-FINAL.pdf.

Morgan, J.P., 2016. Emerging Markets Bond Index Monitor March 2016. J.P. Morgan. Accessed 1 April 2016 at. https://markets.jpmorgan.com/#research.emerging_markets.index.

Keyuraphan, S., Thanarak, P., Ketjoy, N., Rakwichian, W., 2012. Subsidy schemes of renewable energy policy for electricity generation in Thailand. Procedia Eng. 32, 440−448. https://doi.org/10.1016/j.proeng.2012.01.1291.

Leach, G., 1992. The energy transition. Energy Policy 20 (2), 116−123. https://doi.org/10.1016/0301-4215(92)90105-b.

Malik, M., Abdallah, S., 2019. Sustainability initiatives in emerging economies: a socio-cultural perspective. Sustainability 11 (18), 4893.

Manibog, Y., 2017. The challenge of investing in emerging markets clean energy: barriers to structuring viable projects. Renewable Energy Law Policy Rev. 8 (1), 7−18. Retrieved June 29, 2020, from. www.jstor.org/stable/26377516.

McMullen-Laird, L., 2016. BRICS Face $51 Billion Annual Shortfall for Clean Energy. The Diplomat. Retrieved from. https://thediplomat.com/2016/11/brics-face-51billion-annual-shortfall-for-clean-energy/.

MSCI, 2015. MSCI Emerging Market Indexes. Accessed 2 February 2015 at. https://www.msci.com/our-solutions/indexes.

Nathaniel, S., Barua, S., Hussain, H., Adeleye, N., 2020. The determinants and interrelationship of carbon emissions and economic growth in African economies: fresh insights from static and dynamic models. J. Public Aff. https://doi.org/10.1002/pa.2141. Online early view.

Nathaniel, S.P., Barua, S., Ahmed, Z., 2021. What drives ecological footprint in top ten tourist destinations? Evidence from advanced panel techniques. Environ. Sci. Pollut. Res. https://doi.org/10.1007/s11356-021-13389-5.

Nilsen, J., Rovelli, R., 2001. Investor risk aversion and financial fragility in emerging economies. J. Int. Financ. Mark. Inst. Money 11 (3−4), 443−474. https://doi.org/10.1016/s1042-4431(01)00045-2.

OECD, 2011. Towards Green Growth. OECD. Retrieved from. http://mail.icwc-aral.littel.uz/green-growth/files/oecd3.pdf.

OECD, 2017. Green Growth Indicators 2017. OECD Green Growth Studies. OECD Publishing, Paris. https://doi.org/10.1787/9789264268586-en.

OECD, 2020. Green Growth Indicators. OECD. Retrieved 30 June 2020 from. https://stats.oecd.org/Index.aspx?DataSetCode=GREEN_GROWTH.

OECD, 2014. Green Growth Indicators 2014. OECD Green Growth Studies. OECD Publishing. https://doi.org/10.1787/9789264202030-en.

Oliva, P., Alexianu, M., Nasir, R., 2019. Suffocating Prosperity: Air Pollution and Economic Growth in Developing Countries. International Growth Centre. Retrieved from. https://www.theigc.org/wp-content/uploads/2019/12/IGCJ7753-IGC-Pollution-WEB_.pdf.

Polzin, F., Migendt, M., Täube, F., von Flotow, P., 2015. Public policy influence on renewable energy investments—a panel data study across OECD countries. Energy Policy 80, 98−111. https://doi.org/10.1016/j.enpol.2015.01.026.

Quitzow, R., Thielges, S., Goldthau, A., Helgenberger, S., Mbungu, G., 2019. Advancing a global transition to clean energy: the role of international cooperation (2019−48). Econo. Open Access Open Assess. E-J. 13, 1−18.

Rahman, S.M.M., Barua, S., 2016. The design and adoption of green banking framework for environment protection: lessons from Bangladesh". Aust. J. Sustainable Bus. Soc. 2 (1), 1−19.

Rahman, M.M., Rana, R.H., Barua, S., 2019. The drivers of economic growth in South Asia: evidence from a dynamic system GMM approach. J. Econ. Stud. 46 (4) (forthcoming).

Rana, R.H., Barua, S., 2015. Financial development and economic growth: evidence from a panel study on South Asian countries. Asian Econ. Financ. Rev. 5 (10), 1159−1173.

Romei, V., 2020. The Asian Century Is Set to Begin. Financial Times. Retrieved from. https://www.ft.com/content/520cb6f6-2958-11e9-a5ab-ff8ef2b976c7.

Schellnhuber, H., 2001. World in Transition, first ed. Earthscan Publications, London, Sterling, VA, pp. 97−103.

S&P Dow Jones, 2018. S&P Dow Jones Indices 2018 Country Classification Consultation. Standard & Poor's. Accessed 6 March 2019 at. https://www.spice-indices.com/idpfiles/spice-assets/resources/public/documents/725551_spdji2018countryclassificationconsultation6.13.18.pdf.

The World Bank, 2020. World Development Indicators. Retrieved 22 June 2020 from. http://datatopics.worldbank.org/world-development-indicators/.

UNESCAP, 2020a. Energy for Sustainable Development. United Nations ESCAP. Retrieved 26 January 2020, from. https://www.unescap.org/our-work/energy.

UNEP, 2011. Towards a Green Economy: Pathways to Sustainable Development and Poverty Eradication. UNEP, Nairobi, Kenya.

UNESCAP, 2013. Green Growth Indicators: A Practical Approach for the Asia and the Pacific. UNESCAP. Retrieved from. https://www.unescap.org/sites/default/files/publications/GGI_2014.pdf.

United Nations ESCAP, 2018. Energy Transition Pathways for the 2030 Agenda in Asia and the Pacific. United Nations publication, Bangkok. Retrieved from. https://www.unescap.org/sites/default/files/publications/Energy%20Transition%20Pathways-RTR%202018%20Web.pdf.

United Nations ESCAP, 2020b. Green Growth and Green Economy. United Nations ESCAP (2020). Retrieved 26 January 2020, from. https://www.unescap.org/our-work/environment-development/green-growth-green-economy/about.

World Bank, 2012. Inclusive Green Growth: The Pathway to Sustainable Development. World Bank Publications. Retrieved from. https://ycsg.yale.edu/sites/default/files/files/Inclusive_Green_Growth_May_2012.pdf.

World Economic Forum, 2020. Fostering Effective Energy Transition 2020 Edition. Insight Report. Retrieved 14 June 2020 from. http://www3.weforum.org/docs/WEF_Fostering_Effective_Energy_Transition_2020_Edition.pdf.

Zamfir, A., Colesca, S., Corbos, R., 2016. Public policies to support the development of renewable energy in Romania: a review. Renew. Sustain. Energy Rev. 58, 87−106. https://doi.org/10.1016/j.rser.2015.12.235.

Chapter 15

Making green finance work for the sustainable energy transition in emerging economies

Suborna Barua[1], Shakila Aziz[2]
[1]Department of International Business, University of Dhaka, Dhaka, Bangladesh; [2]School of Business & Economics, United International University, Dhaka, Bangladesh

1. Introduction

There is an increasing level of awareness around the world that the global economy is inextricably linked to the environment, and the viability of the one depends on the sustainability of the other. The awareness has led to the incorporation of environmental criteria into all dimensions of economic activities and decisions. The term "Green Finance" has evolved in the global financial, environmental, and regulatory systems that broadly cover financing systems (including products, processes, and organizations) promoting environmentally sustainable economic growth. Green finance entails the simultaneous development of the financial industry, the improvement of the environment, and economic growth, as it mobilizes financial resources to benefit the environment while reducing harm and managing environmental risks (Chartered Banker Institute, 2020; Barua and Chiesa, 2019; Chiesa and Barua, 2019). It is an umbrella term covering sustainable finance, environmental finance, carbon finance, and climate finance (Barua, 2019). It promotes the development and use of financial markets to invest in environmentally sustainable enterprises, technologies, and policies (Noh, 2018).

Green finance covers both public and private investments in environmental goods, commodities, and projects to minimize the damage to the environment (Lindenberg, 2014; Chiesa et al., 2021). Green financing products play a role through different branches of the financial services industry in promoting environmentally sustainable economic development. In retail banking, instruments include green mortgages, green home equity loans, green infrastructure loans, car loans, and cards. There are instruments such as green bonds and climate bonds in securities markets to mobilize environment-friendly financing and investments (Barua and Chiesa, 2019; Chiesa and Barua, 2019). In corporate and investment banking, products include green project finance, green securitization, green venture capital and private equity, green indices, and carbon commodities. The asset management industry offers green fiscal funds, green investment funds, and carbon funds. The insurance industry has green insurance and carbon insurance. There can also be derivative instruments for green technologies and companies (Noh, 2018). As a key to achieving environmentally sustainable economic growth, the evolution of innovative green finance solutions is considered critical in facilitating the gradual switch to clean and renewable energy.

The need for an energy transition through green finance is urgent, particularly in the context of emerging economies—the economies that currently or are likely to grow at a faster than average rate and are expected to be the key drivers of the world economy in the next few decades (McKinsey Global Institute, 2018; HSBC, 2018). The inherent nature of emerging economies is their higher economic growth, which requires a rapidly increasing energy supply level. Almost all emerging economies are either purely or almost entirely reliant on fossil fuels to fuel their economic growth, which continues to degrade the environmental quality at an accelerating rate. Since achieving faster economic growth receives priority, the protection of the environment remains secondary in these economies (Barua, 2021). Furthermore, the lack of financial and economic capacity and dearth of research and technological innovation work as a contributing factor in placing the environment a secondary priority. For the same reasons, emerging economies keep consuming fossil fuel—based energy at an accelerating rate, contributing to the environment's degradation.

This chapter discusses the role and potential of green finance mechanisms in facilitating a sustainable energy transition to achieve sustainable growth in emerging economies. Considering the available lists of emerging economies and Asia's potential to lead the world economy, we focus on five major Asian emerging economies, namely Bangladesh, India, Vietnam, Indonesia, and the Philippines. Our analyses use time-series energy and economic growth data from 1990 to 2019 for the world and the selected economies, taken from the World Bank, the International Renewable Energy Agency, and the British Petroleum statistical review. We first discuss interrelationships between energy consumption, development, and the environment, in the context of the selected economies. We then highlight how the selected economies perform in terms of the energy transition and how the traditional financial markets can be used for enabling a sustainable energy transition through green finance mechanisms. We further discuss the policy and regulatory adaptations required to enable the mechanisms to work. This chapter is organized into eight sections: Section 2 reviews the literature on the energy-environment-economy nexus and the role of green finance, Section 3 highlights the state of the energy transition at the world level, Section 4 elaborates on the globally available green finance mechanisms, Section 5 discusses the role of green finance in emerging economies, Section 6 and seven discuss the energy transition pattern in emerging economies and green finance mechanisms available to support them, respectively, Section 8 outlines some potential green finance mechanisms for emerging economies, and finally Section 9 highlights some policy needs, followed by a conclusion in the final section.

2. The energy-environment-economy nexus and the role of green finance in emerging economies

Energy consumption is an inseparable part of modern industrial economies. A large body of research ascertains the environmental and growth consequences of fossil fuel—based energy consumption in industrial production. Based on a review of some studies, Ozturk (2010) reveals that there is often a link between energy consumption and economic growth, with causation going in one or both directions. There are also findings that energy consumption and economic growth are linked to carbon emission (Waheed et al., 2019; Mardani et al., 2019; Nathaniel et al., 2021; Nathaniel et al., 2020).

The environmental impacts of energy consumption and economic growth necessitate a sustainable energy transition. A review of studies using decomposition methods on different countries shows that energy intensity decreases in most countries globally, particularly in developed countries, and switching to clean energy sources is less prevalent in emerging economies (Xu and Ang, 2013). The activity structure change in the transportation and residential sectors led to higher carbon intensities in developed and developing countries (Xu and Ang, 2013). The global transfer of energy-intensive industries, usually in the manufacturing sector, has decreased energy-related domestic emissions in developed countries but has instead increased it in developing countries. International trade and foreign direct investments particularly have transferred emissions from one country to another and have increased emissions along the supply chain, particularly in emerging economies. Therefore, even if domestic energy footprints may appear to decrease in developed countries, the indirect emissions caused by their consumption lead to increased emissions in emerging economies (Lan et al., 2016; Xu and Dietzenbacher, 2014). This makes it even more important to increase the use of clean energy in developing or emerging economies.

As developing countries grow faster, their energy demand increases, alongside their emissions. The relationship has been extensively studied particularly for the selected economies considered in this study. Several studies suggest that economic growth and energy consumption have contributed to increased emissions in Asian countries over the last few decades (Rahman, 2017; Rahman et al., 2019). Studies suggest that energy consumption is a key driver of economic growth in Asian economies, for example, in South Asia (Rahman et al., 2019). The relationship between energy consumption, financial development, economic growth, and emissions have been extensively studied for India by Shahbaz et al. (2017), Boutabba (2014), Yang and Zhao (2014), and Alam et al. (2011). Energy consumption plays an important role in the economic growth of ASEAN countries like Indonesia, Thailand, and the Philippines (Azam et al., 2015; Shahbaz et al., 2013; Jafari et al., 2012; Nugraha and Osman, 2019; Hwang and Yoo, 2014). Other studies suggest that energy and electricity consumption result in economic growth as well as CO_2 emissions in Bangladesh, as does financial development (Alam et al., 2012; Ahamad and Islam, 2011; Shahbaz et al., 2014; Rana and Barua, 2015; Rahman and Kashem, 2017; Khatun and Ahamad, 2015). Studies suggest that financial development and technological innovation can also affect economic growth and energy intensity (Pan et al., 2019). This indicates that if the financial sector can be realigned with sustainable energy objectives, it can play a significant role in fostering the energy transition and sustainable development.

Studies suggest that energy consumption in the transport sector leads to economic growth in Indonesia and Thailand. Therefore, these countries' transport sector should embrace efficient and green energy by increasing its share of public transport and electric vehicles (Chandran and Tang, 2013). Among others, research on the environmental Kuznets curve hypothesis supports its existence in the case of India (Ahmad et al., 2016), Indonesia (Alam et al., 2016), Bangladesh, and Indonesia (Shahbaz et al., 2016; Yıldırım et al., 2014; Saboori and Sulaiman, 2013; Heidari et al., 2015). However, it is contradicted in the case of Vietnam (Al-Mulali et al., 2015).

The literature review provides a consensus that energy consumption and economic activities affect each other positively. On the other hand, they both affect the environment adversely in most cases, unless both energy consumption and economic growth happen sustainably. Several studies suggest that countries should implement energy efficiency measures without fearing an adverse effect on the economy (Lim et al., 2014). A sound and stable financial system that includes a healthy capital market, fiscal governance, and sound financial fundamentals can contribute to a long-term reduction in environmental degradation and eventually decrease emissions with increasing income (Nasreen et al., 2017). In shifting toward sustainable energy to drive up economic growth, it is necessary to mobilize adequate and targeted financing and investments. While most emerging economies, including the selected ones in this study, have an underdeveloped financial system, they largely depend on public funding, overseas development assistance, and foreign direct investments (Rana and Barua, 2015; Mahembe and Odhiambo, 2019). The literature suggests that financial development encourages both economic growth and energy consumption. Given these findings, if the direction and the nature of financial development can be transformed into sustainability-oriented businesses or aligned with "green" objectives, it can undoubtedly encourage shifting toward sustainable energy production and consumption and economic growth. A green financial system can mobilize the necessary large-scale green finance and investments from the private and international sources toward a sustainable energy transition (Barua, 2019). Given the strong role financial markets traditionally play in an economy, if green financing can be widely operationalized, it can substantially reduce a country's dependency on government or external sources. Since government funding and overseas assistance in many developing economies are limited in supply and come with conditions and structural inefficiencies, the private financial system's immense potential to mobilize green financing and investment can replace them effectively and efficiently (Barua, 2019).

3. Energy transition in the world and the emerging economies

Fig. 15.1 shows the total primary energy consumption (TPEC) at the world level and the share of renewables from 1990 to 2019. While there is a continued and steady rise in TPEC since 1990, renewables consumption patterns look different. The share of total renewable energy consumption remains more or less flat until around 2005, followed by a rising trend, though at a lower rate than TPEC. The pattern indicates that the global call to promote renewables perhaps started to deliver real outputs after 2005. Despite such efforts, the share of renewables in 2019 was just around 13% of the total energy consumed. An important point to note is that the majority of the renewables

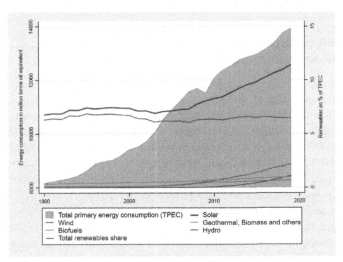

FIGURE 15.1 World energy consumption and the share of renewables. *Source: Author developed based on BP statistical review of world energy data (British Petroleum, 2020). BP Statistical Review of World Energy. British Petroleum. https://www.bp.com/en/global/corporate/energy-economics/ statistical-review-of-world-energy.html; total primary energy consumption includes Oil, Natural Gas, Coal, Nuclear energy, Hydroelectricity, and all other renewables.*

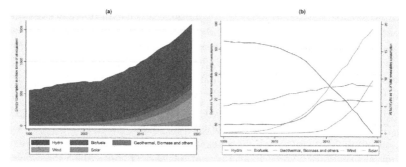

FIGURE 15.2   World renewable energy consumption by energy type. *Source: Author developed based on BP statistical review of world energy data (British Petroleum, 2020). BP Statistical Review of World Energy. British Petroleum. https://www.bp.com/en/global/corporate/energy-economics/statistical-review-of-world-energy.html.*

consumed over the period considered were hydropower, its share floating around 7%—8% of total energy consumed. The other sources of renewables such as solar, wind, biofuels, biomass, geothermal, and others remain below 3% of the world's total energy consumption over the period. Among them, wind supersedes solar since the early 2000s, while both show a rising trend recently. However, within the renewable energy consumption portfolio, global reliance on hydropower declines consistently over the period, while wind and solar energy are increasingly preferred, which could be partly exlplained by their falling porduction cost (Fig. 15.2). The trends and patterns overall indicate that a global transition probably awaits a long walk into the future .

4. Green financing for energy transition: current applications across countries

Given the importance of finance in directing sustainable investments in the economy, it is essential for emerging economies to have dedicated financing sources for the renewable and sustainable energy sector. It requires the development of market infrastructure, the allocation of funds, and policy enforcements. Research shows that the development of the stock market and an increase in private investment, particularly from international sources, can help develop the renewable energy industry in developing economies and ultimately reduce their emissions (Kutan et al., 2018). Therefore, financial markets and instruments can be used as a key channel to promote renewable energy and sustainable development (Rahman and Barua, 2016). The use of private finance through the financial markets and institutions can be defined as "market-based" mechanisms of financing the energy transition. Several market-based financial mechanisms currently mobilize funds for sustainable economic activities worldwide; some of the major ones are highlighted in this section.

4.1 Green bonds

One prominent instrument for renewables is green bonds, where the proceeds are used for renewable energy development. Green bonds are plain vanilla fixed income instruments that enable investors to participate in financing green projects to help countries mitigate or adapt to the effects of climate change and environmental degradation (Barua and Chiesa, 2019). Climate change produces a wide range of effects and economic costs across societies and the economy (Barua, 2019; Barua and Valenzuela, 2018; Barua et al., 2020), and an extensive and continued finance flow is needed to adapt to and mitigate them. Green bonds are different from traditional bonds because of the reaction to market volatility arising from environmental risk. These bonds are popular with different classes of investors who prefer environment-friendly investments. They can be used to finance solar and wind projects, rehabilitate fossil fuel energy systems and invest in energy efficiency in the transport and building sectors (Reichelt, 2010). The top issuers of renewable energy green bonds are corporations, followed by government agencies and financial institutions. The top five countries issuing such bonds are the United States, Germany, Spain, China, and the Netherlands (IRENA, 2020a). The largest underwriters of green bonds include Credit Agricole, BNP Paribas, and HSBC, although the latter leads in issued for emerging markets. Leading exchanges listing green bonds include the Luxembourg Stock Exchange, the German Stock Exchange, and Euronext Paris. About 60% of all global green bonds are for financing energy or green buildings (Climate Bonds Initiative, 2020). The main difference is in the reaction to market volatility, arising from environmental risk, in which case investors seem to prefer bonds of longer maturity. In many emerging Asian economies, the green bond market's growth depends on the growth of the overall bond market (Ng and Tao, 2016). In the context of our emerging economies, the green bonds market can be tapped by greater standardization of the instruments to reduce the cost of issuing and to make it easier for smaller issuers to issue them profitably. Climate-related financing should also be brought into the mainstream instead of being confined to a niche market (Sartzetakis, 2020).

4.2 Institutional investors

Institutional investors like pension funds, insurance companies, sovereign wealth funds, foundations, and endowments have an increasing amount of funds at their disposal, amounting to around USD 85 trillion as of 2020 (IRENA, 2020b). Institutional investors show an increasing appetite for emerging economies as demand for their services and activities rise in new markets (IRENA, 2020b). However, these funds remain untapped as a source of financing for renewable energy. As of now, institutional investors are involved in not more than 2% of renewable energy projects so far, concentrating on solar, wind, and hydropower (IRENA, 2020b). The bond-like returns of renewable energy assets provide institutional investors with a suitable investment avenue. Usually, larger institutional investors invest directly in renewable energy projects, while smaller ones invest in funds linked to renewables. Green bonds and project bonds, with risk mitigation mechanisms, can tap into the institutional investors' funds to finance renewable energy projects (IRENA, 2020b).

4.3 Asset securitizations

Asset securitization is a suitable way of financing renewable energy projects and has been used successfully in several countries. In the United States, issuers include Hannon Armstrong, SunRun, and SolarCity, who have made a number of issues since 2013 (Clean Energy Solutions, 2020). Other countries using asset-backed securities include France, China, Australia, and Brazil (Climate Bonds, 2020). Securitization has been used to finance solar energy in China (Risen Energy, 2019), Brazil (Climate Finance Lab, 2020), Canada, and Australia (Climate Bonds, 2018). Government agencies have issued green asset-backed bonds, residential and commercial mortgage-backed securities. The proceeds have been used for renewable energy projects, energy efficiency in buildings, and transports (Climate Bonds, 2018).

4.4 Green credit and loans

A popular financing model currently available in many economies—either developed or emerging—is green credit or lending by banks. Being the engine of economic growth for many developing economies, banks have a significant power to influence how economic agents—consumers and producers—manage their funds (Barua, 2020). In many countries such as Bangladesh and India, banks have launched "green banking" practice. They provide green credit or loans at concessional rates for environment-friendly private sector projects covering renewable energy, energy efficiency, and pollution abatement (Barua, 2020; Rahman and Barua, 2016). These subsidized rate financings are often supported by the central banks through the supply of additional liquidity and appropriate policy guidelines specifically targeting energy and environment-related projects. Green credits or loans are generally featured with concessional interest rates, extended repayment periods, and strong compliance and monitoring requirements (Barua, 2020).

5. Financing sustainable energy transition in emerging economies

Achieving a sustainable energy transition requires moving gradually from nonrenewable or fossil fuels to renewable energy sources. However, renewable energy's commercial competitiveness relative to fossil fuels is affected by the cost of capital for renewables projects. Evidence suggests that the weighted average cost of capital affects the levelized cost of electricity for renewable technologies like solar and wind (Vartiainen et al., 2020). Emerging economies often have higher capital costs due to the underdeveloped state of financial markets, which makes renewable energy investments uncompetitive and expensive. This is where green finance can play a significant role.

Green financing can flow from market-based sources (e.g., capital markets) or development finance institutions. Although Asian countries enjoy the largest absolute amount of investments, China actually receives the bulk, with other Asian countries like India and Vietnam following far behind. Overall, emerging countries in Asia received the most investment from other Asian countries, while the European Union was the second biggest investor (Bloomberg NEF, 2019). The major institutions investing in emerging market renewables industries include development banks, followed by project developers and utilities (Bloomberg NEF, 2019). The countries making the most investments in renewable energy are the United States, France, Spain, Germany, and Japan, while China makes large investments in other Asian countries under the Belt and Road Initiative. Countries with a nonmonopolized power market structure attract more investments (Bloomberg NEF, 2019).

The East-Asia Pacific region was the biggest destination for renewable energy investments in the last decade, followed by Western Europe, OECD Americas, and Latin America. Most utility-scale renewable energy projects in the wind and photovoltaic (PV) technologies are financed by a mix of debt and equity, with grants and concessional finance making up a negligible amount (IRENA, 2018). The debt to equity ratio usually falls between 60% and 70%, although it can rise to 80% in some cases. The greatest private investments come from the East Asia-Pacific region, whereas Western Europe generates the most public funds. However, in addition to private financing, public resources are allocated to fund supportive policy measures for renewable energy. Such policies include capital subsidies or grants, feed-in-tariffs, public investments, auctions, and tradable Renewable Energy Certificates (RECs). Most countries of the world today have at least some of these policies. Private investment sources include project developers, commercial, financial institutions, corporations, households, institutional investors, private equity, venture capital, infrastructure funds, corporate end-users, and utilities. Of these, project developers are the largest financiers, followed by commercial financial institutions, corporations, and households (IRENA, 2018). Private sources usually make investments within their national border, but cross border financing takes place through public financing sources. Solar and wind projects have received the largest share of funds. The falling cost of renewables has led to greater capacity installed in renewables projects. Therefore, even if the monetary amount of investments has fallen in some years, it represents greater capacity additions (IRENA, 2018).

6. The state of energy transition in selected emerging economies

Consistent with the global awareness and efforts for a greener energy portfolio, emerging economies appear to play a proactive, although insufficient. However, the five countries under consideration appear to be exceptions. Fig. 15.3 shows the renewable energy consumption trend in the selected emerging economies. Up to 2015, there is a steady decline in renewables' share out of total energy consumption in the five economies. The decline rates are faster for Bangladesh and Vietnam compared to others, while the Philippines always had a lower level of renewable consumption share over the years considered. Eventually, all countries except the Philippines converge around 35% of renewables share in their total energy consumption.

Figs. 15.4 and 15.5 clearly show a negative association between the share of renewable energy consumption and economic growth and increases in per

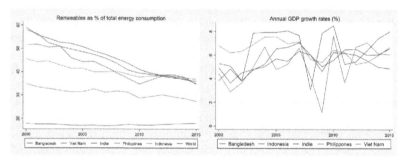

FIGURE 15.3 Renewable energy consumption and economic growth in selected economies. *Source: (The World Bank, 2020). World Development Indicators. https://databank.worldbank.org/.*

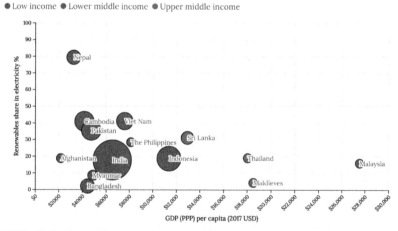

FIGURE 15.4 Renewables share versus economic growth in selected emerging economies. *Source: Author developed based on data from The World Bank, 2020. World Development Indicators. https://databank.worldbank.org/.*

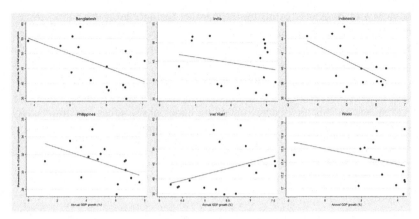

FIGURE 15.5 Renewable energy consumption and economic growth in selected economies. *Source: Author developed based on data from The World Bank, 2020. World Development Indicators. https://databank.worldbank.org/.*

capita income. The only exception is Viet Nam as a single country case. It means as economies keep growing, the share of renewable consumption keeps decreasing; alternatively, increased economic growth is linked to an increased share of nonrenewable energy consumption. The trends are entirely opposite to what one would expect in consideration of the sustainable energy transition. In other words, the selected emerging economies are perhaps moving in a direction opposite to sustainable energy transition and growth. The fact can be validated by looking at the countries' state of energy transition. Table 15.1 shows the 2020 Energy Transition Index (ETI), which reflects the readiness and performance of the emerging economies selected in this discussion. The figures in Table 15.1 suggest a relatively lower level (overall ETI rank

TABLE 15.1 Energy transition readiness, selected economies.

Country	System performance (%)	Transition readiness (%)	2020 ETI score (%)	2020 ETI rank[a]
Philippines	62	49	55.3	57
Vietnam	57	50	53.5	65
Indonesia	61	44	52.4	70
India	54	49	51.5	74
Bangladesh	54	43	48.4	87

[a]*a higher ETI rank indicates poorer performance and vice versa.*
Source. Energy Transition Index 2020 (World Economic Forum, 2020). Fostering Effective Energy Transition 2020 Edition. Insight Report. Retrieved 14 June 2020 from http://www3.weforum.org/docs/WEF_Fostering_Effective_Energy_Transition_2020_Edition.pdf.

above 50) of readiness of all the economies for an energy transition toward nonrenewables. As another manifestation of poor energy transition performance, the table shows that the transition readiness scores in the index for the countries falls either below or just equal 50%.

A lower level of energy transition readiness means the countries are not in a favorable state to achieve a significant degree of energy transition quickly. The reasons for this are diverse, including financial resources constraints, lack of awareness, policy gap, and weak regulatory infrastructure. Given the poor readiness, it is imperative to review the developments related to energy transition in these economies.

6.1 Bangladesh

The transition toward sustainable energy remains a challenge in Bangladesh, as it has long been known as an energy-starved nation. Only about 34% of the total energy consumed in the country originates from renewable sources in the country, while the rest is generated by fossil fuels such as natural gas and coal. Electricity remains the key energy driving the country's economic activities and growth. About 85% of Bangladesh's population has access to electricity, up from 20% in 2000, mostly powered by natural gas and other fossil fuels (IEA, 2020a). Although the shares in solar PV are growing, natural gas accounts for the majority of power generation in the country. The only hydroelectric power plant in Bangladesh is situated in Kaptai, with a current operating capacity of 230 megawatts (MW). The Bangladesh Power Development Board identified two additional sites in Sangu (140 MW) and Matamuhuri (75 MW) for large-scale hydropower plants. However, further use of hydropower appears to be limited due to Bangladesh's flat terrain (GOB, 2008). Bangladesh has only recently introduced renewable energy measures and introduced a Renewable Energy Policy in 2008 to step up the use of renewable energy sources and reduce on-going energy inefficiency (GOB, 2008; GOB, 2013). Some of the recently considered measures by the government include—energy star labeling program, a BRESL project aiming to standardize and label energy efficiency of six different energy efficient equipment, biofuel projects supported by cogeneration energy from Sugarcane Bagasse, providing fiscal incentives to renewable energy project developers, arranging financial assistance for solar mission projects, providing green energy tariff for the consumers of renewable energy sources, and conducting awareness programs (GOB, 2013; Karim et al., 2019).

To diversify its energy portfolio, Bangladesh started a major project in-cooperation with Russia to build the country's first-ever nuclear power plant, known as the Rooppur Nuclear Power Project (RNPP), with 2.4 GW capacity (Sourav, 2018). While RNPP is about to solve much of the country's power crisis, its sustainability contribution remains questionable since the management of the RNPP reactor wastes is yet not clearly articulated, at least in public. If not properly neutralized, the wastes could cause massive harm to the environment and community health and create long-term damage to the ecosystem.

6.2 India

India has set a growth target of about 9% through 2024—25, which will make it the fastest-growing economy in the world. From 2014 to 2020, India has changed its subsidy policy by reducing renewable energy subsidies by 35% and drastically increasing fossil fuel subsidies, which indicates its lack of capability to direct policy funds toward clean energy (IISD and GSI, 2020). India has a strong solar thermal industry, with thermal collector installations in 2017 rising around 26% from 2016. Among the industries, the Indian silk industry has transformed its energy sources from wood or briquettes into solar thermal for about 1500 units (REN21, 2020). India has recently undertaken several targets that include: reducing emission intensity by 33%—35% from 2005 levels, achieving 40% of the combined electricity generated from renewable energy supplies by 2030 supported by technology transition, generating a significant carbon sink of 2.5—3 billion tons of CO_2 equivalent by additional forest and tree covers, and enhancing investments in renewable energy sectors.

Further to the current initiatives, the IEA (2020b) outlines some key medium to long-term targets for India's energy transition; for example, enhancing further growth of distributed renewable energy—especially solar PV, implementing a sustainable bioenergy policy, reducing fossil fuel subsidies and shifting more public resources to clean energy, and targeting consumption subsidies for clean energy access (IEA, 2020b; IISD and GSI, 2020).

6.3 Indonesia

More than half of Indonesia's existing electricity comes from fossil fuels. The move to more green energies is a matter of priority as Indonesia makes a major contribution to global greenhouse gas (GHG) pollution and climate change from fossil fuel combustion (UNESCAP, 2019). However, Indonesia is determined to achieve the sustainable development goals (SDGs) and implement a National Energy Planning devised in 2017. Alongside the on-going policy initiatives, the following targets are currently pursued: increasing electricity consumption to around 2500 kWh per capita by 2025, developing biogas-based infrastructure, preparing a blueprint for the production of biogas at 47.4 million standard cubic feet per day in 2025, achieving the primary renewable energy mix goals of 23% by 2025 and at least 31% by 2050, and reducing national GHG emissions from the energy sector by 34.8% in 2025 and 58.3% in 2050 (UNESCAP, 2019).

The Asia-Pacific Economic Cooperation (APEC, 2017) suggests some further initiatives to facilitate the energy transition in Indonesia, which include prioritizing biomass technology, finding substitutes of natural gas to increase the access of clean cooking fuel, substituting the use of oil with biofuel to meet the energy needs of the transport sector, and including energy efficiency measures to achieve the policy targets.

6.4 The Philippines

Policy initiatives in the Philippines promise to reach a 70% reduction in carbon pollution in the electricity, transport, waste, forestry, and manufacturing sectors by 2030. By the end of 2017, overall installed capacity in the Philippines exceeded the point of 22,728 MW, of which coal remained the key source of power with an installed capacity of 80,049 MW (35.4%), followed by 7079 MW (31.1%) of renewable sources, 4153 MW (18.3%) of oil energy sources, and 3457 MW of gas (15.2%) (ADB, 2018b). The progress so far is encouraging, but the country needs to do more. The Institute for Energy Economics and Financial Analysis (IEEFA, 2019) suggests several steps bring diversity in energy sources and achieve greater access to green energy in the Philippines. Key suggestions include removing diesel subsidy as a form of financing measure, preparing a fossil fuel displacement model, increasing different modes of renewable energy to increase capacity, backing up the capital cost of renewable energy projects, removing the automatic pass-through of fossil fuels through making power supply agreement with suppliers supported by the constant reduction of the cost of renewable energy sources, and channeling an asset risk warning to the financing partners of the country's fossil fuel projects.

6.5 Vietnam

The fast-growing economy of Vietnam is highly energy-intensive and is traditionally heavily fossil fuel–dependent. Moving toward more clean energies would lead to the achievement of SDGs linked to electricity as well as the SDGs concerning poverty, environment, health, energy, and nutrition. With the faster economic growth demanding more restructuring of the energy market, Vietnam has become a net importer of coal, which can trigger long-term energy security risks (UNDP, 2020). In recent years, the hydro and fossil fuel combined power capacity of Vietnam has increased dramatically due to strong government spending and substantial investments in hydropower. However, the country realizes the need for a sustainable energy transition as it has recently taken several initiatives. Some of the key initiatives are— incentivizing investment in green projects, exempting import duty for imports of renewable energy–based capital assets, and corporate tax exemptions or incentives for companies with clean energy business such as the solar power ventures (UNDP, 2018).

7. The use of green finance for energy transition in the selected emerging economies

Financing for energy transition in the selected emerging economies remains mostly reliant on development institutions, with little from the public sources. Fig. 15.6 shows that renewable energy finance flows from key development organizations to emerging economies as of 2020 (IRENA, 2020c). The major financier remains the World Bank Group, followed by Kreditanstalt für Wiederaufbau (KfW) in the form of grants, loans, guarantees, and bonds. India received the largest financing (USD 8400 million), followed by Indonesia (USD 3190 million). The largest financing was made in targeting Hydropower (USD 4539), Multiple technologies (USD 4111), and Solar (USD 3613). However, these financings are limited compared to what is needed to achieve a genuine energy transition.

Given that renewable energy sources from development agencies and governments are limited compared to the need, it is imperative to engage the private financing mechanisms with energy transition objectives. The private financial markets have immense potential to supply the large volume of financing needed. Such engagements could help solve dual problems: fulfilling the domestic energy transition funding gap and minimizing or eliminating external financing reliance that often comes with strict conditions. As a ray of hope, the emerging economies considered in this work have some "market-based" green finance mechanism functioning already. Table 15.2 reports the available green financing instruments, products, and processes currently available in the selected economies. For example, Indonesia is one of the biggest issuers of green bonds in Asia (USD 2.9 billion outstanding), mostly funding energy, buildings, and transport projects. The Philippines is the first in the world

Renewable Energy Financing Flows

Flow direction: Investor > Asset Class > Country>Technology

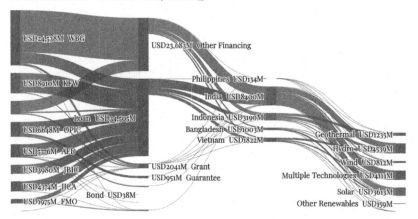

FIGURE 15.6 Renewable energy finance flows and their sources in selected economies. *Source: Author developed based on IRENA (2020).*

TABLE 15.2 Available market-based green financing mechanisms in the selected economies.

Country	Existing financing mechanisms
Bangladesh	- Government climate funds - Development finance - Green bank loans
India	- Loan securitizations - Direct financing - Asset-backed securities - Green bonds - Green bank loans
Vietnam	- Development finance - Private financing and joint ventures - Concessional loans - Viability gap funding - Energy service companies - Green stock market index
The Philippines	- Corporate balance sheet - Project finance - Special bank loans
Indonesia	- Green sovereign Sukuk - Green bonds - Green loans - Green infrastructure investment funds and green index products. - Blended finance instruments, - Credit enhancement services - Viability gap funding, - First loss provisions, - Contingent and asset-backed loan

Source: Author developed based relevant literature.

to issue certified climate bonds (worth USD 630 million) for geothermal power generation. To date, the Philippines has issued USD 2 billion worth of green bonds, all of them entirely or partly related to renewable energy. However, Vietnam has issued only USD 27 million worth of green bonds but supporting water adaptation, not energy. On the other hand, about 33% of ASEAN green bonds are for energy projects, whereas 34% are for low carbon buildings (Rimaud et al., 2019). A country-wise elaboration on the currently available market-based renewable energy financing mechanisms follows.

7.1 Bangladesh

The Bangladesh government has two green funds contributing to the green energy transition. One is the self-financed Bangladesh Climate Change Trust

Fund and the other is the donor financed Bangladesh Climate Change Resilience Fund. Bangladesh Bank (the central bank of Bangladesh) has ruled that the private commercial banks and financial institutions must create climate risk funds and allocate 5% of green finance funds. From this 5%, the banks allocate around 16% to renewable energy projects. The Bangladesh Bank also refinances renewable energy investments. It has launched green banking practices nationwide through which commercial banks supply loans at preferential rates to environment-friendly projects, including renewable energy. The commercial banks, in turn, receive refinancing and subsidies for their loan prices from Bangladesh Bank. The Infrastructure Development Company Limited is the main nonbank financial institution to finance renewable energy technologies and projects in the country, while some venture capital firms also follow the same. The small size of renewable energy projects in Bangladesh usually draws investments from small enterprises, where small borrowers often face trouble in proving their creditworthiness to the lending organizations. They also face higher transaction costs relative to their capital and high policy risk due to their lesser bargaining power with regulators (Hossain, 2018). Bangladesh is yet to enter the green bond market as there is no traditional bond market, although experts estimate that a green bond market has significant potential. The climate-smart investment potential in Bangladesh up to 2030 is around USD 172 billion, indicating a large potential market for green bonds. Bangladesh has an investment gap of USD100 to 130 billion in funding in the electricity sector up to 2040. The investment potential is USD 3.2 billion in renewable energy, USD 118.8 billion in green buildings, and USD 23.7 billion in the transport sector. However, several barriers will have to be overcome to realize this potential. The bond market has to be developed, the green energy projects must compete with other projects, and there should be a broader portfolio of bankable underlying projects. At present, issuers of securities for green energy projects face high taxes and transaction costs, domestic investors struggle to get good credit ratings, and international investors face high country and currency risks (Vivid Economics, 2019).

7.2 India

India has ambitious plans to expand its renewable energy generation to 450 GW by 2030, for which it will need USD 200 billion. A large part of this will be in distributed solar PV. So far, nonbank finance corporations (NBFCs) have securitized small loans for distributed renewables in the country, and channeled the funds through microfinance institutions. There are public sector banks, private sector banks, and government and private NBFCs, providing financing for renewable energy projects. International financing organizations like the World Bank, Asian Development Bank, and KfW have also provided financing for India's solar energy. However, local sources account for more than 85% of renewables financing. Small scale projects are expected to expand the renewable energy share in the Indian market. Still, these projects are at a disadvantage in financing, as they face higher interest rates, lack of funds, and lack of long-term financing. Asset-backed securities (ABSs) have been proposed as a solution to the financing difficulties faced by distributed renewables. India has a sophisticated financial sector that can create a market for ABS and green bonds. As of December 2019, India had issued green bonds of USD 10.3 billion, which made the country the 15th largest green bond issuer in the world. Over 80% of these bonds have financed renewable energy projects. Green ABS has not yet been issued in the country (Vaze et al., 2019). The Reserve Bank of India has identified renewable energy as a priority sector for lending, but banks have not provided the expected amount of funds in the renewable sector, instead favoring fossil fuel generation projects. Green Banks have been created to provide accessible loans to renewable energy projects but have had little success. One source of the problem is classifying renewable energy within the greater power sector without signifying its green qualities. The tenure of debt is short (up to 8 years) compared to the life of renewables projects (15 years), and the cost of debt is too high. Lastly, sometimes institutional investors also perceive renewables projects as riskier (Sarangi, 2019).

7.3 Indonesia

Indonesia has an active capital market and policy landscape to support green financing. In 2018, it issued the world's first green sovereign Sukuk (Islamic asset-backed bonds). In addition to this, there are green bonds and loans, dedicated green infrastructure investment funds, and green index products. Development finance institutions in Indonesia provide additional support by providing blended finance instruments and credit enhancement services like partial credit or risk guarantees, viability gap funding, first loss provisions, contingent loans, and asset-backed loans. These green financing tools are used for climate-resilient projects, including renewable energy generation (Davidson et al., 2019). Some of the challenges of using green bonds for renewable energy in Indonesia include the absence of currency hedging, high cost for small issuers, market illiquidity, low credit ratings, and lack of bankable projects. However, these problems can be addressed by encouraging retail investments, providing credit enhancement services, and hedging services (Climate Bonds Initiative, 2019). Indonesian banks provide limited funds to renewable energy projects, considering them highly risky. However, there is a policy to prevent banks from lending to environmentally damaging projects. Energy finance challenges in Indonesia include high transaction costs due to the small scale of individual projects, high collateral requirements, low tenure (5 years) of bank loans compared to renewables' project life (20 years), and lack of supportive financing regulations (Liebman et al., 2019).

7.4 Vietnam

Vietnam has an infrastructure gap of around USD 100 B in the energy sector up to 2040, creating a scope for green bonds. At present, Vietnam has a small but functional bond market, dominated by sovereign instruments. Vietnam has the biggest renewable energy target for hydropower, followed by solar and wind. So far, the renewable energy projects have relied on development banks, private financing and joint ventures, concessional loans, and energy services companies. Proposed financing for upcoming projects includes mezzanine or subordinated debt and preferred stock, subsidiary, project financing vehicles, YieldCos and investment trusts, but no green bonds (Davidson et al., 2020). The Ha Noi and the Ho Chi Minh City Stock Exchanges are currently working with relevant ministries to encourage green disclosures by listed companies and the listing of green bonds. A green index, called the Green and Sustainable Index, has been listed in the Ho Chi Minh City Stock Exchange and includes 30 top stocks based on green criteria. Financing challenges include the lack of popularity of long-term bonds and lack of expertise in banks about renewable energy projects (ADB, 2018a). The government promotes lending to renewable energy SMEs through public and private banks. However, it also implements a low tariff for electricity, which places real estate (RE) developers at risk due to long-term changes in inflation and interest rates. Moreover, the popular tenure for debt in the financial markets is 10 years, whereas the power purchase agreement contracts of RE companies is usually 20 years (Nguyen et al., 2019).

7.5 The Philippines

In the Philippines, renewable energy projects are financed by corporate balance sheets rather than project finance. Small and new companies find it difficult to obtain bank financing for the early stages of projects. To overcome this hurdle, governmental financial institutions like the Land Bank of the Philippines and the Development Bank of the Philippines provide special loans. These loans cover financing at zero percent interest for preparatory activities like feasibility studies. Financing institutions require equity shares up to 30%, and some companies have difficulty raising domestic equity. Large companies have recourse to project debt financing. The World Bank and the International Finance Corporation provide loans and guarantees through local banks to support renewable energy projects (IRENA, 2017). Project finance funds in the Philippines favor large solar and wind projects with feed-in-tariffs. Financiers include private companies and financial companies. Publicly listed large power companies financed hydropower and biomass energy through corporate finance (Barroco and Herrera, 2019).

8. Financing transition toward renewables: what more can be done?

Table 15.3 shows some of the potential market-based financing mechanisms for the energy transition. While any country could consider these potential mechanisms, they are particularly useful for emerging economies. Given the limited depth and breadth of sustainable energy financing in the countries being studied, introducing the mechanisms outlined in the self-explanatory Table 15.3 could help meet the funding gap in the energy sector.

In addition to the private market-based financing mechanisms available now or to be implemented in the future, other government or development assistance-based financing could be regularized in the selected emerging

TABLE 15.3 Potential market-based products or instruments for financing energy transition.

Financial institution type	Some potential products/instruments
Commercial and investment banks	- Ensuring that commercial banks could offer affordable working capital or long-term loans or credits on a priority basis specifically for companies using or producing sustainable energy. - Enabling investment banks to offer innovative products that encourage and mediate green equities issuances, private equity, and venture capital financing on a priority basis, particularly for companies using or producing sustainable energy. - Allowing both commercial and investment banks to actively encourage financing through loans or investments for start-ups or matured companies that do extensive research on sustainable energy technologies and infrastructure.
Asset managers	- Including more green securities such as green bonds to traditional portfolio investments. - Devising dedicated sustainable energy portfolios for different funds which will include only companies using or producing sustainable energy. - Launching specialized sustainable energy funds such as mutual funds and Exchange Traded Funds (ETFs), which will consider only companies using or producing sustainable energy.

Continued

TABLE 15.3 Potential market-based products or instruments for financing energy transition.—cont'd

Financial institution type	Some potential products/instruments
Securities market	- Introducing green energy equities where the fund raised will only be used for sustainable energy production and distribution purposes. - Using traditional equities with green labeling, where equity Initial Public Offerins (IPO)s traditional companies will be labeled as green energy facilitator if the issuer uses sustainable energy sources for its business operation and/or production. - Introducing dedicated green energy market segment in equity and bond markets, where companies that use or produce sustainable energy sources will only be allowed to be listed for capital financing and their shares will be subsequently traded.
Leasing companies	- There could be "green energy leasing," where individuals, firms, and investors would be able to lease capital machineries (e.g., photovoltaic panels, solar energy storage systems, turbines) required for the energy transition or producing sustainable energy for longer periods. Lease costs and term needs to be more favorable compared to traditional leases.
Insurance companies	- Investments in real estate (RE) are often expensive, and investors consider them riskier. Insurance companies could introduce new sustainable energy insurance, covering investment loss, technology obsolesce or failure, or cash flow dropping below a threshold for new RE investments. - Insurance companies could also offer cheap and affordable insurance for RE equipment and machineries for both firms and individuals at the household levels. This would help cover the RE users and investors for their losses in case of sudden and premature breakdown of equipment or machineries. For example, firms and individuals using PV solar panels may find some of them out of work prematurely; this risk could be well covered by sustainable energy insurance products. - Insurance companies generally invest a large portion of their cash inflows from premiums in real-sector projects. Insurers could prioritize lending or invest in companies using or producing sustainable energy.

Source: Author developed based on literature review.

economies. Almost none of the emerging economies has yet opted for carbon pricing. Carbon pricing measures the explicit cost of GHG pollution. It connects them to their sources, labeling a price for the damages done to the environment, typically in the form of a carbon dioxide (CO_2) tax. While it helps to move the responsibility of GHG pollution back to those liable for pollution, the fund could be redirected under government supervision and management to creation through the implementation of public or privately initiated sustainable and renewable energy projects (World Bank, 2020). Several carbon pricing policies could be implemented. For example, the emissions trading system which limits the overall level of greenhouse gas emissions for the emitter and allows the low-emission industries to trade their extra allowances to larger emitters; a Carbon Tax that effectively sets the price on carbon emissions by defining a specified tax rate on GHG emissions or the carbon content of fossil fuel; and results-based climate finance—a financing strategy in which offers are made to consumers based on actual performance with respect to prespecified environmental results or objectives such as a certain level of energy efficiency or a share of renewables usage in the consumption process.

In addition to carbon pricing, most governments, including emerging economies, apply fiscal supports (e.g., subsidies) to fossil fuels. The emerging economies can gradually withdraw the fossil fuel subsidies down to zero and redirect the supports to renewable and sustainable energy projects. Furthermore, in some developing countries, large infrastructure projects are implemented through public-private partnership (PPP) initiatives. The governments may undertake large renewable and sustainable energy projects and finance them through PPP initiatives. This could be a game-changer since a large fund composition would be easier due to the participation of both the government and the private enterprise, and operation efficiency would be maximized since the private enterprises generally are vested with the duty of regular management of the venture. It is worth noting that even in the PPP cases, the private sector entrepreneurs could be supported with necessary green financing by the market-based mechanisms. On the other hand, the government funding participation could be through regular public sources or redirecting fossil fuel subsidies and/or the fund generated through carbon pricing, as stated earlier.

9. The need for a coordinated policy consideration

Leading a successful energy transition in order to achieve sustainable green growth requires proper policy support and a regulatory framework. Governments across the world, including those of the selected emerging economies, respond to the energy transition with different policies and regulatory measures to green their energy mix. While preparing policy frameworks, it is necessary to set short-term and long-term goals with a specific timeline of

TABLE 15.4 Policies to achieve Renewable energy target overtaken around the world.

Policies	2016	2017
Countries with national/state/provincial renewable energy targets	176	179
Countries with 100% renewable electricity targets	57	57
Countries with 100% renewable heating and cooling targets	1	1
Countries with 100% renewable transport targets	1	1
Countries with 100% renewable energy in primary or final energy targets	1	1
States/provinces/countries with heat obligations/mandates	21	22
States/provinces/countries with biofuel mandates	68	70
States/provinces/countries with feed-in policies	110	113
States/provinces/countries with RPS/quota policies	33	33
Countries with tendering (held in 2017)[a]	34	29
Countries with tendering (cumulative)	73	84

[a]*Data for tendering reflect all countries that have held tenders at any time up through the year of focus.*
Source: Renewable Energy Policy Network for the 21st Century REN21, 2020. Advancing the Global Renewable Energy Transition. REN21. https://www.ren21.net/wp-content/uploads/2019/08/Highlights-2018.pdf.

attainment at the national and subnational levels. Table 15.4 shows the latest available information about the number of countries with particular policy measures. The table shows that existing policy measures are purely energy market-oriented and directly address the energy market dynamics to accelerate the transition and do not consider any policy measures related to financing the energy transition. Given that directing adequate financial resources is critical for a successful and gradual energy transition, policy measures need to consider this. As such, there should be specific policies addressing the issues of adequate, timely, and low-cost financing, either separately in the national energy policy frameworks or appropriately embedded into the pure energy market policies.

As countries are different in terms of economic and social characteristics, there can be no "one size fits for all" approach in formulating policy frameworks, although some commonality may exist. One such case is that of the emerging economies considered in this work, where many of the fast-growing economies share homogeneity in terms of the economy's growth, urbanization, energy consumption patterns, weaker regulatory frameworks, limited economic resources, and so on. Such homogeneity often paves the way for knowledge sharing and a coordinated approach to foster a sustainable energy transition. Therefore, a mix of homogeneity and heterogeneity needs to be applied when considering energy transition policy frameworks in the emerging economies' context. The countries need to adopt a long-term oriented and timeline-wise target-based policy framework. A strict and comprehensive regulatory framework should be developed to aid the policy implementation. The set of policy and regulatory frameworks should have adequate provisions to reward sustainable energy initiatives at the public and private levels and penalize those acting in the opposite direction.

10. Challenges in energy transition in emerging economies

Several challenges lie in the way of an effective and sustainable energy transition in emerging economies, and the selected five countries are also crippled with many of these challenges. Many of the challenges significantly impede private energy financing mechanisms. Some of the key challenges are discussed below.

- In emerging economies like the five selected ones in this study, the energy market is mostly dominated by a monopoly or, at best, a duopoly, often under government ownership and control. The established market and service monopoly may be viewed as a systemic barrier. Such a strongly concentrated and less participatory environment is often inefficient and increases the cost of transition toward renewable energy (Sen and Ganguly, 2017).
- A knowledge gap about sustainability and advanced energy technologies remains a critical challenge for most emerging economies, including the countries considered in this study. There is a lack of awareness about sustainability considerations among energy stakeholders such as investors, consumers, and policymakers. Key stakeholders at the operational and policy levels or the regulatory offices often have limited or no knowledge about new energy technologies, their needs and performances, and the financing mechanisms needed to put them in place (IRENA, IEA and REN21, 2018).
- To bring about a sustainable transition in the energy sector, robust policy and regulatory support are needed, including modifying existing regulations and introducing new regulatory and policy standards to facilitate new and innovative products and projects. Without a supportive, timely, and state-of-the-art policy and regulatory framework, financing the energy transition would very difficult as private financiers and investors may find it uncertain, risky, and financially infeasible.
- The acceptance level of renewable energy technologies is highly dependent on the initial expenditure, production expense, economic standing, and the

existence of subsidies and subventions. Renewable energy's initial capital and installation costs are comparatively higher than that of traditional energy sources, which increases the cost of renewable energy production (Kariuki, 2018). It might discourage potential investors in the sector.
- For many emerging economies, the geographic location and unavailability of geological resources may be an obstacle to renewable energy production; for example, setting up wind turbines may not be feasible in countries with unstable wind speeds.

11. Conclusion

The assessment of five major emerging economies of the world shows that their growth decouples with renewable energy consumption over time, reflecting the countries' secondary priority on the sustainable energy transition. One of the key reasons for lower renewables consumption in these economies is the lack of financial and economic capacity. Emerging economies, including the five countries studied in this work, largely depend on overseas and development assistance with little contribution from public sources. As such, the private market-based financing mechanism makes little to no contribution to the sustainable energy transition in these economies. There is a lot of work needed to be done to redirect the traditional financial markets and institutions toward the energy transition in emerging economies, which requires continuous innovation in products, services, processes, policies, and regulations. All countries, including the emerging ones, need to consider the SDGs and protect the earth for future generations. Considering their high-growth impact, emerging economies need to act now to green both their energy mix and their economic growth. They need to realize that a sustainable energy transition is no longer an option anymore; instead it is a necessity for the global good.

References

ADB, 2018a. Green Financing in Viet Nam: Barriers and Solutions. Asian Development Bank, Tokyo.

ADB, 2018b. Philippines Energy Sector Assessment, Strategy, and Road Map. Asian Development Bank. https://www.adb.org/sites/default/files/publication/463306/philippines-energy-assessment-strategy-road-map.pdf.

Ahamad, M.G., Islam, A.N., 2011. Electricity consumption and economic growth nexus in Bangladesh: revisited evidences. Energy Policy 39, 6145–6150.

Ahmad, A., Zhao, Y., Shahbaz, M., Bano, S., Zhang, Z., Wang, S., Liu, Y., 2016. Carbon emissions, energy consumption and economic growth: an aggregate and disaggregate analysis of the Indian economy. Energy Policy 96, 131–143.

Al-Mulali, U., Saboori, B., Ozturk, I., 2015. Investigating the environmental Kuznets curve hypothesis in Vietnam. Energy Policy 76, 123–131.

Alam, M.J., Begum, I.A., Buysse, J., Rahman, S., Huylenbroeck, G.V., 2011. Dynamic modeling of causal relationship between energy consumption, CO_2 emissions and economic growth in India. Renew. Sustainale Energy Rev. 15, 3243−3251.

Alam, M.J., Begum, I.A., Buysse, J., Huylenbroeck, G.V., 2012. Energy consumption, carbon emissions and economic growth nexus in Bangladesh: cointegration and dynamic causality analysis. Energy Policy 45, 217−225.

Alam, M.M., Murad, M., Noman, A.H., Ozturk, I., 2016. Relationships among carbon emissions, economic growth, energy consumption and population growth: testing Environmental Kuznets Curve hypothesis for Brazil, China, India and Indonesia. Ecol. Indicat. 70, 466−479.

APEC, 2017. Energy Efficiency Finance in Indonesia Current State, Barriers and Potential Next Steps. Asia-Pacific Economic Cooperation. https://www.apec.org/.

Azam, M., Khan, A.Q., Bakhtyar, B., Emirullah, C., 2015. The causal relationship between energy consumption and economic growth in the ASEAN-5 countries. Renew. Sustainale Energy Rev. 47, 732−745.

Barroco, J., Herrera, M., 2019. Clearing barriers to project finance for renewable energy in developing countries: a Philippines case study. Energy Policy 135, 1110008.

Barua, S., 2019. Financing Sustainable Development Goals: A Review of Challenges and Mitigation Strategies. Business Strategy and Development, pp. 1−17. https://doi.org/10.1002/bsd2.94 (Early view).

Barua, S., 2020. The Principles of Green Banking: Managing Environmental Risk and Sustainability. De Gruyter, Boston (MA).

Barua, S., 2021. Human Capital, Economic Growth, and Sustainable Development Goals: An Evaluation of Emerging Economies. The Dynamics of Intellectual Capital in Current Era. Springer, Singapore, pp. 129−148. https://doi.org/10.1007/978-981-16-1692-1_6.

Barua, S., Chiesa, M., 2019. Sustainable financing practices through Green Bonds: what affects the funding size? Bus. Strat. Environ. 28 (6), 1131−1147.

Barua, S., Colombage, S., Valenzuela, E., 2020. Climate Change Impact on Foreign Direct Investment Inflows: A Dynamic Assessment at the Global, Regional and Economic Level. SSRN Electronic J. https://doi.org/10.2139/ssrn.3674777.

Barua, S., Valenzuela, E., 2018. Climate Change Impacts on Global Agricultural Trade Patterns: Evidence from the Past 50 Years. Proceedings of the 6th International Conference on Sustainable Development 2018. United Nations Sustainable Development Solutions Network and Columbia University, USA. https://ssrn.com/abstract=3281550.

Bloomberg NEF, 2019. Emerging Markets Outlook 2019: Energy Transition in the World's Fastest Growing Economies. Bloomberg Finance.

Boutabba, M.A., 2014. The impact of financial development, income, energy and trade on carbon emissions: evidence from the Indian economy. Econ. Modell. 40, 33−41.

British Petroleum, 2020. BP Statistical Review of World Energy. British Petroleum. https://www.bp.com/en/global/corporate/energy-economics/statistical-review-of-world-energy.html.

Chandran, V., Tang, C.F., 2013. The impacts of transport energy consumption, foreign direct investment and income on CO_2 emissions in ASEAN-5 economies. Renew. Sustainale Energy Rev. 24, 445−453.

Chartered Banker Institute, 2020. What Is Green Finance? https://www.charteredbanker.com/uploads/assets/uploaded/6e89f43e-6a3b-41c7-a2a65d41deeee960.pdf. (Accessed 30 June 2020).

Chiesa, M., Barua, S., 2019. The surge of impact borrowing: the magnitude and determinants of green bond supply and its heterogeneity across markets. J. Sustainable Finance Invest. 9 (2), 138−161. https://doi.org/10.1080/20430795.2018.1550993.

Chiesa, M.A., Ben, M., Barua, S., 2021. Does a Company's Environmental Performance Influence Its Price of Debt Capital? Evidence from the Bond Market. The Journal of Impact and ESG Investing (Spring), jesg.2021.1.015. https://doi.org/10.3905/jesg.2021.1.015.

Clean Energy Solutions, 2020. Asset-Backed Securities. Clean Energy Solutions. https://cleanenergysolutions.org/instruments/asset-backed-securities.

Climate Bonds, 2018. Green Securitisation Unlocking Finance for Small-Scale Low Carbon Projects. Climate Bonds.

Climate Bonds, 2020. Securitisation as an Enabler of Green Asset Finance in India. Climate Bonds.

Climate Bonds Initiative, 2019. Barriers and Solutions to Green Bond Issuance in Indonesia. Climate Bonds Initiative.

Climate Bonds Initiative, 2020. 2019 Green Bond Market Summary. Climate Bonds Initiative.

Climate Finance Lab, 2020. Green Receivables Fund (Green FIDC). https://www.climatefinancelab.org/project/green-receivables-fund-green-fidc/.

Davidson, K., Gunawan, N., Filkova, M., Giorgi, A., 2019. Green Infrastructure Investment Opportunities Indonesia Update Report. Climate Bonds Initiative.

Davidson, K., Nguyễt, P.T., Gunawan, N., 2020. Green Infrastructure Investment Opportunities Vietnam 2019 Report. Climate Bonds Initiative.

GOB (Government of Bangladesh), 2008. Renewable Energy Policy of Bangladesh. Policy Report. Ministry of Power, Energy and Mineral Resources, Government of the People's Republic of Bangladesh. Retrieved 22 June 2020 from. http://www.sreda.gov.bd/d3pbs_uploads/files/policy_1_rep_english.pdf.

GOB (Government of Bangladesh), 2013. 500MW Solar Program (2012-2016). An Initiative to Promote Renewable Energy Program in Bangladesh. Power Division, Ministry of Power Energy and Mineral Resources, Government of the People's Republic of Bangladesh.

Heidari, H., Katircioğlu, S.T., Saeidpour, L., 2015. Economic growth, CO_2 emissions, and energy consumption in the five ASEAN countries. Electr. Power Energy Syst. 64, 785−791.

Hossain, M., 2018. Green finance in Bangladesh: policies, institutions, and challenges. In: Sachs, J.D., Woo, W.T., Yoshino, N., Taghizadeh-Hesary, F. (Eds.), Handbook of Green Finance: Energy Security and Sustainable Development. Springer, Singapore, pp. 513−537.

HSBC, 2018. The World in 2030. HSBC. Retrieved from. https://enterprise.press/wp-content/uploads/2018/10/HSBC-The-World-in-2030-Report.pdf.

Hwang, J.-H., Yoo, S.-H., 2014. Energy consumption, CO_2 emissions, and economic growth: evidence from Indonesia. Qual. Quantity 48, 63−73.

IEA, 2020a. Bangladesh Country Profile. International Energy Agency. https://www.iea.org/countries/bangladesh.

IEA, 2020b. India 2020-Energy Policy Review. International Energy Agency. https://www.iea.org/reports/india-2020.

IEEFA, 2019. The Philippine Energy Transition Building a Robust Power Market to Attract Investment, Reduce Prices, Improve Efficiency and Reliability. Institute for Energy Economics and Financial Analysis. http://ieefa.org/wp-content/uploads/2019/03/The-Philippine-Energy-Transition_March-2019.pdf.

IISD, GSI, 2020. Mapping India's Energy Subsidies 2020: Fossil Fuels, Renewables, and Electric Vehicles. https://www.iisd.org/sites/default/files/publications/india-energy-transition-2020.pdf.

IRENA, 2017. Renewables Readiness Assessment: The Philippines. International Renewable Energy Agency, Abu Dhabi.

IRENA, 2018. Global Landscape of Renewable Energy Finance. International Renewable Energy Agency, Abu Dhabi.

IRENA, 2020a. Renewable Energy Finance: Green Bonds. International Renewable Energy Agency, Abu Dhabi.

IRENA, 2020b. Renewable Energy Finance: Institutional Capital. International Renewable Energy Agency, Abu Dhabi.

IRENA, 2020c. Renewable Energy Finance Flows. Retrieved from IRENA International Renewable Energy Agency. https://www.irena.org/Statistics/View-Data-by-Topic/Finance-and-Investment/Renewable-Energy-Finance-Flows.

IRENA, IEA, REN21, 2018. Renewable Energy Policies in a Time of Transition. IRENA, OECD IEA, REN21. https://www.irena.org/-/media/Files/IRENA/Agency/Publication/2018/Apr/IRENA_IEA_REN21_Policies_2018.pdf.

Jafari, Y., Othman, J., Nor, A.H., 2012. Energy consumption, economic growth and environmental pollutants in Indonesia. J. Pol. Model. 34, 879−889.

Karim, M.E., Karim, R., Islam, M.T., Muhammad-Sukki, F., Bani, N.A., Muhtazaruddin, M.N., 2019. Renewable energy for sustainable growth and development: an evaluation of law and policy of Bangladesh. Sustainability 11 (5774), 1−30. https://doi.org/10.3390/su11205774. www.mdpi.

Kariuki, D., 2018. Barriers to Renewable Energy Technologies Development. De Gruyter. https://doi.org/10.1515/energytoday-2018-2302.

Khatun, F., Ahamad, M., 2015. Foreign direct investment in the energy and power sector in Bangladesh. Renew. Sustainale Energy Rev. 52, 1369−1377.

Kutan, A.M., Paramati, S.R., Ummalla, M., Zakar, A., 2018. Financing renewable energy projects in major emerging market economies: evidence in the perspective of sustainable economic development. Emerg. Mark. Finance Trade 54 (8), 1761−1777.

Lan, J., Malik, A., Lenzen, M., McBain, D., Kanemoto, K., 2016. A structural decomposition analysis of global energy footprints. Appl. Energy 163, 436−451.

Liebman, A., Reynolds, A., Robertson, D., Nolan, S., Argyriou, M., Sargent, B., 2019. Green finance in Indonesia: barriers and solutions. In: Sachs, J.D., Woo, W.T., Yoshino, N., Taghizadeh-Hesary, F. (Eds.), Handbook of Green Finance: Energy Security and Sustainable Development. Springer, Singapore, pp. 557−586.

Lim, K.-M., Lim, S.-Y., Yoo, S.-H., 2014. Oil consumption, CO_2 emission, and economic growth: evidence from the Philippines. Sustainability 6, 967−979.

Lindenberg, L., 2014. Definition of Green Finance. German Development Institute.

Mahembe, E., Odhiambo, N.M., 2019. Foreign aid, poverty and economic growth in developing countries: A dynamic panel data causality analysis. Cogent Econ. Finance 7 (1), 1626321. https://doi.org/10.1080/23322039.2019.1626321.

Mardani, A., Streimikiene, D., Cavallaro, F., Loganathan, N., Khoshnoudi, M., 2019. Carbon dioxide (CO_2) emissions and economic growth: a systematic review of two decades of research from 1995 to 2017. Sci. Total Environ. 649, 31−49.

McKinsey Global Institute, 2018. Outperformers: High-Growth Emerging Economies and the Companies that Propel Them. Mckinsey & Company. Retrieved 15 June 2020 from. https://www.mckinsey.com/~/media/McKinsey/Featured%20Insights/Innovation/Outperformers%20High%20growth%20emerging%20economies%20and%20the%20companies%20that%20propel%20them/MGI-Outperformers-Full-report-Sep-2018.pdf.

Nasreen, S., Anwar, S., Ozturk, I., 2017. Financial stability, energy consumption and environmental quality: evidence from South Asian economies. Renew. Sustainale Energy Rev. 67, 1105−1122.

Nathaniel, S.P., Barua, S., Ahmed, Z., 2021. What drives ecological footprint in top ten tourist destinations? Evidence from advanced panel techniques. Environ. Sci. Pollut. Res. https://doi.org/10.1007/s11356-021-13389-5.

Nathaniel, S., Barua, S., Hussain, H., Adeleye, N., 2020. The determinants and interrelationship of carbon emissions and economic growth in African economies: fresh insights from static and dynamic models. J. Publ. Aff. https://doi.org/10.1002/pa.2141. Online early view.

Ng, T.H., Tao, J.Y., 2016. Bond financing for renewable energy in Asia. Energy Policy 95, 509−517.

Nguyen, T.C., Chuc, A.T., Dang, L.N., 2019. Green finance in Viet Nam: barriers and solutions. In: Sachs, J.D., Woo, W.T., Yoshino, N., Taghizadeh-Hesary, F. (Eds.), Handbook of Green Finance: Energy Security and Sustainable Development. Springer, Singapore, pp. 675−705.

Noh, H.J., 2018. Financial Strategy to Accelerate Green Growth. Asian Development Bank Institute, Tokyo.

Nugraha, A.T., Osman, N.H., 2019. CO_2 emissions, economic growth, energy consumption, and household expenditure for Indonesia: evidence from cointegration and vector error correction model. Int. J. Energy Econ. Pol. 9 (1), 291.

Ozturk, I., 2010. A literature survey on energy−growth nexus. Energy Policy 340−349.

Pan, X., Uddin, M.K., Han, C., Pan, X., 2019. Dynamics of financial development, trade openness, technological innovation and energy intensity: evidence from Bangladesh. Energy 171, 456−464.

Rahman, M.M., 2017. Do population density, economic growth, energy use and exports adversely affect environmental quality in Asian populous countries? Renew. Sustainale Energy Rev. 77, 506−514.

Rahman, S.M.M., Barua, S., 2016. The design and adoption of green banking framework for environment protection: lessons from Bangladesh. Aust. J. Sustainable Bus. Soc. 2 (1), 1−19.

Rahman, M.M., Kashem, M.A., 2017. Carbon emissions, energy consumption and industrial growth in Bangladesh: empirical evidence from ARDL cointegration and Granger causality analysis. Energy Policy 110, 600−608.

Rahman, M.M., Rana, R.H., Barua, S., 2019. The drivers of economic growth in South Asia: evidence from a dynamic system GMM approach. J. Econ. Stud. 46 (4) (forthcoming).

Rana, R.H., Barua, S., 2015. Financial development and economic growth: evidence from a panel study on South Asian countries. Asian Econ. Financ. Rev. 5 (10), 1159−1173.

Reichelt, H., 2010. Green Bonds: A Model to Mobilise Private Capital to Fund Climate Change Mitigation and Adaptation Projects. The World Bank, Washington.

REN21, 2020. Advancing the Global Renewable Energy Transition. REN21. https://www.ren21. net/wp-content/uploads/2019/08/Highlights-2018.pdf.

Rimaud, C., Siva, H., Almeida, M., Whiley, A., Tukiainen, K., 2019. ASEAN Green Finance State of the Market 2019. Climate Bonds Initiative.

Saboori, B., Sulaiman, J., 2013. CO2 emissions, energy consumption and economic growth in Association of Southeast Asian Nations (ASEAN) countries: a cointegration approach. Energy 55, 813−822.

Sarangi, G.K., 2019. Green finance in India: barriers and solutions. In: Sachs, J.D., Woo, W.T., Yoshino, N., Taghizadeh-Hesary, F. (Eds.), Handbook of Green Finance: Energy Security and Sustainable Development. Springer, Singapore, pp. 539−556.

Sartzetakis, E.S., 2020. Green bonds as an instrument to finance low carbon. Econ. Change Restruct. 1−25.

Sen, S., Ganguly, S., 2017. Opportunities, barriers and issues with renewable energy development − a discussion. Renew. Sustainale Energy Rev. 69, 1170−1181. https://doi.org/10.1016/j.rser. 2016.09.137.

Shahbaz, M., Hye, Q.M., Tiwari, A.K., Leitão, N.C., 2013. Economic growth, energy consumption, financial development, international trade and CO_2 emissions in Indonesia. Renew. Sustainale Energy Rev. 25, 109−121.

Shahbaz, M., Uddin, G.S., Rehman, I.U., Imran, K., 2014. Industrialization, electricity consumption and CO2 emissions in Bangladesh. Renew. Sustainale Energy Rev. 31, 575−586.

Shahbaz, M., Mahalik, M.K., Shah, S.H., Sato, J.R., 2016. Time-varying analysis of CO_2 emissions, energy consumption, and economic growth nexus: statistical experience in next 11 countries. Energy Policy 98, 33−48.

Shahbaz, M., Hoan, T.H., Mahali, M.K., Roubau, D., 2017. Energy consumption, financial development and economic growth in India: new evidence from a nonlinear and asymmetric analysis. Energy Econ. 63, 199−212.

Sourav, R., 2018. The dilemma of the energy law and policy triangle in recent energy laws and policies in Bangladesh. Hydro Nepal J. Water Energy Environ. 22, 10−15. https://doi.org/10.3126/hn.v22i0.18991.

The World Bank, 2020. World Development Indicators. https://databank.worldbank.org/.

UNDP, 2018. Private Funding Opportunities for Renewable Energy and Energy Efficiency Investments in Viet Nam. UNDP. https://www.vn.undp.org/content/vietnam/en/home/library/environment_climate/private-funding-opportunities-for-renewable-energy-and-energy-ef.html.

UNESCAP, 2019. National Energy Policy in Indonesia and It's Alignment to Sustainable Development Goals 7 (SDG7) and Paris Agreement (NDC). ESCAP, Bangkok. https://www.unescap.org/sites/default/files/UN%20ESCAP%20Workshop%20Indonesia%20Mr.%20Budi%20Cahyono.pdf.

Vartiainen, E., Masson, G., Breyer, C., Moser, D., Medina, E.R., 2020. Impact of weighted average cost of capital, capital expenditure, and other parameters on future utility-scale PV levelised cost of electricity. Wiley Photovoltaics 28 (6), 439−453.

Vaze, P., Bhattacharya, S., Kumar, N., Filkova, M., Giuliani, D., 2019. Securitisation as an Enabler of Green Asset Finance in India. Climate Bonds Initiative.

Vivid Economics, 2019. Green Bonds Development in Bangladesh - A Market Landscape. Vivid Economics Limited, London.

Waheed, R., Sarwar, S., Wei, C., 2019. The survey of economic growth, energy consumption and carbon emission. Energy Rep. 5, 1103−1115.

World Bank, 2020. World Development Indicators. http://datatopics.worldbank.org/world-development-indicators/. (Accessed 30 June 2020).

World Economic Forum, 2020. Fostering Effective Energy Transition 2020 Edition. Insight Report. Retrieved 14 June 2020 from. http://www3.weforum.org/docs/WEF_Fostering_Effective_Energy_Transition_2020_Edition.pdf.

Xu, X., Ang, B., 2013. Index decomposition analysis applied to CO_2 emission studies. Ecol. Econ. 93, 313−329.

Xu, Y., Dietzenbacher, E., 2014. A structural decomposition analysis of the emissions embodied in trade. Ecol. Econ. 101, 10−20.

Yang, Z., Zhao, Y., 2014. Energy consumption, carbon emissions, and economic growth in India: evidence from directed acyclic graphs. Econ. Modell. 38, 533−540.

Yıldırım, E., Sukruoglu, D., Aslan, A., 2014. Energy consumption and economic growth in the next 11 countries: the bootstrapped autoregressive metric causality approach. Energy Econ. 44, 14−21.

Chapter 16

A revisit of the globalization and carbon dioxide emission nexus: evidence from top globalized economies

Ali Syed Raza[1], Nida Shah[1], Arshian Sharif[2], Muhammad Shahbaz[3]

[1]*Business Administration, IQRA University, Karachi, Pakistan;* [2]*Othman Yeop Abdullah Graduate School of Business, University Utara Malaysia, Malaysia;* [3]*School of Management and Economics, Beijing Institute of Technology, Beijing, China*

1. Introduction

In the last two decades, globalization being an international fact which affects each individual in every step of their life. Globalization is described as the process of reducing the barriers in order to achieve liberalized flow of capital, finance, goods, services, commodities and labor across the globe (Lal 2000; Bowles 2005; Bhensdadia and Dana, 2004). Globalization aims to accomplish uniform socioeconomic and political system throughout the world's economies. Globalization generally connects all the economies by foreign direct investment (FDI) and trade along with its various consequences. Through which economies enhance their degree of real per capita income, financial development, openness, and environmental qualities. Along with this every economy requires to reach the optimal ratio of income per capital through investment and trade. The process of increasing growth with the help of urbanization and industrialization also enhances some unattractive and adverse external fact, like pollution which ultimately degrades the environmental quality of country. Industrial production activity plays a vital role in the growth of any economy. In addition, Sharif et al. (2017), Raza et al. (2016), and Sharif, Raza (2016) confirmed that energy consumption has a significant input in the industrial production and economic growth, but it has some side effects like environmental pollution in the form of carbon dioxide (CO_2). These two gases are the reason for ecological imbalances and global climate change and it is the source for massive economic and welfare harms for the people on the globe. Hence, the usage of energy consumption and production based upon its total effect on economy whether it is good or not.

Energy-Growth Nexus in an Era of Globalization. https://doi.org/10.1016/B978-0-12-824440-1.00002-3

More the growth in terms of trade and FDI (both are the proxy of globalization) of any economy means enhanced the outer competition and solid connection of an economy in investment and trade with the world, which ultimately enhance the growth of the economy. During this process of investment and trade, a huge amount of CO_2 releases from the consumption of energy. An economy has to maintain a reduced rate of openness, reduced rate of economic growth, and reduce the rate of industrialization because of no searching of alternative energy which will discharge lesser carbon emission. Therefore, the impact of globalization based on the impact of openness on economic growth and the impact on energy too. This is due to their inborn relationship with each other. It is confirmed from the theory and past studies that economic growth is connected with higher energy consumption and its impact on the environment, except any variable deals in the energy demand model. It is hard to separate the impact of economic growth on carbon emission and similarly the impact of energy consumption on economic growth in carbon and economic growth estimation model, respectively. There is a chance of attaining biased estimation about their vigorous relationship of energy-growth carbon emission variables. Along with this, the amount of openness is also based on liberalization methods which adopted by apprehensive economies by favor to their investments and trade.

On the contrary, those who remain stiff achieved slow economic growth. Therefore, the economies of the world has been divided over the socioeconomic consequences of globalization and so the literature. The literature on growth-globalization nexus mainly uses trade liberalization, and FDI as the key measures of globalization to check its impact on countries' economic growth (Feenstra and Hanson, 1997; Gaston and Nelson, 2002; Mah, 2003; Kai and Hamori, 2009; Celik and Basdas, 2010). Based on the findings of both the empirical and theoretical studies, the neoclassical growth theory supports globalization. This theory explains that globalization enhances efficiency and encourages growth through upgraded technology transfer and resource distribution. It is revealed that globalization allows an increase in the mobilization of deposits and also increases the size of FDI and exports. Simultaneously, FDI is also an essential part of economic growth in any country. FDI-growth nexus based on exogenous and endogenous growth theories has been discussed widely in empirically studies (Borenszteinet al., 1998; Ghazali, 2010; Rachdi and Saidi, 2011).

According to the exogenous growth theory, FDI can affect the economic growth by providing the technological development in the host country. On other hand, endogenous growth theory argues that FDI can affect economic growth by providing the positive and significant spillover effects in human capital and gross domestic product (Borensztein et al., 1998, Stanisic, 2015). In developing countries, these indications help to increase in economic advancement, income, and employment and a diminution in inequality. According to Heckschler-Ohlin-Samuelson model, developed countries export skill intensive product in which they have comparative advantage, but the developing countries, mainly export labor intensive product in which they have a relative advantage. Thus, this enhances the demand for lower skill labor and reduces inequality in developing countries.

There is a convincible theoretical explanation in the favor of lesser carbon releasing is depends on the cause endowments hypothesis. This hypothesis emphasizes that factor endowment explains a county's polluting industry and comparative advantages are characteristically capital intensive. Consequently, polluting industries are more expected to be focused on capital, ample developed nations irrespective of their change in the environmental policy (Copeland and Taylor, 2004). However, the past studies related to this issue are also very limited and the increasing wealth of globalization as a source by joining the economy has shaped an inclusive interest from both globalization policy analyst and academicians. The past studies on this topic provide very contrasting evidence and the harmony is yet to reemerge. So, our contribution of this study is as follows.

In today's world, energy consumption plays a vital role in economic development of any country. In the last 30 years, energy consumption has risen for more than 45% with more than 40% of CO_2 emission (Energy Information Administration). Nowadays, one of the most important production inputs is energy not only as a most important economic indicator but also use as a strategic commodity for world's politics and its importance is increasing day by day. A serious problem facing by the world today is the lack of strict policies and its enforcement regarding CO_2 emission resulting in environmental pollution. This dilemma of global warming is effecting badly on the environment in especially poor or undeveloped countries. The emission is continuously rising from 1980 to 2005 and so on. Change in climate and energy security, increase in temperature that average (global warming) is mainly caused by the increase in carbon emission. Country manufacturing, economic activities, and growth increases with the inflow of FDI ultimately increase energy consumption as well.

Present study makes various significant contributions to the globalization, energy-growth relationship between globalized economy in four different aspects. The majority of the past studies has focused on the panels of different countries in describing the vigorous relationship between globalization and environmental degradation. As per our knowledge, the main denunciation on the past studies is the selection of panels. The panel of countries selected for a past studies has an issue of heterogeneity and also might be an issue of cross-sectional dependent throughout the panel. Present study overcomes this issue by applying recent heterogeneous panel analysis techniques along with cross-sectional dependence (CD) test. This is essential, as globalization policies making and set by the global level can also affect single countries all together, along with the additional exogenous shocks. The current study is the pioneering attempt to focus on globalization and environmental degradation by opting heterogeneous panel analysis for most globalized countries.

Second, relating to other studies, this study is different because the criteria for selecting a panel are not random. For the selection of the panel, we apply the KOF Index of Globalization introduced by KOF Swiss Economic Institute. The KOF ranking is based on three factors that are economic globalization index, social globalization index, and political globalization index. We select top 50 globalized countries, according to the ranking of KOF index. Third, past studies usually used trade or globalization index in their models, we deploy all the proxy of globalization including globalization index, FDI, and trade so that we can explain the comparative effect of these in the economic growth route.

Finally, present study examines the long run out elasticities related to every type of globalization for panel and separate countries. These elasticities reflect both the cross-sectional and time dimensional nature of the panel and give, the more substantial result as compared to the time series techniques. Along with this, present study advances earlier studies in the environmental and globalization literature by opting new established econometric techniques. For example, the Pesaran (2004) analyzed CD test in order to identify the cross-sectional dependence between the variables, whereas the Westerlund (2008) bootstrap cointegration is opted to identify the long-run relationship between the considered variables. Further, the fully modified ordinary least square (FMOLS) is used to investigate the long-run coefficients and the causal relationship between the variables. These techniques are valuable and beneficial for the policy makers for making the policies bases on the result of long-run demand for globalization in the carbon emission process for considering countries.

The remaining of the paper is ordered as follows. Section 2 presents the literature review of globalization, energy, and growth nexus hypotheses based on the past empirical studies. Section 3 presents a brief statistics of globalization in different countries. Section 4 explains the model, data analysis, and empirical findings. In Section 5, we finalize the conclusions and deliver policy implications.

2. Literature review

2.1 Theoretical background

The connection between environmental degradation and economic growth was reflected by environmental Kuznets curve (EKC) which was investigated by Simon Kuznets (1960). He revealed initially economic growth is positively associated with environmental degradation, and it worsens environmental quality until it reaches a turning point where it starts decreasing because of many changes that come from new investment in the pollution-free economy and clean industrialization that environmental degradation rises correspondently as one with economic activities. He found that there is a negative relation existing between income level and CO_2 emissions, with the increase in income level pollution will decrease, a similar study was conducted on G-7 countries by Shahbaz et al. (2019) examined biomass energy consumption led to having more CO_2 emissions he further investigated that the EKC hypothesis is more valid in G-7 countries where capitalization has decreased CO_2 emissions. Environmental quality has improved with the increase in the level of financial development, similarly institutional quality, urbanization impedes environmental quality, and trade openness also helps to decrease the degradation of the environment whereas globalization increases pollution.

2.2 Panel countries

In earlier studies, the globalization is measured with the proxy of trade openness, FDI. After the development of the KOF global index, it was used to examine its linkage with CO_2 emission.

Atici (2009) investigated the trade openness (globalization) and CO_2 emission association in the central and east European economies. It is concluded that globalization measured through trade openness variable does not have any effect on the CO_2 emission in the countries. The association between the trade, FDI, and CO_2 emission is also tested in ASEAN countries by Atici (2012). It is concluded that export acts as the main contributor of CO_2 emission, whereas the FDI has a negative effect on the CO_2 emission. Ertugrul et al. (2016) analyzed the impact of trade openness on the CO_2 emission in top 10 CO_2 emitting economies. The data comprised of the years 1971−2011. They reported that trade openness acts as an important contributor of the CO_2 emission in the economies. The causal association between the trade openness and CO_2 emission in newly industrialized countries is tested in the work of Ahmed et al. (2016). They reported a unidirectional causal relationship between the trade openness and CO_2 emission, i.e., trade openness causes the CO_2 emission in the examined countries.

Al-Mulali (2012) also examined the role of FDI, gross domestic product (GDP), and total trade on CO_2 emission in 12 Middle East countries by using the panel data comprise of the years 1990−2009. It is concluded that the FDI, GDP, and total trade increases the CO_2 emission in the examined countries. Another study on the FDI and CO_2 emission is done in the context of G20 countries by Lee (2013). The result concluded that FDI reduces the CO_2 emission in the economies by using the energy-efficient equipment. Pao and Tsai (2011) reported the bidirectional causal association exists between the FDI and CO_2 emission in BRIC countries. The nexus between the FDI and CO_2 emission is also tested in 54 countries using the dynamic simultaneous-equation by Omri et al. (2014). They reported that bidirectional causality exists between the FDI and CO_2 emission.

Leitão and Shahbaz (2013) examined the globalization and CO_2 emission relationship in 18 countries. The dataset comprised of the years 1990−2010 and the generalized method of moments (GMMs) technique was applied. It is concluded from the outcome that globalization has a positive effect on the CO_2 emission. They further reported that globalization acts as an important engine to improve the productivity of an economy by using the resources effectively and efficiently. Lee and Min (2014) also analyzed the role of globalization in reducing the CO_2 emission by using the data of 255 countries comprised of years 1980−2011. They reported that the globalization plays an important role in minimizing the CO_2 emission. Shahbaz et al. (2016) uses the autoregressive-distributed lag (ARDL) technique to analyze the globalization and CO_2 emission nexus in 19 African countries. They used the panel data comprised of years 1971−2012. The result concluded that all the variables are co-integrated in the long run, and the positive association exists between them. However, the impact of globalization on CO_2 emission varies from country to country.

Bu et al. (2016) examined the globalization and CO_2 emission association by using the data of 166 countries comprised of years 1990−2009. They concluded that all the three types of globalization, i.e., social, political, and economic, increases the CO_2 emission. However, the impact of the globalization varies between Organization for Economic Co-operation and Development (OECD) and non-OECD countries. Paramati et al. (2017) reported that political globalization has a negative effect on the CO_2 emission in European unions, G20, and OECD countries.

Kastratovic (2019) investigates the relationship between FDI and greenhouse gas emission in the agriculture sector of 63 developing countries. This research used panel data for the period 2005 to 2014 and applied technique system GMMs. The finding of the research shows that FDI has a positive impact on greenhouse gas emission in the agriculture sector of developing countries, and the result shows that the increase in FDI of 10% could increase in CO_2 emission intensity in agriculture by 0.17%−0.18%. Furthermore, the study recommends that there is a need to more emphasize environmental policies in the agriculture sector and to coordinate foreign investment promotion.

2.3 Time series literature

Ang (2009) studied the CO_2 emission and trade openness association in China by using the data comprised of the years 1953−2006. It is concluded that trade openness increases the CO_2 emission in the region. Another work on China is done by Jalil and Feridun (2011) and reported that trade openness has a significant effect on the CO_2 emission. Ren et al. (2014) also uses the data of 18 industries of China to explore the impact of FDI and international trade on CO_2 emission. The result of the two-step GMM estimation shows that both the variables have a significant effect on CO_2 emission. Zhang and Zhou (2016) also analyzed whether the FDI inflows reduce the CO_2 emission in China. They used the data from 1995 to 2010 and applied the STIRPAT model.

They reported that FDI helps to reduce the CO_2 emission in the region and shows the acceptance of the pollution halo hypothesis.

In the context of Turkey, Ozturk and Acaravci (2013) examine the trade openness and CO_2 emission and reported the positive association. Lau et al. (2014) examined the role of trade and FDI on CO_2 emission in Malaysia. They applied bounds testing approach and Granger causality techniques and reported that FDI and trade increase the CO_2 emission. Gökmenoğlu and Taspinar (2016) also examine the FDI and CO_2 emission nexus in turkey and concluded that FDI significantly contributes to the CO_2 emission.

Shahbaz et al. (2013) examined the association between the globalization and CO_2 emission in Turkey and reported the existence of positive association among the variables. Shahbaz et al. (2015) also examined the globalization and CO_2 emission in the context of India and reported that globalization has a positive effect on CO_2 emission. This implies that increase in any type of globalization, i.e., social, economic, or political enhances the CO_2 emission in the region. Shahbaz et al. (2016) restudied the globalization and CO_2 emission in the context of Indian economy and reported the negative association exists between the examined the variables. An increase in the globalization activities minimizes the CO_2 emission in the region. The same association (globalization and CO_2 emission) is tested by Shahbaz et al. (2016) in China by using the time series data comprised of the years 1970–2012. They reported that all the variables are co-integrated in the long run. They further concluded that negative association exists between the globalization and CO_2 emission indicating that increase in globalization decreases the CO_2 emission in the Chinese economy. The granger causality test shows that the unidirectional causal relationship exists between the examined variables.

Yu and Xu (2019) conducted this research to examine the relationship between FDI and research as well as development (R&D) on industrial CO_2 emission reduction. The different economic techniques applied to study, for instance, Panel-corrected standard error model was constructed based on Chinese provincial panel data from 2000 to 2017, and GM-model was applied to predict the growth trend of CO_2 emissions at the national level. They used panel data of a sample size of 30 provinces in China. The result of the study suggests that FDI and R&D are all negatively correlated with CO_2 emission intensity. Industrial CO_2 emission in four regions of China shows different results. The eastern region has shown the highest rank in CO_2 emission followed by the western region, the central region, and the northeastern region. The growth trend of CO_2 emission after 2010 in four regions slowed down. Moreover, the industrial sector played a dominating role in CO_2 emission while the change of industrial structure affects the increase of CO_2 emission of a different region.

Chin et al. (2018) investigated this research to identify the determinants of CO_2 emission in Malaysia to enhance sustainable economic growth without a decline in environmental quality and employed the ordinary least squares

(OLS)—based ARDL method. The result of the study suggests that the bound test shows that their relationship exists among CO_2 emission, FDI outflow, GDP, and vertical intra-industry trade (VIIT). It indicates that the ARDL method is appropriate to treat the data. And the Ramsey test result shows that CO_2 emission is higher in the industrial production sector also the positive coefficient implies that VIIT between the two nations brings the scale effect and the composition effect to Malaysia and technique does not affect. Moreover, the positive and significant coefficient of FDI at the 5% level shows that the home country has not given the latest technology to the host country.

Khan et al. (2019) examined the effect of globalization, economic factors, and energy consumption on CO_2 emissions in Pakistan. The study used CO_2 emission as a dependent variable and energy consumption, economic factor and globalization are independent variables. They employed ARDL model to examine both long-run and short-run impacts and used time series data from the year 1971—2016. The result of the research indicates that financial development, energy consumption, trade, political globalization, social globalization, and economic globalization have a positive and significant impact on CO_2 emissions in Pakistan. While innovation, urbanization, have a negative effect in the long run. However, the result of the study in the short-run shows that urbanization, economic growth, energy consumption, financial development, economic globalization, social globalization, and political globalization have a positive impact on CO_2 emissions while FDI, trade, and innovation have a negative impact on CO_2 emissions.

Hanif et al. (2019) investigated this research to examine the relationship between fossil fuel consumption, FDI, and economic growth on carbon emission in 15 developing Asian countries. They used panel data for the period from 1990 to 2013 and applied ARDL to the data. The result of the study, in the long run, suggests that economic growth is increasing carbon emissions significantly when GDP increases 1% of the increase in carbon emission by 0.22%. The relation between fossil fuels and carbon emission is directly and significantly a 1% increase in fossil fuels increase in carbon emission by 0.29%. FDI is also a positive and significant relationship with carbon emissions. One percent increases in FDI the increase in carbon emission by 0.12%. The result of population growth and carbon emission is positive and insignificant. One percent increases in population growth the increase in carbon emission by 0.13%. The cointegration analysis shows that 0.513% of carbon emission in a given period is related to a 1% increase in carbon emission in the previous period besides in the short run only fossil fuel makes a positive and significant contribution to carbon emissions. By results, 1% increase in fossil fuel consumption contribution to 0.029% carbon emissions in the short run. Economic growth and FDI make no significant short-run contribution to carbon emissions. The result of the study suggests that fast deterioration of the region's environment has been due to the large consumption of fossil fuel to benefit economic growth. The overall study recommends that to establish empirical development in the developing countries of Asia, there is a need to empower civil society as regards major development and environmental policies, to anticipate the prevailing pollution.

3. Methodology

Current study taken the data of per capita of real GDP, per capita of CO_2 emission and three dimensions of globalization which are social globalization, political globalization, and economic globalization in order to confirm the presence of EKC for top 50 globalized economy in the globe. The data of real GDP (dollars), CO_2 emission (metric tons) and energy consumption (millions tons), FDI (% of GDP) and trade (summing exports and imports of goods and services) have been collected from World Bank (world development indicators). The data of globalization is collected from KOF index website. This globalization index is created from three different subindices (economic, social, and political globalization).[1] Past studies either use globalization index or FDI or trade along with the function of carbon emission. In this study, we combined all the proxies of globalization, economic growth, and total energy consumption as a determinant of CO_2 emission. Present study covers the time period from 1980 to 2014 for top 50 globalized countries, according to the KOF globalization index. The CO_2 emission function for our model is as follows:

$$CE_{it} = f\left(EC_{it}, Y_{it}, Y_{it}^2, G_{it}\right)$$

$$CE_{it} = f\left(EC_{it}, Y_{it}, Y_{it}^2, F_{it}\right)$$

$$CE_{it} = f\left(EC_{it}, Y_{it}, Y_{it}^2, T_{it}\right)$$

where CE_{it} is a CO_2 emission per capita, EC_{it} is total energy consumption. Y_{it} $\left(Y_{it}^2\right)$ is a real GDP per capita and (square of real GDP per capita), and G_{it} is a the KOF index of globalization, which consists of three different subindices, i.e., social, economic, and political globalization. F_{it} is a foreign direct investment and T_{it} is explain and trade which calculated by the sum of imports and exports of goods and services.

3.1 Estimation techniques

In the present study, we examine the long-run relationship between the variables by using a panel cointegration technique. Likewise, current study investigates the long-run effect of globalization of environmental degradation by using FMOLS approach. Lastly, we use a heterogeneous panel noncausality test to investigate the short-run causal relationship between globalization and environmental degradation.

---

1. See in details http://globalization.kof.ethz.ch/.

3.2 Cross-sectional dependence and CIPS test

Initially, we find that whether the data we have taken has the characteristic of cross-sectional independence or dependence. Pesaran's (2004) CD test is used to fulfill this purpose. This is the major issue which needs to be solved before approaching to panel unit root test. The old unit root test has low power and are not effective when they are used on the panel series, which already has a CD issue (Paramati et al., 2016; Bhattacharya et al., 2016). Consequently, in the present study, we employ Pesaran (2007) cross-sectionally augmented IPS (CIPS) unit root test which is based on the hypothesis of CD. This panel unit root test is the essential for the panel cointegration models. This test is used to examine the order of incorporation of the variables. If entire variables are integrated of the equally level, i.e., $I(1)$, then this is an indication that entire of the data set has a unit root problem at level and are stationary at first difference. So, it can be concluded all the variables in the data set may have a relationship with long-run equilibrium.

3.3 Panel cointegration techniques

In the current study, we apply bootstrap panel cointegration proposed by Westerlund (2007) to examine the long-run relationship between the variables throughout the complete sample of top 50 globalized economies. This analysis is more beneficial if the time series component of every cross-section is smaller. Owing to these features, research scholars have to newly adopt the bootstrapping panel cointegration technique to investigate the long-run relationship between the variables (Lee and Brahmasrene, 2013). The old traditional techniques have accepted the null hypothesis of no cointegration even in case where cointegration is strongly proposed by theories. Irrespective of traditional techniques, Westerlund (2007) has newly established a recent panel cointegration test focused on structural instead of residual dynamics. The outcomes disclose that these tests have restricted normal distributions, and they are more reliable in term of consistency. Westerlund (2007) explains that the outcomes of the new test give decent size accuracy and are extra powerful than the residual based tests by Pedroni (2004). Based on this evidence, the current study will analyze the effect of globalization on environmental degradation.

This study employs the panel cointegration test. This test was given by Westerlund (2007) and Persyn and Westerlund (2008). The hypothesis of cointegration is evaluated by using the two different tests (i) group mean test (ii) panel test. Based on the error correction model, Westerlund (2007) developed four test statistics, Ga, Gt, Pa, and Pt and all these four statistics are normally distributed. The Gt, Pt are calculated in the standard way, by using the standard error parameters of the Error Correction model. Whereas, Ga, Pa are based on the standard errors given by Newey and West (1994), which are adjusted from autocorrelations and heteroskedasticity.

In order to run this test, all the variables are assumed to be stationary at first difference $I(1)$. This test evaluates the absence of cointegration by determining whether error correction is present for the whole group and also in individual panel members. On the basis of the existence of cointegration, the long-run parameters are estimated. In a cross-sectional analysis, the error variance varies across the groups which affect the consistency of the estimators. To overcome this issue, the generalized least squares method can be used. But the variance variability still exists, such as the correlation of the squared residuals with the regressors in each group. Within the group, there are two sources which cause the heteroskedasticity problem, it could either by differences in the variance of the residual terms conditioned on the regressors or by differences in the unconditional variance of the residual terms. So, in order to control both the sources causing the heteroskedasticity, the FMOLS is used.

3.4 Long-run elasticities

Based on a panel data set, the use of OLS is considered to bias and its distribution focus upon an annoyance constraint. Pedroni (2001a,b) reasons that in the case of regression result, annoyance constrains could result for the existence of endogeneity and serial correlation issue between the regressors. So, to solve these problems, we apply the FMOLS model. This technique focuses on the nonparametric method in order to resolve the issue of serial correlation and endogeneity. Therefore, we use FMOLS technique to examine the long-run equilibrium relationship.

3.5 Heterogeneous panel causality test

We investigate the short-run bivariate causal relationship between the variables by selecting a framework that supports the heterogeneity of the models throughout the cross sections. This panel causality technique has been introduced by Dumitrescu and Hurlin (2012). This technique is suitable for stationary data by using the fixed coefficients in the vector autoregressive (VAR) model. The implication of this approach is that it is taking different log structure and similarly heterogeneous measurement throughout the cross-section under both assumptions. Firstly, the null hypothesis of no causal relationship is tested and then the alternative hypothesis for testing the causal relationship at least for few cross sections. Finally, the Wald statistics are computed to each of the cross sections individually for testing Granger noncausality. Dumitrescu and Hurlin (2012) reason that this panel causality test meets to a normal distribution in homogeneous noncausality hypothesis when T indicates to infinity fist and then N represents to infinity.

TABLE 16.1 Descriptive statistics of the variables.

	G	C	E	Y	F	T
Mean	70.30	227.78	3974.64	1.80	3.63	350.00
Median	73.15	55.55	3177.39	2.05	1.67	154.00
Minimum	29.98	−532.41	264.08	−24.57	−43.46	4900.00
Maximum	92.63	6023.76	18177.25	22.72	198.31	2.25
Std. Dev.	14.41	826.54	2909.81	3.82	8.67	575.00
Observations	35	35	35	35	35	35

C explains carbon dioxide emission, E stands for energy consumption, F indicates foreign direct investment, G represents globalization, T represents trade, and Y explains gross domestic product.

4. Data analysis and findings

4.1 Descriptive statistics

At first, we explore the summary statistics of all the under-examined countries. The result related to the statistics is reported in Table 16.1. As seen from the table, the highest value of globalization is 29.98 and the minimum value is 92.63. The minimum carbon emission by the sample countries is −532.41 and the highest emissions from these countries are 6023.76. Similarly, the average energy consumption is 3974.64 and the highest consumption is 18177.25. In the full sample, the minimum gross domestic product is −24.57 and the maximum gross domestic product is 22.72. Likewise, the minimum value of FDI is −43.46 and the maximum value is 198.31. While, the minimum value of trade is 2.25 billion and the maximum value is 490 billion.

4.2 The cross-sectional dependence and unit root tests

Table 16.2 explains the outcomes for the CD test CIPS unit root test. The CD test outcomes for the most globalized countries show that the rejection of the null hypothesis of cross-sectional independence at the 1% level of significance for all variables, signifying indication of CD. We used newly established CIPS unit root test instead of conventional methods of unit root test. This test discusses for CD in the data series.The outcomes of the CIPS unit root test show that the rejection of the null hypothesis for all variables at first-order differentials. Thus, the results of CIPS unit root propose that all the variables are showing nonstationary behavior at the level and showing stationary behavior at the first-order difference. In simple words, entire variables are integrated of order $I(1)$. So, there must be an evidence of cointegration relationship between the variables in long run.

TABLE 16.2 Results of cross-sectional dependence and CIPS unit root test.

Variable	CD test	P-value	CIPS test Level	CIPS test First difference
C	63.672	0.0000	−1.228	−4.884***
Y	39.101	0.0000	−1.000	−3.296***
E	51.759	0.0000	−1.410	−4.278***
G	152.067	0.0000	−1.152	−6.881***
F	38.753	0.0000	−1.071	−6.071***
T	106.874	0.0000	−1.678	−4.440***

***, **, and * indicates statistical significance at 1%, 5%, and 10%.
Source: Authors' estimation.

4.3 Findings of panel cointegration tests

Table 16.3 explains the results of Pedroni panel cointegration test results. These tests reject the null hypothesis of no cointegration at the 1% level of significance because four test of the within measurement (Panel v-statistics, Panel rho-statistics, Panel PP statistics, and Panel ADF statistics) and three tests of the between dimension (group rho, PP statistics, and group ADF statistics) support this rejection. Therefore, all seven tests reveal that the variables move together in the long-run equilibrium in globalization and trade model. Whereas, six tests out of seven tests (except Panel v-statistics) confirm the presence of long-run relationship with FDI model. The cointegration among the variables is also examined by using the Kao test. As seen from the results reported in Table 16.4, the null hypothesis is rejected and the alternative hypothesis is accepted, i.e., cointegration exists among the variables in all globalization, FDI, and trade models.

The cointegration among the variables is also analyzed by using the second generation cointegration test. The results related to the bootstrap panel cointegration are mentioned in Table 16.5. Both with dimensions and within dimensions results are reported. The result confirms the acceptance of alternative hypothesis and the rejection of the null hypothesis. Thus, the second generation test also confirms that the under-examined variables are co-integrated in the long run in all the three models of globalization, trade, and FDI.

4.4 Findings from fully modified ordinary least square

The long-run association among the variables is examined by using the FMOLS technique. The FMOLS technique was given by Phillips and Hansen (1990) and then further modified by the Pedroni (2004). We choose this

TABLE 16.3 Results of Pedroni (Engle-Granger based) panel cointegration.

Estimates	Stats.	Prob.
$C = f(Y + Y^2 + E + G)$		
Panel v-statistic	−6.897	0.000
Panel rho-statistic	−13.241	0.000
Panel PP statistic	−20.340	0.000
Panel ADF statistic	−21.054	0.000
Alternative hypothesis: individual AR coefficient		
Group rho-statistic	−1.560	0.059
Group PP statistic	−10.364	0.000
Group ADF statistic	−10.979	0.000
$C = f(Y + Y^2 + E + F)$		
Panel v-statistic	−1.959	0.975
Panel rho-statistic	−1.438	0.075
Panel PP statistic	−6.003	0.000
Panel ADF statistic	−4.351	0.000
Alternative hypothesis: individual AR coefficient		
Group rho-statistic	−1.312	0.095
Group PP statistic	−8.667	0.000
Group ADF statistic	−7.324	0.000
$C = f(Y + Y^2 + E + T)$		
Panel v-statistic	−2.986	0.001
Panel rho-statistic	−3.907	0.000
Panel PP statistic	−10.646	0.000
Panel ADF statistic	−9.267	0.000
Alternative hypothesis: individual AR coefficient		
Group rho-statistic	−2.790	0.003
Group PP statistic	−13.737	0.000
Group ADF statistic	−14.080	0.000

The null hypothesis of Pedroni's (2004) panel cointegration procedure is no cointegration.
Source: Authors' estimation.

TABLE 16.4 Results of Kao (Engle-Granger based) panel cointegration.

Estimates	Stats.	Prob.
$C = f(Y + Y^2 + E + G)$		
Panel ADF statistic	−2.013	0.022
$C = f(Y + Y^2 + E + F)$		
Panel ADF statistic	−3.368	0.000
$C = f(Y + Y^2 + E + T)$		
Panel ADF statistic	−4.236	0.000

The null hypothesis of Kao Residual Cointegration panel cointegration procedure is no cointegration.
Source: Authors' estimation.

TABLE 16.5 Results of Westerlund (2007) bootstrap panel cointegration.

Statistic	Value	Z value	P value	Robust P value
$C = f(Y + Y^2 + E + G)$				
Gt	−8.603	−3.995	0.000	0.000
Ga	−10.560	−3.909	0.000	0.023
Pt	−18.979	−3.613	0.000	0.028
Pa	−15.512	−2.777	0.003	0.030
$C = f(Y + Y^2 + E + F)$				
Gt	−2.767	−2.326	0.010	0.050
Ga	−9.494	−4.112	0.000	0.000
Pt	−13.432	−3.926	0.000	0.050
Pa	−15.685	−2.530	0.006	0.000
$C = f(Y + Y^2 + E + T)$				
Gt	−2.884	−1.133	0.129	0.000
Ga	−9.306	−4.240	0.000	0.042
Pt	−15.598	−2.238	0.013	0.009
Pa	−18.153	−4.305	0.000	0.000

The null hypothesis of Westerlund (2007) panel cointegration procedure is no cointegration. Using the boot strap approach of Westerlund (2007) to account for cross-sectional dependence, the number of replications is 400. The P-values are for a one-sided test based on normal distribution. The robust P-values are for a one-sided test based on 400 bootstrap replications.
Source: Authors' estimation.

TABLE 16.6 Results of long run analysis through fully modified ordinary least square.

Variable	Fully modified ordinary least square		
	Coeff.	t-stats.	Prob.
Model with globalization index			
Y	0.138	2.500	0.013
Y²	−0.001	−1.922	0.055
E	1.322	22.452	0.000
G	0.038	2.737	0.006
Model with FDI			
Y	0.105	0.315	0.753
Y²	−0.001	−2.353	0.019
E	1.428	33.015	0.000
F	0.016	1.762	0.078
Model with trade			
Y	0.122	0.803	0.422
Y²	−0.001	−2.495	0.013
E	1.103	17.996	0.000
T	0.152	6.961	0.000

Source: Authors' Estimations.

technique because they account for endogeneity and autocorrelation problems and give robust results.

We investigate the long-run estimates by taking the FMOLS coefficients. The results of FMOLS have been reported separately for each model in Table 16.6. The long-run beta value calculated using two different approaches which are very much similar and significant at the 10% level. The EKC hypothesis that undertakes an inverted U-shaped association among CO_2 emission and GDP is confirmed for all models.

Table 16.6 explains that, for the model with globalization and for the FMOLS method, the results of panel estimate conclude that the long-run impact of CO_2 emission by GDP is nearly equivalent to 0.138; a 1% rise in energy consumption will increase the CO_2 emission by 1.322%; a 1% rise in globalization index will increase the CO_2 emission by 0.038%. For the model with FDI by using FMOLS technique: the results of panel estimate conclude that the long-run impact of CO_2 emission by GDP is nearly equivalent to

0.105%; a 1% rise in energy consumption will increase the CO_2 emission by 1.428%; a 1% rise in FDI will increase the CO_2 emission by 0.016%. Finally, the model with trade the results of panel estimate conclude that the long-run impact of CO_2 emission by GDP is nearly equivalent to 0.122; a 1% rise in energy consumption will increase the CO_2 emission by 1.103%; a 1% rise in trade will increase the CO_2 emission by 0.152%.

For all models, it is explained that globalization increase the CO_2 emission in the long run. Therefore, there is a need to introduce environmental friendly technologies when the collaboration between two countries is taken place. We also examined the presence of EKC curve hypothesis between the economic growth and CO_2 emission in most globalized countries. In order to examine this, we construct a new variable square rate of economic growth represented by (y^2) and added in all models. It can be seen from the results report in Table 16.6 that the economic growth has a positive value whereas the square rate of economic growth shows the negative value. This confirms the existence of the Kuznets curve hypothesis, i.e., the inverted U-shape association between the economic growth and CO_2 emission. This implies that at the initial stage, the economic growth increases the CO_2 emission, but after reaching a certain level, it starts minimizing the CO_2 emission. This finding is sensible in a manner that now the countries that are going for economic development via globalization are more inclined to join hands with those countries which are opting for environmentally friendly technologies. Thus, the more development in the examined countries will help the government to decrease the CO_2 emission.

4.5 Heterogeneous Panel Causality Test

The causal relationship between the globalization and CO_2 emission is analyzed by the heterogeneous panel causality test. The result is reported in Table 16.7. The result shows that a unidirectional causal relationship exists between the globalization and CO_2 emission, and the causality is running from globalization to CO_2 emission. Similarly, in the trade model, result indicates an unidirectional causal relationship exists from trade to CO_2 emission and reverse is not possible. While, in the case of FDI, results reveal that there is a bidirectional causal relationship existing between FDI and CO_2 emission. This implies that the globalization causes the CO_2 emission in all the proxies.

4.6 Discussion

The overall combined results of cointegration estimation have confirmed the long-run relationship among globalization, CO_2 emission, FDI, and GDP. Furthermore, FMOLS results are strongly supporting the existence of the EKC hypothesis which is an inverted U-shaped relationship between CO_2 emission and the economic growth of selected countries. But the outputs of the

TABLE 16.7 Results of heterogeneous panel causality test.

Null hypothesis	Stats.	Prob.
G does not homogenously cause C	5.039	0.000
C does not homogenously cause G	1.308	0.258
F does not homogenously cause C	2.898	0.004
C does not homogenously cause F	5.452	0.000
T does not homogenously cause C	22.054	0.000
C does not homogenously cause T	1.538	0.124

*** indicates statistical significance at 1%.
Source: Authors' estimation.

heterogeneous panel causality test are presenting the unidirectional causal relationship between the globalization and CO_2 emission; the same results are found in case of trade and CO_2 emission except for the bidirectional relationship between FDI and CO_2 emission. Previously many studies have been conducted on different countries and economies with numerous variables, similar results were found in these studies. Shahbaz et al. (2015) and Ertugrul et al. (2016).

5. Conclusion and policy recommendation

Current study investigate the role of globalization in environmental degradation by using a multivariate structure and panel data sets for most globalized economies from 1980 to 2014. The outcomes of the current study are the pioneering attempt to investigate the relationship between economic growth environmental degradation and globalization. The outcomes of this study support the fundamental observation in the literature that globalization has a positive and significant impact on the overall economy, and this is possible because it has significant impacts on foreign exchange reserves, tax revenues, income level, and employment opportunity. In short, globalization is probably increasing the environmental degradation of the world and present study also supports this.

The enhance in the environmental degradation by globalization-related activities has significant policy implications. Furthermore, our outcomes bases EKC, propose that after attaining a certain point the impact of economic growth on environmental degradation will significantly decrease. Base on this evidence, we recommend the policy makers to adopt and improve the environmental friendly policies for reducing and controlling the environmental degradation. With the help of this policy, countries can have a positive economic impact of globalization, however avoiding of its worse effects.

Moreover, based on the results, this study gives some valuable recommendations to the policy makers and the government of the economies. The government should take the keenest interest in investing in clean technology, as this will help to protect the environment and help the economies to increase their competitiveness. The government should develop strict environmental regulations and ISO standards related to the technology transfer from the entrant country. They should ensure and allow the entry of new country in the market only if they come up with cleaner environmental technologies. The strict implementations of the environmental standards also induce the domestic operating companies to opt to go-green products and switch themselves to renewable energy.

The policy makers should take a keen interest in reducing the CO_2 emission; they should initiate clean energy projects and should use the FDI inflows and stock markets money in creating such projects. As the CO_2 emission reduction is at its early stage, the government should give safe and healthy business environment to the investors, and encourage the domestic and international investors both to develop clean energy projects. The tax benefits should also be provided to the investors that build clean energy industries.

The environment quality could only improve while countries have policies related to FDI and the standard to be followed when they allow investors. Host countries have to assess the environmental impact of FDI before introducing foreign investors into the country. Moreover, high-emissions countries should improve the level of FDI. The amount of fossil fuel consumption could only reduce when projects related to alternate energy started. High-emissions countries could benefit the most from increasing the levels of economic growth and population size.

The economies should also invest in research and development activities, as this will help them to identify new ways through which the environmental quality can be improved. The economies should collaborate with each other, as this collaboration will help them to share ideas, financial resources, and technological innovations on the common problem and should bring some common solution for the CO_2 emission.

References

Ahmed, K., Shahbaz, M., Kyophilavong, P., 2016. Revisiting the emissions-energy-trade nexus: evidence from the newly industrializing countries. Environ. Sci. Pollut. Control Ser. 23 (8), 7676–7691.

Al-mulali, U., 2012. Factors affecting CO_2 emission in the Middle East: a panel data analysis. Energy 44 (1), 564–569.

Ang, J.B., 2009. CO_2 emissions, research and technology transfer in China. Ecol. Econ. 68 (10), 2658–2665.

Atici, C., 2009. Carbon emissions in Central and Eastern Europe: environmental Kuznets curve and implications for sustainable development. Sustain. Dev. 17 (3), 155–160.

Atici, C., 2012. Carbon emissions, trade liberalization, and the Japan–ASEAN interaction: a group-wise examination. J. Jpn. Int. Econ. 26 (1), 167–178.

Bhattacharya, M., Paramati, S.R., Ozturk, I., Bhattacharya, S., 2016. The effect of renewable energy consumption on economic growth: evidence from top 38 countries. Appl. Energy 162, 733—741.

Bhensdadia, R.R., Dana, L.P., 2004. Globalisation and rural poverty. Int. J. Enterpren. Innovat. Manage. 4 (5), 458—468.

Borensztein, E., De Gregorio, J., Lee, J.W., 1998. How does foreign direct investment affect economic growth? J. Int. Econ. 45 (1), 115—135.

Bowles, P., 2005. Globalization and neoliberalism: a taxonomy and some implications for anti-globalization. Can. J. Dev. Stud. Rev. 26 (1), 67—87.

Bu, M., Lin, C.T., Zhang, B., 2016. Globalization and climate change: new empirical panel data evidence. J. Econ. Surv. 30 (3), 577—595.

Çelik, S., Basdas, U., 2010. How does globalization affect income inequality? A panel data analysis. Int. Adv. Econ. Res. 16 (4), 358—370.

Chin, M.-Y., Puah, C.-H., Teo, C.-L., Joseph, J., Malaysia, L., Rahman, T.A., 2018. The determinants of CO_2 emissions in Malaysia: a new aspect. Int. J. Energy Econ. Pol. 8 (1), 190—194. http://econjournals.com.

Copeland, B.R., Taylor, M.S., 2004. Trade, growth, and the environment. J. Econ. Lit. 42 (1), 7—71.

Dumitrescu, E.I., Hurlin, C., 2012. Testing for Granger non-causality in heterogeneous panels. Econ. Modell. 29 (4), 1450—1460.

Ertugrul, H.M., Cetin, M., Seker, F., Dogan, E., 2016. The impact of trade openness on global carbon dioxide emissions: evidence from the top ten emitters among developing countries. Ecol. Indicat. 67, 543—555.

Feenstra, R.C., Hanson, G.H., 1997. Foreign direct investment and relative wages: evidence from Mexico's maquiladoras. J. Int. Econ. 42 (3), 371—393.

Gaston, N., Nelson, D., 2002. Integration, foreign direct investment and labour markets: microeconomic perspectives. Manch. Sch. 70 (3), 420—459.

Ghazali, A., 2010. Analyzing the relationship between foreign direct investment domestic investment and economic growth for Pakistan. Int. Res. J. Fin. Econ. 47, 123—131.

Gökmenoğlu, K., Taspinar, N., 2016. The relationship between CO_2 emissions, energy consumption, economic growth and FDI: the case of Turkey. J. Int. Trade Econ. Dev. 25 (5), 706—723.

Hanif, I., Faraz Raza, S.M., Gago-de-Santos, P., Abbas, Q., 2019. Fossil fuels, foreign direct investment, and economic growth have triggered CO_2 emissions in emerging Asian economies: some empirical evidence. Energy 171, 493—501. https://doi.org/10.1016/j.energy.2019.01.011.

Jalil, A., Feridun, M., 2011. The impact of growth, energy and financial development on the environment in China: a cointegration analysis. Energy Econ. 33 (2), 284—291.

Kai, H., Hamori, S., 2009. Globalization, financial depth and inequality in Sub-Saharan Africa. Econ. Bull. 29 (3), 2025—2037.

Kastratović, R., 2019. Impact of foreign direct investment on greenhouse gas emissions in the agriculture of developing countries. Aust. J. Agric. Resour. Econ. 63 (3), 620—642. https://doi.org/10.1111/1467-8489.12309.

Khan, M.K., Teng, J.Z., Khan, M.I., Khan, M.O., 2019. Impact of globalization, economic factors, and energy consumption on CO_2 emissions in Pakistan. Sci. Total Environ. 688, 424—436. https://doi.org/10.1016/j.scitotenv.2019.06.065.

Kuznets, S., 1960. Economic growth of small nations. In Economic consequences of the size of nations. Palgrave Macmillan, London, pp. 14—32.

Lal, D., 2000. The Challenge of Globalization: There Is No Third Way. Global Fortune: The Stumble and Rise of World Capitalism. Cato Institute, New York, pp. 29–42.

Lau, L.S., Choong, C.K., Eng, Y.K., 2014. Investigation of the environmental Kuznets curve for carbon emissions in Malaysia: do foreign direct investment and trade matter? Energy Policy 68, 490–497.

Lee, J.W., 2013. The contribution of foreign direct investment to clean energy use, carbon emissions and economic growth. Energy Policy 55, 483–489.

Lee, J.W., Brahmasrene, T., 2013. Investigating the influence of tourism on economic growth and carbon emissions: evidence from panel analysis of the European Union. Tourism Manage. 38, 69–76.

Lee, K.H., Min, B., 2014. Globalization and carbon constrained global economy: a fad or a trend? J. Asia Pac. Bus. 15 (2), 105–121.

Leitão, N.C., Shahbaz, M., 2013. Carbon dioxide emissions, urbanization and globalization: a dynamic panel data. Econ. Res. Guard. 3 (1), 22.

Mah, J.S., 2003. A Note on globalization and income distribution – the case of Korea, 1975–1995. J. Asian Econ. 14, 157–164.

Newey, W.K., West, K.D., 1994. Automatic lag selection in covariance matrix estimation. Rev. Econ. Stud. 61 (4), 631–653.

Omri, A., Nguyen, D.K., Rault, C., 2014. Causal interactions between CO_2 emissions, FDI, and economic growth: Evidence from dynamic simultaneous-equation models. Econ. Model. 42, 382–389.

Ozturk, I., Acaravci, A., 2013. The long-run and causal analysis of energy, growth, openness and financial development on carbon emissions in Turkey. Energy Econ. 36, 262–267.

Pao, H.T., Tsai, C.M., 2011. Multivariate Granger causality between CO_2 emissions, energy consumption, FDI (foreign direct investment) and GDP (gross domestic product): evidence from a panel of BRIC (Brazil, Russian Federation, India, and China) countries. Energy 36 (1), 685–693.

Paramati, S.R., Apergis, N., Ummalla, M., 2017. Financing clean energy projects through domestic and foreign capital: the role of political cooperation among the EU, the G20 and OECD countries. Energy Econ. 61, 62–71.

Paramati, S.R., Ummalla, M., Apergis, N., 2016. The effect of foreign direct investment and stock market growth on clean energy use across a panel of emerging market economies. Energy Econ. 56, 29–41.

Pedroni, P., 2001a. Fully modified OLS for heterogeneous cointegrated panels. Adv. Econom. 15, 93–130.

Pedroni, P., 2001b. Fully modified OLS for heterogeneous cointegrated panels. In: Nonstationary Panels, Panel Cointegration, and Dynamic Panels. Emerald Group Publishing Limited, pp. 93–130.

Pedroni, P., 2004. Panel cointegration: asymptotic and finite sample properties of pooled time series tests with an application to the PPP hypothesis. Econ. Theory 20 (3), 597–625.

Persyn, D., Westerlund, J., 2008. Error-correction-based cointegration tests for panel data. STATA J. 8 (2), 232–241.

Pesaran, M., 2004. General Diagnostic Tests for Cross Section Dependence in Panels. Cambridge Working Papers in Economics No. 0435. Cambridge University, Cambridge.

Pesaran, M.H., 2007. A simple panel unit root test in the presence of cross-section dependence. J. Appl. Econom. 22 (2), 265–312.

Phillips, P., Hansen, B., 1990. Statistical inference in instrumental variables regression with I(1) processes. Rev. Econ. Stud. 57, 99–125.

Rachdi, H., Saidi, H., 2011. The impact of foreign direct investment and portfolio investment on economic growth in developing and developed economies. Interdiscipl. J. Res. Bus. 1 (6), 10−17.

Raza, S.A., Sharif, A., Wong, W.K., Karim, M.Z.A., 2016. Tourism development and environmental degradation in the United States: evidence from wavelet-based analysis. Curr. Issues Tourism 1−23. https://doi.org/10.1080/13683500.2016.1192587.

Ren, S., Yuan, B., Ma, X., Chen, X., 2014. International trade, FDI (foreign direct investment) and embodied CO_2 emissions: a case study of Chinas industrial sectors. China Econ. Rev. 28, 123−134.

Shahbaz, M., Khan, S., Ali, A., Bhattacharya, M., 2016. The impact of globalization on CO_2 emissions in China. Singapore Econ. Rev. 62 (3), 1−29.

Shahbaz, M., Mallick, H., Mahalik, M.K., Loganathan, N., 2015. Does globalization impede environmental quality in India? Ecol. Indicat. 52, 379−393.

Shahbaz, M., Mallick, H., Mahalik, M.K., Sadorsky, P., 2016. The role of globalization on the recent evolution of energy demand in India: implications for sustainable development. Energy Econ. 55, 52−68.

Shahbaz, M., Ozturk, I., Afza, T., Ali, A., 2013. Revisiting the environmental Kuznets curve in a global economy. Renew. Sustainable Energy Rev. 25, 494−502.

Shahbaz, M., Solarin, S.A., Ozturk, I., 2016. Environmental Kuznets Curve hypothesis and the role of globalization in selected African countries. Ecol. Indicat. 67, 623−636.

Shahbaz, M., Balsalobre, D., Shahzad, S.J.H., 2019. The influencing factors of CO_2 emissions and the role of biomass energy consumption: statistical experience from G-7 countries. Environ. Model. Assess. 24 (2), 143−161.

Sharif, A., Raza, S.A., 2016. Dynamic relationship between urbanization, energy consumption and environmental degradation in Pakistan: evidence from structure break testing. J. Manage. Sci. 3 (1), 01−21.

Sharif, A., Afshan, S., Nisha, N., 2017. Impact of tourism on CO_2 emission: evidence from Pakistan. Asia Pac. J. Tourism Res. 1−14. https://doi.org/10.1080/10941665.2016.1273960.

Stanisic, N., 2015. Do foreign direct investments increase the economic growth of Southeastern European transition economies? S. E. Eur. J. Econ. 6 (1), 29−38.

Westerlund, J., 2007. Testing for error correction in panel data. Oxf. Bull. Econ. Stat. 69 (6), 709−748.

Westerlund, J., 2008. Panel cointegration tests of the Fisher effect. J. Appl. Econ. 23 (2), 193−233.

Yu, Y., Xu, W., 2019. Impact of FDI and R&D on China's industrial CO_2 emissions reduction and trend prediction. Atmos. Pollut. Res. 10 (5), 1627−1635. https://doi.org/10.1016/j.apr.2019.06.003.

Zhang, C., Zhou, X., 2016. Does foreign direct investment lead to lower CO_2 emissions? Evidence from a regional analysis in China. Renew. Sustainable Energy Rev. 58, 943−951.

Further reading

Cheng, C., Ren, X., Wang, Z., Yan, C., 2019. Heterogeneous impacts of renewable energy and environmental patents on CO 2 emission - evidence from the BRIICS. Sci. Total Environ. 668, 1328−1338. https://doi.org/10.1016/j.scitotenv.2019.02.063.

Pedroni, P., 2001. Purchasing power parity tests in cointegrated panels. Rev. Econ. Stat. 83 (4), 727−731.

Stock, J., Watson, M., 1993. A simple estimator of cointegrating vectors in higher order integrated systems. Econometrica 61 (4), 783−820.

Chapter 17

Is there an asymmetric causality between renewable energy and energy consumption in BIC countries?

Yusuf Muratoglu[1], Devran Sanli[2], Mehmet Songur[3]
[1]*Department of Economics, FEAS, Hitit University, Corum, Turkey;* [2]*Department of Economics, FEAS, Bartın University, Bartın, Turkey;* [3]*Department of Economics, FEAS, Dicle University, Diyarbakır, Turkey*

1. Introduction

For all practical purposes, energy sources are classified into two categories as renewable energy sources (RESs) and nonrenewable energy sources (N-RESs). N-RESs are a type of energy that is in the static energy stores of nature and remains underground unless it is released by human interaction, which includes fossil resources such as nuclear fuels, coal, oil, and natural gas. Nonrenewable contain energy potential in isolation and are necessary for the external human factor to supply energy. N-RESs are also called finite sources or brown energy sources (Twidell and Weir, 2015).

Renewable energy refers to the process that converts these fuels into useable forms of energy (electricity, heat, chemical, or mechanical power) using resources that are constantly renewed by nature (solar, wind, water, Earth's internal heat, and biomass). RESs are abundant in nature, and related technologies that turn these sources into end-use are continuously evolving. There are many different ways to use renewable energy resources. The primary known RESs are hydroelectric, bioenergy, geothermal energy, solar energy, wind energy, hydrogen, and ocean energy (Boyle, 2004). This form of energy passes from nature as a current or flow independently of man and tool. RESs are also called green energy or sustainable energy. The main feature of RES is that it is obtained from permanent energy flows and is in repetitive periods (Twidell and Weir, 2015).

Hydroelectric power plants convert the energy in flowing water into electricity. The most common form of hydroelectric power is to build a dam on

Energy-Growth Nexus in an Era of Globalization. https://doi.org/10.1016/B978-0-12-824440-1.00007-2

405

the river to provide a large water tank, and water is released into the electric turbines to generate power. Bioenergy is generated from organic substances such as plants based on biomass. Solar energy technologies produce heat by using infinite power and energy of the sun. Besides that, wind energy refers to the energy harvesting from the wind by turbines. Hydrogen is one of the most abundant and simplest elements in the universe. As a result of its use as fuel, only water is released as waste and this is called negative emission. However, hydrogen production, consumption, storage, and transportation require complex technologies and therefore its cost is relatively high. Ocean energy systems convert the kinetic and potential energy found in natural oscillations of ocean waves into electricity. There are several mechanisms for the use of this energy source. Geothermal energy is mainly used for hot springs containing hot spheres, geysers, hot magma, etc., that the earth sphere contains. These formations are used to generate electricity and provide heating (Boyle, 2004; EWEA, 2009; Scheer, 2013; Khaligh and Onar, 2017; Rosen and Koohi-Fayegh, 2017; Zohuri, 2018; Yan, 2018).

Today, the world's energy supply is largely based on fossil fuels. These energy sources have proven to be the main cause of environmental problems, and this cannot be sustained forever. RESs will inevitably dominate the world's energy supply system in the long run. The reason for this includes simplicity due to necessity: There is no alternative energy source to RES (EREC, 2010). Energy-based environmental problems were serious and widespread in the past, but these problems were rarely universal and could be solved in any way. But now climate change and global warming from human activity have become an accepted scientific fact.

Since carbon dioxide (CO_2) emissions from human activities originate from the combustion of fossil fuels used for energy production, the energy issue is at the center of the debate on climate change. Therefore, creating new environmentally friendly and sustainable energy sources has been the focus of scientists and governments (Uğurlu, 2019).

The increase in energy demand is due to the rapidly growing emerging economies. China, India, and other developing Asian countries account for about two-thirds of energy consumption growth. Nearly half of the increasing energy demand arise from only China and India (BP, 2017). Over the 1980–2017 period, the average annual growth rate of real gross domestic product (GDP) in China and India was approximately 9.5% and 6.5%, respectively.

Ongoing economic growth and the increase in population costs the nature and human beings greenhouse gas emissions and climate change, firstly. Table 17.1 provides an overview of the CO_2 emissions by sectors and fuels in the period 2000–17 and 5 years projections. While the International Energy Agency (IEA) expects the total CO_2 emissions to fall below 2000 in 2040 with a very optimistic scenario, the increase in CO_2 emissions between 2000 and 2017 is around 41%. Although fossil fuels account for about 80% of the global

TABLE 17.1 CO$_2$ emissions.

| | \multicolumn{12}{c}{CO$_2$ emissions (million tonnes and share)} |
	2000		2017		2025		2030		2035		2040	
By sector												
Power	9305	0.40	13,587	0.42	10,656	0.36	7839	0.31	5127	0.24	3292	0.19
Industry	3922	0.17	6154	0.19	6273	0.21	5936	0.23	5481	0.26	5081	0.29
Transport	5757	0.25	7986	0.25	7932	0.27	7326	0.29	6373	0.30	5563	0.32
Buildings	2714	0.12	2997	0.09	2767	0.09	2593	0.10	2367	0.11	2202	0.12
Other	1424	0.06	1856	0.06	1907	0.06	1788	0.07	1633	0.08	1510	0.09
By fuel												
Coal	8951	0.39	14,448	0.44	11,335	0.38	8335	0.33	5577	0.27	3855	0.22
Oil	9620	0.42	11,339	0.35	10,657	0.36	9501	0.37	8032	0.38	6886	0.39
Gas	4551	0.20	6795	0.21	7543	0.26	7645	0.30	7373	0.35	6906	0.39
Total	23,123	1.00	32,581	1.00	29,535	1.00	25,482	1.00	20,982	1.00	17,647	1.00

Note: Revised from World Energy Outlook (2018).

primary energy demand, RESs have changed more than double in the last 17 years (IEA, 2018). The main drivers of the rapid increase in renewables are the enhancement in the global environmental conscience, policy support, and related state financial commitments and cost reductions. However, this increase has limited by the technology and high initial costs required for renewable energy in developing economies.

Even though these limitations, significant cost reductions in electricity generation from renewable projects have been realized. Global-weighted average cost of electricity for solar photovoltaic projects decreased from USD 0.36 to USD 0.10/kWh by 72%, concentrated solar power from USD 0.33 to USD 0.22/kWh by 33%, onshore wind from USD 0.08 to USD 0.06/kWh by 25%, and followed by 18% reduction offshore wind from USD 0.17 to USD 0.14/kWh between 2010 and 2017 (IRENA, 2018). According to BP (2018), renewables costs reducing by approximately 24% with every doubling of cumulative capacity. Moreover, emerging and developing economies out-performed developed countries in renewable energy investments. Excluding hydropower, China accounted for a record 45%, India 4%, and Brazil 2% of global investment in renewables (REN21, 2018). Considering that China and Brazil are in the first two places with hydropower generation of 312.7 and 100.3 GW (global share is 28% and 7%, respectively), the importance of investments in other renewable resources is emerging. Brazil has the greenest energy mix of the world, and almost 45% of the total energy consumption will be supplied by RESs in 2023 (IEA, 2018).

Numerous studies and policy reports have provided important information on shifting from fossil fuels to RESs (solar, wind, biofuels, tides, waves, geothermal, renewable municipal waste, etc), have a key role and critical in reducing greenhouse gas emission and diversifying energy supply sources (Apergis and Payne, 2010a,b; BP, 2017, 2018; Yildirim, 2014).

Renewable energy is expected to be the fastest growing source of the energy supply over the period 2015–40 at an average annual growth rate of 7% (BP, 2018). Thus, renewable energy generation and technologies have become the main components of the environment, sustainable development, and energy policies.

Table 17.2 above shows some of the main characteristics of the historical development of renewable energy. What stands out in the table is that there is a significant increase in the share of renewable energy in both world and Brazil, India, and China (BIC) countries. Coal is still dominant in China and India energy consumption, but this is slowly changing.

From 1980 to 2014, while the world's total energy consumption increased by 1.93 times, renewable energy consumption is 2.79; natural gas consumption 2.38; coal consumption increased by 2.18 and oil consumption by 1.42 times. The share of renewable energy in the world has increased by 4% in the last 35 years.

TABLE 17.2 Energy consumption (Quad Btu) 1980–2014.

	1980	1985	1990	1995	2000	2005	2010	2011	2012	2013	2014
Brazil											
Consumption	3.97	4.59	6.17	7.19	8.71	9.48	11.72	12.17	12.45	12.82	13.14
Coal	0.20	0.37	0.38	0.42	0.49	0.48	0.54	0.58	0.58	0.63	0.69
Natural gas	0.06	0.10	0.11	0.17	0.35	0.70	0.94	0.94	1.13	1.33	1.41
Petroleum	2.35	2.20	3.39	3.76	4.49	4.55	5.70	5.86	6.10	6.37	6.65
Renewables	1.37	1.92	2.29	2.85	3.40	3.75	4.54	4.80	4.64	4.49	4.39
Renewables (%)	*0.34*	*0.42*	*0.37*	*0.40*	*0.39*	*0.40*	*0.39*	*0.39*	*0.37*	*0.35*	*0.33*
China											
Consumption	19.83	24.30	31.05	37.42	42.94	74.57	109.36	120.82	129.41	135.04	136.90
Coal	14.53	18.53	24.30	27.90	30.34	54.44	78.15	87.10	92.18	94.99	93.45
Natural gas	0.58	0.53	0.57	0.68	0.94	1.73	3.94	4.83	5.30	6.02	6.69
Petroleum	4.12	4.29	4.91	7.05	9.65	14.28	19.04	20.49	21.56	22.51	23.56
Renewables	0.59	0.95	1.27	1.79	2.00	4.12	8.23	8.41	10.37	11.52	13.20
Renewables (%)	*0.03*	*0.04*	*0.04*	*0.05*	*0.05*	*0.06*	*0.08*	*0.07*	*0.08*	*0.09*	*0.10*
India											
Consumption	3.75	5.26	7.46	9.88	12.57	15.94	21.70	22.81	25.11	25.71	27.31
Coal	1.82	2.64	3.76	5.02	6.08	7.96	10.93	11.62	13.77	14.28	15.64

Continued

TABLE 17.2 Energy consumption (Quad Btu) 1980–2014.—cont'd

	1980	1985	1990	1995	2000	2005	2010	2011	2012	2013	2014
Natural gas	0.06	0.15	0.46	0.72	0.74	1.32	2.43	2.41	2.24	1.98	1.93
Petroleum	1.35	1.88	2.44	3.27	4.71	5.24	6.60	6.67	7.06	7.14	7.36
Renewables	0.52	0.58	0.81	0.87	1.03	1.42	1.75	2.10	2.03	2.31	2.38
Renewables (%)	*0.14*	*0.11*	*0.11*	*0.09*	*0.08*	*0.09*	*0.08*	*0.09*	*0.08*	*0.09*	*0.09*
World											
Consumption	293.46	317.43	358.96	367.86	402.36	465.13	523.73	538.70	550.85	561.77	567.03
Coal	79.18	91.28	101.15	91.73	98.37	128.08	154.15	163.47	169.53	172.95	172.25
Natural gas	53.87	63.37	75.12	81.32	89.60	102.61	119.42	122.99	125.80	127.65	128.15
Petroleum	132.11	123.60	136.48	142.03	156.93	171.70	178.63	179.87	182.92	184.80	187.54
Renewables	28.31	39.18	46.21	52.77	57.46	62.75	71.52	72.38	72.60	76.37	79.10
Renewables (%)	*0.10*	*0.12*	*0.13*	*0.14*	*0.14*	*0.13*	*0.14*	*0.13*	*0.13*	*0.14*	*0.14*

Source: www.eia.gov.

When we examine the developments in the country-specific trends, it is clear that the most important enlargements in terms of renewable energy were experienced in China. For the Chinese economy, renewable energy consumption has increased by approximately 22.3 times in quantity, while it has increased proportionally by 7%. These increases were 3.21 times for Brazil and 4.59 times for India. However, the point to be noted for Brazil is that although the total renewable energy consumption has increased more than three times, the share of proportionally renewable energy in the total has decreased by 1% point. The most remarkable increases for India and Brazil are in natural gas consumption.

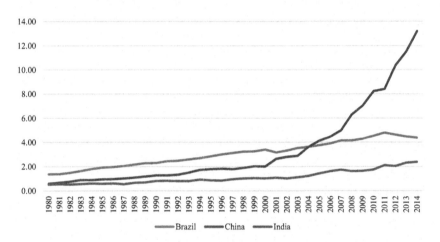

Renewable energy consumption (Quad Btu) 1980–2014.

The graph above illustrates the line of renewable energy consumption of BICs. When the graphic is analyzed, the remarkable point is the breakdown of renewable energy consumption in China after 2000. The progress of the other two countries seems to be more stable. In the past 20 years, China has included scaling power plants and reducing renewable energy costs in its development plans. By reaching these goals, it has made significant progress in green energy. The rapid growth of wind and solar energy technologies both in China and the world has reduced costs for green energy (CNREC, 2019).

The relationship between energy consumption and economic growth and the policies that are the extension of this relationship have tested by researchers with a series of hypotheses (Narayan, 2016; Apergis and Payne, 2009, 2012; Bowden and Payne, 2010; Payne, 2010; Ewing et al., 2007; Lee, 2006).

The first one is the growth hypothesis. Energy consumption, directly and indirectly, affects the growth process in a complementary relationship with other production factors. Under this hypothesis, there is a one-way causality from energy consumption to economic growth. In the case of the validity of this hypothesis, energy saving or negative energy supply shock will diminish economic growth.

According to the conservation hypothesis, policies designed to reduce energy consumption and energy waste have no negative impact on economic growth. The conservation hypothesis is supported if an increase in economic growth causes enhanced in energy consumption. The existence of a one-way causality from economic growth to energy consumption indicates the validity of this hypothesis.

The neutrality hypothesis refers to energy consumption is a small component of the total output of an economy and therefore has little or no impact on the growth process. If there is no causality between energy consumption and economic growth when the neutrality hypothesis is valid.

The feedback hypothesis indicates a bidirectional causality between two variables. It means that energy consumption and economic growth complement each other. Feedback hypothesis posits a two-sided connection between energy policies and economic growth.

The purpose of this paper is to extend the empirical literature on the asymmetric causal relationship between renewable energy consumption and GDP and determining short and long-run cointegration coefficients in the case of China, India, and Brazil. The countries and periods included in the study are not randomly selected. They are chosen from BRICS countries considering the data quality and availability. Another reason for choosing these countries as the sample is that China and India are among the top three countries that consume the most energy and produce CO_2 emissions. In addition, they have the fastest economic growth rate and large renewable energy potential.

When the current literature is reviewed, many studies have been conducted on the relationship between renewable energy and economic growth. It is concluded that these studies have two distinctive features. Firstly, the existing evidence does not include consensus on the relationship between renewable energy and economic growth. The relationship between energy consumption and economic growth has tested by researchers with growth, conservation, neutrality, and feedback hypothesis. The results of the studies differed in terms of hypotheses. The aim of this study is to present which hypothesis is validated. Second, although BRIC countries have made rapid progress in the renewable energy sector, only Venkatraja (2020) has examined the effects of renewable energy on economic growth in these countries.

Firstly, the stationarity process of the series is investigated by augmented Dickey–Fuller (ADF) and Phillips–Perron (PP) unit root tests. To analyze the time series data for countries and in order to get cointegration coefficients separately, autoregressive distributed lag (ARDL) approach is used. For each individual country the test used that the causality approach of Hatemi-J (2012) which is taking into account asymmetric behavior in causality. Moreover, the relationship between renewable energy and GDP is modeled using a production function to avoid omitted variable. A novelty of the paper is that it investigates the role of asymmetric casual behavior renewable energy in the growth process for sample countries over the chosen period.

The paper is constructed as follows. The next section discusses briefly related literature dealing with the nexus between renewables and/or non-renewables and GDP for one or more individual countries, or a panel of countries. The following section describes the methodology and the data used while the empirical results are presented. Finally, concluding remarks given in the last section.

2. Literature review

Economists have been trying to explain the source of economic growth for many years. In this respect, it is important to reveal the relationship between economic growth and energy consumption. For this purpose, a number of studies have focused on these issues in different countries and time periods using different econometric methods. At first, the relationship between energy consumption and economic growth may seem obvious. However, the empirical results of studies investigating the relationship between the two variables are sometimes inconsistent. There may be several reasons for this. For example, using different data sets, applying different econometric methods, and the unique characteristics of countries may be the cause of this contradiction.

In this study, the literature review is divided into two sections as single country-specific studies and multi-countries studies. The first part of the literature review includes single country-specific studies.

Bowden and Payne (2010) used the Toda-Yamamoto causality test (Toda and Yamamoto, 1995) to analyze the causal relationship between renewable and nonrenewable energy consumption by sector to real GDP in the United States over the time span of 1949—2006. The results showed that there is not causality from renewable energy consumption in the commercial and industrial sectors to real GDP. In addition there is unidirectional causality between residential energy consumption and real GDP. Otherwise, the results indicate bidirectional causality between nonrenewable energy consumption in the commercial and residential sectors and real GDP. There is unidirectional causality between industrial nonrenewable energy consumption and real GDP.

Fang (2011) used multivariate OLS to investigate the impact of renewable energy consumption and renewable energy consumption share on GDP and GDP per capita for China over the time period of 1978—2008. The empirical findings showed that renewable energy consumption positively effects real GDP and GDP per capita. The impact of renewable energy consumption share (SREC) on GDP and GDP per capita is not significant.

Payne (2012) examined the causality between renewable energy consumption, real GDP, carbon emissions, and real oil prices in the United States over the period of 1949—2009. In this study, author used the Toda-Yamamoto causality test. The Toda-Yamamoto causality test result showed that renewable energy legislation and policies had a positive impact on renewable energy consumption since 1978 and it was statistically significant. There is not a causal effect of real GDP, carbon emissions, and real oil prices on renewable energy consumption. However, the result indicated that the unexpected shocks to real GDP and carbon emissions are statistically significant and these shocks have an impact on renewable energy consumption over time.

Lin et al. (2013) used linear and nonlinear causality test to examine the causality relation between the industrial renewable energy consumption sector, the residential and commercial renewable energy consumption sector, the transportation renewable energy consumption sector, the electric power renewable energy consumption sector, and real GDP for the United States over the period of 1989−2008. According to the linear causality test results, there is bidirectional causality between the industrial renewable energy consumption sector and real GDP. Moreover, there is not causality between the residential and commercial renewable energy consumption sector, the transportation renewable energy consumption sector, the electric power renewable energy consumption sector, and real GDP. The nonlinear causality test results showed that there is bidirectional causality between the electric power renewable energy consumption sector and real GDP. Also, the results indicate the unidirectional causality among real GDP and the residential and commercial renewable energy consumption sector and the transportation renewable energy consumption sector.

Lin and Moubarak (2014) attempted to investigate the relationship between renewable energy consumption and economic growth in China. The authors selected the period 1977−2011 and used ARDL approach, Johansen cointegration test, and Granger causality test. According to the results, there is cointegration among the variables, and the causality test results suggest bidirectional long-run causality between renewable energy consumption and economic growth.

Shahbaz et al. (2015) examined the relationship between renewable energy consumption and economic growth for Pakistan over the period of 1972−2011. They used ARDL model and rolling window approach for cointegration. They examined VECM Granger causality test for an analyzed causal relationship between series and used an innovative accounting approach for testing the robustness of the causality test results. The empirical findings showed that there is cointegration among the variables. The results showed that renewable energy consumption has a positive effect on economic growth. The rolling window results indicated renewable energy consumption, capital, and labor positively effect the economic growth, except for a few quarters. The causality test result suggested the feedback effect between renewable energy consumption and economic growth.

Alper and Oğuz (2016) used asymmetric causality test approach and ARDL approach to investigate the causal relationship between variables and the impact of renewable energy consumption on economic growth for eight new EU member countries (Bulgaria, Cyprus, Czech Republic, Estonia, Hungary, Poland, Romania, Slovenia) over the period of 1990−2009.

According to test results, the impact of renewable energy consumption on economic growth is positive and statistically significant in Bulgaria, Estonia, Poland, and Slovenia. There is no causal relationship between variables in Cyprus, Estonia, Hungary, Poland, and Slovenia. The results support the neutrality hypothesis in Cyprus, Estonia, Hungary, Poland, and Slovenia. There is causality from economic growth to renewable energy consumption in the Czech Republic. The conservation hypothesis was supported in the Czech Republic. Also, there is causality from renewable energy consumption to economic growth in Bulgaria. The growth hypothesis was supported in Bulgaria.

Shakouri and Yazdi (2017) examined the causality between economic growth, renewable energy consumption, energy consumption, capital, and trade openness using ARDL bound testing approach and Granger causality test in South Africa from 1971 to 2015. The result showed that the variables are cointegrated and there is a long-run relationship between the variables. When the results of Granger causality test are examined, there is a bidirectional causality relation from renewable energy consumption and trade openness to economic growth. The causality test result suggested the feedback hypothesis between renewable energy consumption and trade openness to economic growth.

Amri (2017) analyzed economic growth-energy consumption nexus in Algeria over the 1980−2012 period. ARDL approach, Gregory−Hansen, and Johansen cointegration tests, and VECM Granger causality test was used in this study. Gregory−Hansen and Johansen cointegration tests were performed to confirm the ARDL. According to Gregory−Hansen and Johansen cointegration tests, there is a long-run relationship between variables. The short-run and long-run ARDL result showed that the nonrenewable energy and capital positively impact economic growth whereas renewable energy is not significant. In addition, the causality test result showed that the feedback hypothesis exists between nonrenewable energy consumption and output in both the short and long run. Also, there is a one-way causality from renewable energy to GDP in the long run. However, there is not causality from renewable energy to GDP in the short run.

Wang et al. (2018) evaluated the relationship between renewable energy consumption, economic growth, and human development index with two-stage least square method in Pakistan over the time span of 1990−2014. This study showed that the impact of renewable energy consumption on the human development index is negative and the CO_2 emission improves human development index. In addition, trade openness negatively affects human development index. The causality test results suggest a feedback hypothesis between an environmental factor and human development in the long run.

This part of the literature review includes multi-countries studies.

Apergis and Payne (2010a,b) used a multivariate panel data approach for analyzed the causality between renewable energy consumption and economic growth in 13 countries in Eurasia over the time span of 1992−2007. The result of the heterogeneous panel cointegration test showed that there is a long-run relationship between all variables. The error correction models showed that there is bidirectional causality between renewable energy consumption and economic growth. Thus, the empirical findings lend support for the feedback hypothesis of the interdependent relationship between renewable energy consumption and economic growth. Test results suggest a feedback hypothesis.

Menegaki (2011) utilized multivariate panel methods for analyzed the causal relationship between economic growth and renewable energy in 27 European countries in 1997−2007. The result indicated that there is not a causal relationship between energy consumption and GDP.

Apergis and Payne (2012) tested the causality between renewable and nonrenewable energy consumption and economic growth in 80 countries. Authors selected the period 1990−2007 and they used a multivariate panel framework. The result showed that there is bidirectional causality between renewable and nonrenewable energy consumption and economic growth in the short and the long-run. Test results suggest a feedback hypothesis.

Salim et al. (2014) used the panel cointegration and causality technique for a test to the relationship between renewable energy, nonrenewable energy consumption, industrial output, and economic growth for OECD countries in the 1980−2011 period. The cointegration tests' results showed that there is a long-run relationship between the variables. The result of the causality test indicated a bidirectional causality among industrial output and renewable and nonrenewable energy consumption. In addition, the authors found short-run bidirectional relationship between economic growth and nonrenewable energy consumption and they indicated unidirectional causality between economic growth and renewable energy consumption.

Apergis and Payne (2014) examined the determinants of renewable energy consumption using panel framework in 25 OECD countries over the period of 1980−2011. For this purpose, renewable energy consumption per capita, real GDP per capita, CO_2 emissions per capita, and real oil prices variables are used in the study. The result of panel cointegration and error correction model indicated that there is long-run relationship between all variables. According to the panel error correction model, there is a feedback relationship between the variables.

Apergis and Payne (2015) examined the relationship between renewable energy consumption per capita, GDP per capita, CO_2 emissions per capita, and real oil prices. The authors used panel cointegration and error correction model for 11 South American countries (Argentina, Bolivia, Brazil, Chile, Colombia, Ecuador, Paraguay, Peru, Suriname, Uruguay, and Venezuela) over the time of 1980−2010. The test result showed that there is a long-run equilibrium relation between renewable energy consumption per capita and all independent variables. All independent variables have a positive impact on renewable energy consumption per capita. In addition, the authors found a bidirectional causal relationship between renewable energy consumption per capita and real

GDP per capita in the short and long-run. The results of the paper indicated that there is a feedback relationship between the variables.

Koçak and Sarkgüneşi (2017) the relationship between renewable energy consumption and economic growth using Pedroni panel cointegration approach and Dumitrescu and Hurlin heterogeneous panel causality approach in nine Black Sea and Balkan countries (Albania, Bulgaria, Georgia, Greece, Macedonia, Romania, Russian Federation, Turkey, and Ukraine) over the period of 1990−2012. The result showed that there is a long-run relationship between renewable energy consumption and economic growth. The result of panel causality test indicated that the feedback hypothesis was supported in panel. When countries are examined, growth hypothesis was supported in Bulgaria, Greece, Macedonia, Russia, and Ukraine, the feedback hypothesis was supported in Albania, Georgia, and Romania, the neutrality hypothesis was supported in Turkey.

3. Data and methodology

The relationship between renewable energy consumption and economic growth for Brazil, China, and India was investigated using annual data for the period 1980−2014. Labor and capital are the primary inputs used in a production process. Cobb−Douglas production function has augmented with renewable energy input as the third factor of production. Therefore, in the model established under the Cobb−Douglas production function, capital, labor, and renewable energy consumption are defined as a function of GDP. The model established in this framework can be shown as follows:

$$\ln Y_t = \beta_0 + \beta_1 \ln K_t + \beta_2 \ln L_t + \beta_3 \ln REN_t + \varepsilon_t \qquad (17.1)$$

where Y_t is the real GDP of constant 2011 US dollars, K_t is the real gross fixed capital formation, L_t is number of persons engaged, and REN_t is renewable energy consumption (quadrillion BTU). In addition, while creating the labor series, the employment number of each country was multiplied by the human capital index included in the Penn World Table. The aim here is to prevent the labor from taking place homogeneously in the established model. Thus, labor differences among countries were also included in the model. The data of real GDP and real gross fixed capital are sourced from World Development Indicators. The data of number of persons engaged and human capital index are sourced from Penn World Table 9.0. The data of renewable energy consumption are sourced from Energy Information Administration. In addition, all variables are in natural logarithms.

3.1 Cointegration analysis

This study utilizes cointegration analysis and causality test to analyze the relationship among variables. In this context, ARDL bounds testing approach developed by Pesaran et al. (2001) is used in order to examine the long-run relationship that exists among variables. ARDL bounds test is the most suitable approach when the order of integration of variables is different. However, in the ARDL bounds test, the order of integration of the variables is expected to be I(0) or I(1). If the order of integration of one of the variables is I(2), the ARDL bounds test cannot be used (Pesaran et al., 2001; Narayan, 2005). Unlike other cointegration tests, ARDL boundary test is also suitable for small time series. In addition, ARDL bounds test generally provides unbiased estimates of the long-run model and valid t-statistics even when some of the regressors are endogenous.

In this context, the unit root properties of variables should be investigated to determine the order of integration of variables. Therefore, ADF and PP unit root tests were used in this study. Dickey and Fuller (1979) developed a unit root test for cases where error terms were uncorrelated. Later, Dickey and Fuller also modified the unit root test for cases where error terms were correlated. This test is referred to as ADF unit root test. This study used ADF test which was proposed by Said and Dickey (1984). The null hypothesis in ADF test is "series contains unit root," thus series is not stationary. The calculated test statistic is compared to the MacKinnon critical values. In general, null hypothesis can be rejected if the calculated test statistic is negatively less than the MacKinnon critical values. PP (Phillips and Perron, 1988) test is more flexible than ADF test according to the hypothesis about error term. For ADF test, error term is independent and with constant variance. The asymptotic distribution of the PP t-statistic is the same as the ADF t-statistic critical values.

After examining the unit root characteristics of the series, it is necessary to investigate the existence of a long-run relationship between the series. In this context, ARDL bounds testing approach is used in this study. One of the most important advantages of the ARDL bounds test approach is that it analyzes both short and long-run relationships among variables. In addition, Pesaran and Shin (1998) argue that ARDL bounds testing approach has consistent results against autocorrelation. The ARDL version of the model is as follows:

$$\Delta \ln Y_t = \gamma_1 + \sum_{i=1}^{k1} \alpha_{1i} \Delta \ln Y_{t-i} + \sum_{i=0}^{l1} \phi_{1i} \Delta \ln K_{t-i} + \sum_{i=0}^{m1} \eta_{1i} \Delta \ln L_{t-i}$$

$$+ \sum_{i=0}^{n1} \theta_{1i} \Delta \ln REN_{t-i} + \delta_1 Y_{t-1} + \delta_2 K_{t-1} + \delta_3 L_{t-1} + \delta_4 REN_{t-1} + \varepsilon_{1t}$$

$$(17.2)$$

where ε_{1t} is error term and Δ is the first difference operator. The bounds testing procedure is based on the joint F-statistic (or Wald statistic) for cointegration analysis among variables. The null hypothesis of no cointegration among the variables in Eq. (17.2) is $(H_0 : \delta_1 = \delta_2 = \delta_3 = \delta_4 = 0)$ against the alternative hypothesis $(H_1 : \delta_1 \neq \delta_2 \neq \delta_3 \neq \delta_4 \neq 0)$. Pesaran et al. (2001) report two sets of critical values for a given significance level. Two sets of critical values for a given significance level that are reported. Pesaran et al. (2001) provide critical value bounds for all classifications of the regressors into purely I(1), purely I(0) or mutually cointegrated. If the computed F-statistics exceeds the upper critical bounds value, then the null hypothesis is rejected. If the F-statistics value lies below the lower bounds critical value, there is no cointegration relationship among the variables. However, if the F-statistics value lies between the upper and lower bound critical values, the null hypothesis is inconclusive.

In the ARDL bounds testing approach, if there is a long-run relationship among the variables, the coefficients of short-run and long-run models can be estimated by the following equations.

$$
\ln Y_t = \gamma_2 + \sum_{i=1}^{k2} \alpha_{2i} \ln Y_{t-i} + \sum_{i=0}^{l1} \phi_{2i} \ln K_{t-i} + \sum_{i=0}^{m2} \eta_{2i} \ln L_{t-i}
$$
$$
+ \sum_{i=0}^{n2} \theta_{2i} \ln REN_{t-i} + \varepsilon_{2t} \tag{17.3}
$$

$$
\Delta \ln Y_t = \gamma_3 + \sum_{i=1}^{k3} \alpha_{3i} \Delta \ln Y_{t-i} + \sum_{i=0}^{l3} \phi_{3i} \Delta \ln K_{t-i} + \sum_{i=0}^{m3} \eta_{3i} \Delta \ln L_{t-i}
$$
$$
+ \sum_{i=0}^{n3} \theta_{3i} \Delta \ln REN_{t-i} + \psi ECT_{t-1} + \varepsilon_{3t} \tag{17.4}
$$

where ψ is the coefficient of error correction term (ECT). It is expected to have a statistically significant coefficient with a negative sign.

3.2 Causality analysis

Granger causality test and Toda-Yamamoto causality test are used frequently to investigate the causality relationship between variables. However, the causality relationships between positive and negative shocks of variables cannot be determined using these tests. Granger and Yoon (2002) stated that the cointegration relationship among the variables may differ when examined separately for positive and negative shocks. On the other hand, Hatemi-J (2012) developed the asymmetric causality test by indicating that the positive and negative shocks may also differ in the causality relationship.

We can show random walk processes for lnY and lnREN which take into account the positive and negative components, respectively.

$$
\ln Y_t = \ln Y_{t-1} + e_{1t} = \ln Y_0 + \sum_{i=1}^{t} e_{1i} \tag{17.5}
$$

$$
\ln REN_t = \ln REN_{t-1} + e_{2t} = \ln REN_0 + \sum_{i=1}^{t} e_{2i} \tag{17.6}
$$

In line with Granger and Yoon (2002), positive and negative shocks can be shown as following, respectively:

$$e_{1i}^+ = \max(e_{1i}, 0) \qquad (17.7)$$

$$e_{1i}^- = \max(e_{1i}, 0) \qquad (17.8)$$

$$e_{2i}^+ = \max(e_{2i}, 0) \qquad (17.9)$$

$$e_{2i}^- = \max(e_{2i}, 0) \qquad (17.10)$$

Within the framework of these equations, the equalities of $\ln Y$ and $\ln REN$ can be expressed by arranging as follows:

$$\ln Y_t = \ln Y_{t-1} + e_{1t} = \ln Y_0 + \sum_{i=1}^{t} e_{1i}^+ + \sum_{i=1}^{t} e_{1i}^- \qquad (17.11)$$

$$\ln REN_t = \ln REN_{t-1} + e_{2t} = \ln REN_0 + \sum_{i=1}^{t} e_{2i}^+ + \sum_{i=1}^{t} e_{2i}^- \qquad (17.12)$$

The negative and positive shocks of each variable in cumulative form can be shown as follows:

$$\ln Y_t^+ = \sum_{i=1}^{t} e_{1i}^+, \ \ln Y_t^- = \sum_{i=1}^{t} e_{1i}^-, \ln REN_t^+ = \sum_{i=1}^{t} e_{2i}^+, \ln REN_t^- = \sum_{i=1}^{t} e_{2i}^-$$

$$(17.13)$$

Under the assumption that the causality relationship between positive shocks of renewable energy consumption and GDP $\left[y_t^+ = \left(\ln Y_t^+, \ \ln REN_t^+ \right) \right]$ is examined, the causality relationship between the two variables can be expressed with the vector autoregressive (VAR_p) model as follows:

$$y_t^+ = v_t + A_1 y_{t-1}^+ + \ldots + A_p y_{t-p}^+ + \ldots + A_{p+d} y_{t-p-d}^+ + \varepsilon_t^+ \qquad (17.14)$$

where d represents the maximum degree of integration. This model can be abbreviated as in Eq. (17.15):

$$Y = DZ + \delta \qquad (17.15)$$

where,

$$Y: = \left(y_1^+, \ldots, y_T^+ \right) (n \times T) \text{ matrix}, \qquad (17.16)$$

$$D: = \left(v, A_1, \ldots, A_p, \ldots, A_{p+d} \right) (n \times (1 + n(p + d))) \text{ matrix}, \qquad (17.17)$$

$$Z_t: = \begin{bmatrix} 1 \\ y_t^+ \\ y_{t-1}^+ \\ \vdots \\ y_{t-p+1}^+ \end{bmatrix} ((1+np) \times 1) \text{ matrix, for } t = 1, ..., T, \qquad (17.18)$$

$$Z: = (Z_0, ..., Z_{T-1})((1+np) \times T)\text{matrix and,} \qquad (17.19)$$

$$\delta: = \left(\varepsilon_1^+, ..., \varepsilon_T^+\right)(n \times T)\text{matrix.} \qquad (17.20)$$

The null hypothesis of non-Granger causality, $H_0 : C\beta = 0$, is tested by the following test method:

$$\text{MWald} = (C\beta)' \left[C\left((Z'Z)^{-1} \otimes S_U\right)C' \right]^{-1} (C\beta), \qquad (17.21)$$

where $\otimes$ represents the Kronecker product, S_U is the variance-covariance matrix of the unrestricted VAR model, and C is a $p \times n(1 + np)$ indicator matrix with elements ones for restricted parameters and zeros for the rest of the parameters. The MWALD test statistic has asymptotically χ^2 distribution. It is also assumed that the error term is normally distributed. Hatemi-J (2012) in asymmetric causality analysis. Three important situations are outstanding: determining the lag length of the VAR model, determining the additional lag length to be added to the model, and obtaining critical values for the Wald test statistics. This study, the optimum lag length is determined by Hatemi-J criterion developed by Hatemi-J (2003).

4. Empirical findings

Results of the ADF unit root and PP unit root tests are presented in Table 17.1. In both unit root tests, both intercept and intercept and trend models were used. For three countries included in the analysis in this study, some variables are stationary in the level. In Brazil, the ln REN is stationary in the level in both the ADF and PP test in the intercept model. In China, the ln CAP is stationary in the level in the intercept and trend model in ADF test. The ln LAB is stationary in the level in the intercept model in the PP test for China and India. At the same time, all variables are stationary in the first differences. In this context, it would be more appropriate to use the ARDL methodology to investigate the long-run relationship among variables (Table 17.3).

In the study, Schwartz Bayesian criteria are used to determine the appropriate lag lengths in ARDL bounds testing approach. In this context, the F-statistic values obtained from the ARDL bounds test are presented in Table 17.4. The appropriate ARDL model for Brazil is (3,3,2,1). The F-statistic

TABLE 17.3 Unit root test results.

| | Augmented Dickey–Fuller test | | | | | |
| | Brazil | | China | | India | |
	Intercept	Intercept and trend	Intercept	Intercept and trend	Intercept	Intercept and trend
lnY	0.884	−2.806	−0.041	−1.962	2.295	−1.040
lnCAP	0.395	−3.138	−0.103	−4.505***	0.745	−1.864
lnLAB	−0.898	−1.761	−2.188	−0.795	−1.749	−0.274
lnREN	−2.972**	−0.986	1.733	−0.295	0.341	−2.673
Δ lnY	−5.112***	−4.263***	−3.798***	−3.712**	−4.816***	−5.607***
Δ lnCAP	−4.913***	−4.957***	−4.609***	−4.528***	−5.607***	−5.643***
Δ lnLAB	−4.415***	−4.367***	−3.344**	−3.431*	−2.625*	−3.253*
Δ lnREN	−4.227***	−5.459***	−5.660***	−6.386***	−6.792***	−6.924***

| | Phillips-Perron test | | | | | |
| | Brazil | | China | | India | |
	Intercept	Intercept and trend	Intercept	Intercept and trend	Intercept	Intercept and trend
lnY	0.740	−3.004	−0.369	−2.442	0.999	−0.823
lnCAP	0.395	−3.144	0.413	−2.087	0.749	−1.892

lnLAB	−0.773	−2.112	−9.007***	−2.128	−3.334**	−0.029
lnREN	−3.843***	−0.448	1.999	−0.295	2.508	−2.468
Δ lnY	−5.263***	−5.019***	−3.407**	−3.395*	−4.812***	−6.764***
Δ lnCAP	−4.899***	−4.938***	−5.173***	−4.853***	−5.615***	−5.643***
Δ lnLAB	−4.476***	−4.443***	−3.248**	−3.497*	−2.698*	−3.359*
Δ lnREN	−4.227***	−5.458***	−5.681***	−6410***	−7.840***	−16.573***

***, **, and * represent 1%, 5% and 10% of significance levels, respectively. The optimal lag length in ADF test is based on SBC, and in PP test, the bandwidth is based on Newey–West bandwidth.

TABLE 17.4 Estimated autoregressive distributed lag models and bounds F-test for cointegration.

Country	Model	F-stat	ECM(-1)	Critical values					
				1%		5%		10%	
				I(0)	I(1)	I(0)	I(1)	I(0)	I(1)
Brazil	(3,3,2,1)	4.289**	−1.002 [0.000] ***	3.65	4.66	2.79	3.67	2.37	3.20
China	(3,0,1,1)	16.239***	−0.243 [0.000]***	3.65	4.66	2.79	3.67	2.37	3.20
India	(3,0,0,1)	4.669***	−0.250 [0.000]***	3.65	4.66	2.79	3.67	2.37	3.20

***, **, and * represent 1%, 5%, and 10% of significance levels, respectively. Critical values are obtained from Pesaran et al. (2001). Numbers in brackets are P-values.

is 4.289, and it is above the upper bound critical value of 10% significance obtained from Pesaran et al. (2001). Similarly, appropriate ARDL model for China and India are (3,0,1,1) and (3,0,0,1), respectively. Furthermore, the F-statistics are 16.239 and 4.669, respectively. China's F-statistic is above the upper bound critical value of 1% significance, while India's F-statistic is above the upper bound critical value of 5% significance. Therefore, since the absence hypothesis is rejected in all three countries, it can be stated that there is a long-run cointegration relationship among the variables. Finally, the error correction coefficient [ECM(-1)] is both negative and statistically significant as expected in all three countries.

In the study, the stability of the short-run and long-run coefficients are investigated with the cumulative sum (CUSUM) and the cumulative sum of squares (CUSUMQ) tests. According to the findings, the estimated coefficients are stable for the period considered. On the other hand, diagnostic test results are presented in Table 17.5. When we look at Jarque–Bera normality test results, we can say that the residuals obtained from the estimates are normally distributed. Breusch–Godfrey LM test results confirm the absence of serial correlation in error terms. The results of the Breusch–Pagan–Godfrey test show that residuals are homoscedastic. Finally, the results of the Ramsey-Reset test confirm that there is no specification error for the functional form in all countries.

What stands out in the table is long-run coefficients obtained from the estimates, renewable energy consumption has a positive effect on growth in Brazil and China. However, the coefficient of renewable energy in India is not statistically significant. Additionally, the impact of gross fixed capital and total employment on economic growth is positive in all countries. In this context, we can say that the contribution of gross fixed capital investment and total employment to economic growth in all countries is more than the contribution of renewable energy consumption.

When we look at the short-run forecast findings in Table 17.5, we can state that renewable energy consumption has a positive effect on economic growth in Brazil. At the same time, one period-lag renewable energy consumption has a positive impact on economic growth in China. In India, renewable energy consumption has no statistically significant effect on economic growth. On the other hand, while capital has a positive effect on economic growth in all countries, total employment has a positive effect only on economic growth in Brazil and China.

The findings of the asymmetric causality relationship between the renewable energy consumption and the positive and negative shocks of economic growth presents in Table 17.6. All calculated test statistics are smaller than the critical values obtained from 10,000 bootstrap replications. Therefore, the null hypothesis could not be rejected. In this context, the findings show that there is no causal relationship between renewable energy consumption and economic growth in Brazil, China, and India in both positive shocks and negative shocks.

TABLE 17.5 The results of long and short run.

Dep. Variable: ln Y	Brazil	China	India
Long run			
Intercept	10.723 [0.000]***	−8.568 [0.053]*	4.969 [0.000]***
lnCAP	0.279 [0.000]***	0.309 [0.025]**	0.767 [0.000]***
lnLAB	0.437 [0.000]***	1.165 [0.000]***	0.188 [0.086]*
lnREN	0.115 [0.000]***	0.297 [0.004]***	−0.076 [0.722]
Short run			
Intercept	10.741 [0.002]***	−2.098 [0.137]	1.293 [0.018]**
lnY(−1)	0.582 [0.019]**	1.172 [0.000]***	0.536 [0.005]***
lnY(−2)	−0.070 [0.814]	−0.714 [0.004]***	−0.147 [0.453]
lnY(−3)	−0.514 [0.005]***	0.297 [0.015]**	0.351 [0.028]**
lnCAP	0.235 [0.000]***	0.076 [0.057]*	0.199 [0.000]***
lnCAP(−1)	−0.119 [0.099]*	−	−
lnCAP(−2)	0.021 [0.793]	−	−
lnCAP(−3)	0.143 [0.012]**	−	−
lnLAB	0.443 [0.000]***	−1.933 [0.009]***	0.049 [0.143]
lnLAB(−1)	−0.231 [0.120]	2.218 [0.003]***	−
lnLAB(−2)	0.225 [0.092]*	−	−
lnREN	0.195 [0.002]***	−0.0039 [0.229]	0.054 [0.236]
ln REN(−1)	−0.080 [0.064]*	0.113 [0.004]***	−0.074 [0.107]
Diagnostic tests			
Serial correlation[a]	2.643 [0.101]	1.570 [0.231]	0.554 [0.583]
Heteroscedasticity[b]	0.891 [0.570]	0.761 [0.639]	0.963 [0.480]
Normality[c]	0.055 [0.973]	1.989 [0.370]	0.451 [0.798]
Functional form[d]	0.995 [0.333]	0.806 [0.429]	0.807 [0.428]
CUSUM	Stable	Stable	Stable
CUSUMQ	Stable	Stable	Stable

***, **, and * represent 1%, 5%, and 10% of significance levels, respectively. Numbers in brackets are P-values.
[a]The Breusch–Godfrey LM test statistic for no serial correlation.
[b]The Breusch–Pagan–Godfrey test statistic for homoscedasticity.
[c]The Jarque–Bera statistic for normality.
[d]The Ramsey's reset test statistic for regression specification error.

TABLE 17.6 The results of asymmetric causality test.

Null hypothesis	Wstat	Critical values		
		%1	%5	%10
Brazil				
$lnY^+ \nRightarrow lnREN^+$	0.255	9.460	4.495	2.984
$lnREN^+ \nRightarrow lnY^+$	0.632	8.488	4.456	2.956
$lnY^- \nRightarrow lnREN^-$	0.292	7.933	4.231	2.899
$lnREN^- \nRightarrow lnY^-$	0.001	8.065	4.396	2.953
China				
$lnY^+ \nRightarrow lnREN^+$	0.000	8.502	4.162	2.844
$lnREN^+ \nRightarrow lnY^+$	1.664	9.580	4.665	2.975
$lnY^- \nRightarrow lnREN^-$	0.404	10.074	4.637	2.925
$lnREN^- \nRightarrow lnY^-$	0.100	9.361	4.358	2.918
India				
$lnY^+ \nRightarrow lnREN^+$	0.138	8.327	4.258	2.981
$lnREN^+ \nRightarrow lnY^+$	0.001	8.289	4.263	2.890
$lnY^- \nRightarrow lnREN^-$	0.001	7.658	4.037	2.832
$lnREN^- \nRightarrow lnY^-$	1.196	7.543	4.252	2.889

***, **, and * represent 1%, 5%, and 10% of significance levels, respectively. Critical values are obtained from 10,000 bootstrap replications.

5. Results and policy implications

The use of fossil fuels in the economic growth process of countries causes environmental degradation. The increase in renewable energy consumption contributes to the reduction of global warming and other environmental problems. This study aims to examine the impact renewable energy consumption on the economic growth process of BIC countries, and it purposes to explore the long-run relationship between renewable energy and economic growth.

This study examines renewable energy consumption and economic growth by using an alternative methodology based on asymmetric causality, and ARDL cointegration techniques using data from BIC. According to the findings while there are existing long-run relations among renewable energy and GDP. The most striking result to emerge from the study is that renewable energy plays a significant and positive effect to improve the economic growth process in Brazil and China, but not in India. Moreover, research has also shown that there is no causality running from renewable energy to GDP in

BIC. These results suggesting evidence of the neutrality hypothesis which is meaning that the consumption of renewable energy has a minor role in the determination of GDP for BICs.

The economic growth of the BIC countries is not only affected by the increase in renewable energy factors but also by country-specific factors. The BIC countries differ from each other based on structural characteristics. Besides the differences in countries' monetary and fiscal policies, there are also differences in terms of demography, productivity, technology, and natural resources. These differences between countries lead to differences in their economic growth. The GDP per capita is much higher in China, lower in Brazil than China, and much lower in India. study provides statistical evidence that country-specific factors play an important role in the growth of the country's economy besides of renewable energy factors.

References

Alper, A., Oguz, O., 2016. The role of renewable energy consumption in economic growth: evidence from asymmetric causality. Renew. Sustain. Energy Rev. 60, 953−959.

Amri, F., 2017. The relationship amongst energy consumption (renewable and non-renewable), and GDP in Algeria. Renew. Sustain. Energy Rev. 76, 62−71.

Apergis, N., Payne, J.E., 2009. Energy consumption and economic growth in Central America: evidence from a panel cointegration and error correction model. Energy Econ. 31 (2), 211−216.

Apergis, N., Payne, J.E., 2010. Renewable energy consumption and economic growth: evidence from a panel of OECD countries. Energy Policy 38 (1), 656−660.

Apergis, N., Payne, J.E., 2010. Renewable energy consumption and growth in Eurasia. Energy Econ. 32 (6), 1392−1397.

Apergis, N., Payne, J.E., 2012a. A global perspective on the renewable energy consumption-growth nexus. Energy Sour. B Energy Econ. Plann. 7 (3), 314−322.

Apergis, N., Payne, J.E., 2012b. Renewable and non-renewable energy consumption-growth nexus: evidence from a panel error correction model. Energy Econ. 34 (3), 733−738.

Apergis, N., Payne, J.E., 2014. The causal dynamics between renewable energy, real GDP, emissions and oil prices: evidence from OECD countries. Appl. Econ. 46 (36), 4519−4525.

Apergis, N., Payne, J.E., 2015. Renewable energy, output, carbon dioxide emissions, and oil prices: evidence from South America. Energy Sour. B Energy Econ. Plann. 10 (3), 281−287.

Bowden, N., Payne, J.E., 2010. Sectoral analysis of the causal relationship between renewable and non-renewable energy consumption and real output in the US. Energy Sour. B Energy Econ. Plann. 5 (4), 400−408.

Boyle, Godfrey, 2004. Renewable energy. Oxford University Press.

BP, 2017. Energy Outlook-2017. BP, UK.

BP, 2018. Energy Outlook-2018. BP, UK.

CNREC, 2019. China Renewable Energy Outlook-2019. https://www.dena.de/fileadmin/dena/Publikationen/PDFs/2019/CREO2019_-_Executive_Summary_2019.pdf.

Dickey, D.A., Fuller, W.A., 1979. Distribution of the estimators for autoregressive time series with a unit root. J. Am. Stat. Assoc. 74 (366a), 427−431.

EREC, European Renewable Energy Council, 2010. In: Renewable Energy in Europe: Markets, Trends and Technologies. Routledge.

EWEA, 2009. Wind Energy-The Facts. A Guide to the Technology, Economics and Future of Wind Power. Earthscan, London.

Ewing, B.T., Sari, R., Soytas, U., 2007. Disaggregate energy consumption and industrial output in the United States. Energy Policy 35 (2), 1274–1281.

Fang, Y., 2011. Economic welfare impacts from renewable energy consumption: the China experience. Renew. Sustain. Energy Rev. 15 (9), 5120–5128.

Granger, C.W., Yoon, G., 2002. Hidden Cointegration. U of California, Economics Working Paper.

Hatemi-J, A., 2003. A new method to choose optimal lag order in stable and unstable VAR models. Appl. Econ. Lett. 10 (3), 135–137.

Hatemi-J, A., 2012. Asymmetric causality tests with an application. Empir. Econ. 43 (1), 447–456.

IEA, 2018. World Energy Outlook 2018. International Energy Agency, Paris.

IRENA, 2018. Renewable Power Generation Costs in 2017. International Renewable Energy Agency, Abu Dhabi.

Khaligh, A., Onar, O.C., 2017. Energy Harvesting: Solar, Wind, and Ocean Energy Conversion Systems. CRC Press.

Koçak, E., Sarkgüneşi, A., 2017. The renewable energy and economic growth nexus in Black Sea and Balkan countries. Energy Policy 100, 51–57.

Lee, C.C., 2006. The causality relationship between energy consumption and GDP in G-11 countries revisited. Energy Policy 34 (9), 1086–1093.

Lin, B., Moubarak, M., 2014. Renewable energy consumption–economic growth nexus for China. Renew. Sustain. Energy Rev. 40, 111–117.

Lin, H.P., Yeh, L.T., Chien, S.C., 2013. Renewable energy distribution and economic growth in the US. Int. J. Green Energy 10 (7), 754–762.

Menegaki, A.N., 2011. Growth and renewable energy in Europe: a random effect model with evidence for neutrality hypothesis. Energy Econ. 33 (2), 257–263.

Narayan, P.K., 2005. The saving and investment nexus for China: evidence from cointegration tests. Appl. Econ. 37 (17), 1979–1990.

Narayan, S., 2016. Predictability within the energy consumption–economic growth nexus: some evidence from income and regional groups. Econ. Modell. 54, 515–521.

Payne, J.E., 2010. Survey of the international evidence on the causal relationship between energy consumption and growth. J. Econ. Stud. 37 (1), 53–95.

Payne, J.E., 2012. The causal dynamics between US renewable energy consumption, output, emissions, and oil prices. Energy Sour. B Energy Econ. Plann. 7 (4), 323–330.

Pesaran, M.H., Shin, Y., 1998. An autoregressive distributed-lag modelling approach to cointegration analysis. Econom. Soc. Monogr. 31, 371–413.

Pesaran, M.H., Shin, Y., Smith, R.J., 2001. Bounds testing approaches to the analysis of level relationships. J. Appl. Econom. 16 (3), 289–326.

Phillips, P.C., Perron, P., 1988. Testing for a unit root in time series regression. Biometrika 75 (2), 335–346.

REN21, 2018. Renewables 2018 Global Status Report. REN21 Secretariat, Paris.

Rosen, M.A., Koohi-Fayegh, S., 2017. Geothermal Energy: Sustainable Heating and Cooling Using the Ground. John Wiley & Sons.

Said, S.E., Dickey, D.A., 1984. Testing for unit roots in autoregressive-moving average models of unknown order. Biometrika 71 (3), 599–607.

Salim, R.A., Hassan, K., Shafiei, S., 2014. Renewable and non-renewable energy consumption and economic activities: further evidence from OECD countries. Energy Econ. 44, 350–360.

Scheer, H., 2013. The Solar Economy: Renewable Energy for a Sustainable Global Future. Routledge.

Shahbaz, M., Loganathan, N., Zeshan, M., Zaman, K., 2015. Does renewable energy consumption add in economic growth? An application of auto-regressive distributed lag model in Pakistan. Renew. Sustain. Energy Rev. 44, 576–585.

Shakouri, B., Khoshnevis Yazdi, S., 2017. Causality between renewable energy, energy consumption, and economic growth. Energy Sour. B Energy Econ. Plann. 12 (9), 838–845.

Toda, H.Y., Yamamoto, T., 1995. Statistical inference in vector autoregressions with possibly integrated processes. J. Econom. 66 (1–2), 225–250.

Twidell, J., Weir, T., 2015. Renewable Energy Resources. Routledge.

Uğurlu, E., 2019. Renewable energy strategies for sustainable development in the European union. In: Kurochkin, D., Shabliy, E., Shittu, E. (Eds.), Renewable Energy. Palgrave Macmillan, Cham.

Venkatraja, B., 2020. Does renewable energy affect economic growth? Evidence from panel data estimation of BRIC countries. Int. J. Sustain. Dev. World Ecol. 27 (2), 107–113.

Wang, Z., Zhang, B., Wang, B., 2018. Renewable energy consumption, economic growth and human development index in Pakistan: evidence form simultaneous equation model. J. Clean. Prod. 184, 1081–1090.

Yan, J., 2018. Negative-emissions hydrogen energy. Nat. Clim. Change 8 (7), 560–561.

Yildirim, E., 2014. Energy use, CO_2 emission and foreign direct investment: is there any inconsistence between causal relations? Front. Energy 8 (3), 269–278.

Zohuri, B., 2018. Hydrogen Energy: Challenges and Solutions for a Cleaner Future. Springer.

Further reading

Moe, E., 2016. Renewable Energy Transformation or Fossil Fuel Backlash: Vested Interests in the Political Economy. Springer.

Chapter 18

Energy consumption, financial development, globalization, and economic growth in Poland: new evidence from an asymmetric analysis

Yılmaz Toktaş[1], Agnieszka Parlinska[2]

[1]*Department of Economics, Merzifon Faculty of Economics and Administrative Sciences, Amasya University, Amasya, Turkey;* [2]*Institute of Economics and Finance, Warsaw University of Life Sciences WULS — SGGW, Warsaw, Poland*

1. Introduction

In parallel with the development of endogenous growth theories, economic literature started to explore the links between globalization, financial development, and economic growth with energy consumption. Globalization, financial development, and economic growth impacts on energy consumption in various combinations have become the important research area in last decades.

The effect of globalization on energy consumption can be seen in three different ways: the scale effect, the technical effect, and the composition effect (Shahbaz et al., 2018b). Financial development can also affect energy consumption in several ways: consumer effect, business effect, and the wealth effect (Chang, 2015; Sadorsky, 2010, 2011a). The most effective and direct being that it facilitates the purchasing of big-ticket items (BTIs). Financial development also makes it easier for the business world to acquire financial capital; and, it makes it possible to obtain additional funds through stock market development (Sadorsky, 2011a). Finally, economic growth has a direct impact on energy consumption. For example, increases in industrial production can cause higher energy demand. Financial development can also affect energy consumption through economic growth. While some economists consider the importance of financial development for economic development, some economists do not believe in this importance. Economists have

Energy-Growth Nexus in an Era of Globalization. https://doi.org/10.1016/B978-0-12-824440-1.00010-2
431

surprisingly different views on the importance of the financial system for economic growth (Levine, 1999). Depending on the efficiency of economic growth, it can have a positive or negative impact on energy consumption. Economic efficiency results in less energy use for more or the same level of output (Komal and Abbas, 2015).

It should be also mentioned that reducing energy consumption and waste is increasingly important for the European Union. In 2007, EU leaders set a goal to reducing the EU annual energy consumption by 20%−2020. The analysis shows that in Poland after a gradual decrease in the years 2007−2014, energy consumption in recent years has started to increase and is now slightly higher than the linear trajectory for the 2020 goals. This is due to weather fluctuations, especially the colder 2015 and 2016, but also the increase in economic activity and low oil prices. The energy intensity of industry in 2005−2017 was steadily improving by 22%, and energy savings actually offset some of the impact of these increases (EU, 2019). According to Fig. 18.1, 2000−2005 is the lowest period of energy consumption for Poland. After the democratization process, Poland's energy consumption has been significantly decreased.

In Poland, energy efficiency has increased above EU average, if we take as a reference the relation of generally stable energy consumption in recent years to the gross domestic product (GDP) growing at that time. In the years 2006−2016, unit energy consumption decreased (Fig. 18.1-including the deflator provided by Eurostat) by 27.3% (EU, 2019).

Poland has one of the fastest-growing economies among EU countries. Transaction to the market-based economic system in 1990 has changed Poland's economy and linked the Polish system to the global system. The EU

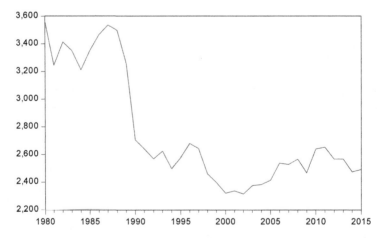

FIGURE 18.1 Energy consumption in Poland. *Source: WorldBank, 2020. Global Financial Development Report. Retrieved from https://www.worldbank.org/en/publication/gfdr/gfdr-2016/ background/financial-development.*

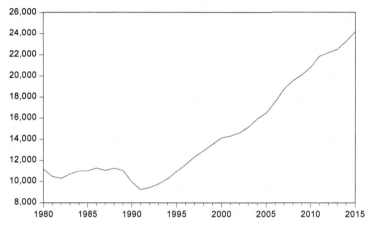

FIGURE 18.2 Gross domestic product per capita in Poland (US$). *Source: IMF, 2019. World Economic Outlook (WEO). October 2019. Retrieved from https://www.imf.org/external/pubs/ft/ weo/2020/01/weodata/index.aspx.*

membership helped Poland's economic growth potential, which continued the execution of investments related to European funds (Fig. 18.2).

Between 1990 and 2011, the Polish economy underwent a transformation from centrally planned to the market-led system. As seen on the figure, the globalization index of Poland has a structural change that democratization date of 1990 (Gorynia and Wolniak, 2002) (Fig. 18.3).

The remainder of the chapter is organized as follows: Part 2 presents an overview of the current literature. The description of the data and model is

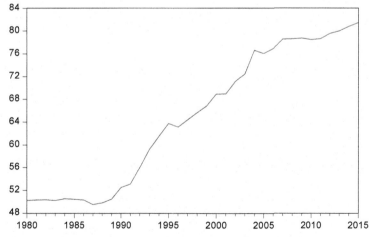

FIGURE 18.3 Globalization index in Poland. *Source: Gygli, S., Haelg, F., Potrafke, N., Sturm, J.-E., 2019. The KOF globalisation index—revisited. Rev. Int. Organ. 14(3), 543—574.*

included in Part 3. In Part 4, the methodological framework was presented. Part 5 shows results of the conducted analysis and discuss the findings. Last part presents the conclusion and implications.

2. Literature review

2.1 Economic growth and energy consumption

The connection between economic growth and energy consumption has already been theoretically and empirically studied and is well documented. From the 1980s to the present day, many different authors analyzed the relationship between energy consumption and economic growth by implementing different econometric techniques in the single country or country group studied, according to the period or the econometric method used. From this point of view, a common decision on the relationship between energy consumption and economic growth has not been reached in the literature.

Öztürk (2010) and Jakovac (2018) presented a detailed literature review on the relationship between energy consumption and economic growth, revealing the dependency between economic growth and energy consumption in their studies. The directions of causal relationship between energy consumption and economic growth nexus can be categorized in four groups:

1. No causality EC $\neq$ EG that refers to the "neutral hypothesis" means that no causality exists between energy consumption and growth.
2. Unidirectional causality EC $\rightarrow$ EG that refers to the "conservation hypothesis" means that the causality runs from energy consumption to growth.
3. Unidirectional causality EG $\rightarrow$ EC that refers to the "growth hypothesis" means that the causality runs from growth to energy consumption.
4. Bidirectional causality EC $\leftrightarrow$ EG that refers to the "feedback hypothesis" means that bidirectional causality exists between energy consumption and growth (Jakovac, 2018; Öztürk, 2010).

In Table 18.1, list of the selected empirical literature on the connection between energy consumption and economic growth by author, time range, country/region/group of countries, methodology, and empirical results was presented. The presented examples include most countries of the world. Over time, various empirical studies concentrated on different countries (sometimes many different authors analyzed single country), or groups of countries, time frame, main variables (or their substitutes), and quantitative methods.

The autoregressive distributed lags bounds testing, vector autoregressive regression, Granger causality, nonlinear autoregressive distributed lags bounds testing, generalized method of moments, and variance decomposition were used most often by researchers. But also, the dynamic ordinary least squares, fully modified ordinary least square and multivariate cointegration, and correlation analysis could be noticed within the applied methodology.

TABLE 18.1 Summary of the selected studies for economic growth-energy consumption nexus.

Authors	Period	Country	Methodology	Results
Shahbaz et al. (2017)	1960–2015	India	NARDL bounds test, AC	$EC^- \rightarrow EG^-$ Positive shock in EC increase EG
Streimikiene and Kasperowicz, 2016	1995–2012	V4 countries and 14 EU old member states	Panel FMOLS and DOLS	$EC \rightarrow EG$ Energy consumption has a positive impact on economic growth and impact of energy consumption in V4 countries is higher than EU countries
Kasperowicz (2014)	2000–2012	Poland	GC	$EG \leftrightarrow EC$
Smiech and Papiez (2014)	1993–2011	EU member states	Panel GC	Greece, Bulgaria, Latvia $EG \leftrightarrow EC$ France, Romania, Slovakia $EG \rightarrow EC$ Poland $EC \rightarrow EG$
Sbia et al. (2014)	1975–2011	UAE	ARDL bounds test, GC	Economic growth has a positive impact on energy demand. In short-run $EG \rightarrow EC$
Saboori et al. (2014)	1960–2008	OECD countries	FMOLS	Bidirectional relationship between road energy consumption and economic growth
Belke et al. (2011)	1981–2007	25 OECD member Countries	Panel cointegration, Panel GC	In the long-run $EG \leftrightarrow EC$
Öztürk et al. (2010)	1971–2005	51 countries	Pedroni, cointegration, Panel GC	Energy consumption and GDP are cointegrated Low-income countries: $EG \rightarrow EC$ Middle-income countries: $EG \leftrightarrow EC$
Zhang and Cheng (2009)	1960–2007	China	GC	$EG \rightarrow EC$

Continued

TABLE 18.1 Summary of the selected studies for economic growth-energy consumption nexus.—cont'd

Authors	Period	Country	Methodology	Results
Belloumi (2009)	1971–2004	Tunisia	GC, VECM	In the long-run: EC ↔ EG In the short-run: EC → EG
Payne (2009)	1949–2006	USA	Toda–Yamamoto causality test	EG ≠ EC
Halıcıoğlu (2009)	1960–2005	Turkey	GC, ARDL bounds test	EG ≠ EC
Narayan and Prasad (2008)	1960–2002	30 OECD countries	bootstrapped causality testing approach	Various results depending of the country Approximately 80% of the OECD countries, EG ≠ EC
Lee et al. (2008)	1960–2001	22 OECD countries	Panel cointegration, Panel VECM	EC ↔ EG
Huang et al. (2008)	1972–2002	82 Low-, middle-, and high-income countries	Panel GC	Middle- and high-income countries: EG → EC Low-income countries: EG ≠ EC
Erdal et al. (2008)	1970–2006	Turkey	Pair-wise GC, Johansen cointegration	EC ↔ EG
Ang (2007)	1960–2000	France	GC, Johansen cointegration, ARDL bounds test	In the long-run: EG → EC In the short-run: EC → EG
Lee and Chang (2005)	1954–2003	Taiwan	Johansen–Juselius cointegration, VECM	EC → EG
Ghali and El-Sakka (2004)	1961–1997	Canada	cointegration, VECM, GC	EC ↔ EG
Altınay and Karagöl (2004)	1950–2000	Turkey	Hsiao's version of GC	EG ≠ EC

Study	Period	Countries	Method	Results
Soytaş and Sarı (2003)	1950–1992	G-7 countries and Top 10 emergin markets exclude China	Johansen-Juselius, cointegration and GC	EC ↔ EG (Argentina); EG → EC (Italy, Korea); EC → EG (Turkey, France, Japan, Germany); EG ≠ EC (Poland)
Hondroyiannis et al. (2002)	1960–1996	Greece	VECM, GC	EC ↔ EG
Fatai et al. (2002)	1960–1999	New Zealand	GC, ARDL, Toda and Yamamoto causality test	EG ≠ EC
Soytaş et al. (2001)	1960–1995	Turkey	Johansen-Juselius cointegration, GC	EC → EG
Erol and Yu (1987)	1952–1982	6 industrialized countries	GC	EC ↔ EG (Japan); EG → EC (Italy, Germany); EC → EG (Canada); EG ≠ EC (France, UK)
Yu and Choi (1985)	1950–1976	5 countries	GC	EG ≠ EC (UK, USA, Poland); EG → EC (Philippines); EG → EC (Korea)
Akarca and Long (1980)	1950–1970	USA	Sim's technique	EG ≠ EC
Kraft and Kraft (1978)	1947–1974	USA	GC	EG → EC

ARDL, autoregressive distributed lags bounds testing; CEEC, Central and Eastern European countries; DOLS, dynamic ordinary least squares; EC≠EG means that no causality exists between energy consumption and growth; EC→EG means that the causality runs from energy consumption to growth; EC↔EG means that bidirectional causality exists between energy consumption and growth; EG→EC means that the causality runs from growth to energy consumption; EU, European Union; FMOLS, fully modified ordinary least square; GC, Granger causality; GMM, generalized method of moments; GVD, generalized variance decomposition; IR, impulse response; NARDL, nonlinear autoregressive distributed lags bounds testing; OECD, Organization for Economic Cooperation and Development; V4, Visegrad countries; VAR, vector autoregression model; VD, variance decomposition; VECM, vector error correction model.
Source: Own Estimation.

Kraft and Kraft (1978) used US data for the period 1947 to 1974 to examine the relationship between energy consumption and economic growth using the Granger causality test as a result of the study; they identified one-way causality from economic growth to energy consumption. It can be noticed that some empirical results give evidence of the growth hypothesis that energy consumption leads to economic growth (Lee and Chang, 2005; Soytaş et al., 2001; Streimikiene and Kasperowicz, 2016) while others prove the opposite hypothesis that causality runs from growth to energy consumption (Kraft and Kraft, 1978; Öztürk, 2010; Sbia et al., 2014; Zhang and Cheng, 2009). The research of Bekhet et al. (2017), Erdal et al. (2008), Ghali and El-Sakka (2004), Hondroyiannis et al. (2002), Kasperowicz (2014), and Lee and Chang (2008) support the feedback hypothesis that the bidirectional causality exists between energy consumption and growth. While research from the last group prove a natural hypothesis that the causality does not exist between energy consumption and growth (Akarca and Long, 1980; Altınay and Karagöl, 2004; Fatai et al., 2002; Halıcıoğlu, 2009; Payne, 2009); Smiech and Papiez (2014). Also, some researchers emphasize the differences between individual countries, regions, or created groups (Erol and Yu, 1987; Huang et al., 2008; Öztürk et al., 2010; Soytaş and Sarı, 2003; Yu and Choi, 1985). The results of those research are often contradictory, and the lack of consensus on this matter could result in inadequate selection and implementation of economic and energy policies.

Considering the research for Poland, similar trends and a variety of results can be observed. It is connected with the time frame, main variables (or their substitutes) as well as variety of quantitative methods implemented in the research. The research of (Narayan and Prasad, 2008; Smiech and Papiez, 2014; Soytaş and Sarı, 2003; Yu and Choi, 1985) have proved the lack of causality between energy consumption and growth in Poland during the examined time frame. While other studies (Belke et al., 2011; Streimikiene and Kasperowicz, 2016) pointed out bidirectional causality exists between energy consumption and growth. Kasperowicz (2014) on the basis of the causality results estimated that economic growth of Poland is electricity-dependent.

2.2 Financial development and energy consumption

The last decades have been the time when scientists also to pay increasing attention to the relationship between financial development and energy consumption nexus. As was mentioned Chang (2015), Sadorsky (2010, 2011a) theoretically explains the three effects channels of financial development on energy demand, i.e.,

- Consumer effect as a direct effect where a development of financial system provides chances for consumers to spend their savings for energy-consuming BTIs which can affect a country's total demand for energy;

- Business effect- where a development of financial system provides opportunities for investment and innovation activities which can cause a business demand for energy increases along with the financial development of the economy;
- Wealth effect- where increased stock market activity is reflected a main economic growth and prosperity index, but it also creates wealth effect in terms of affecting confidence among consumers and business firms.

Nevertheless, other scientists pointed out that financial development can cause decreasing energy consumption by providing easily accessible and low-cost capital to access to energy-friendly technologies (Chang, 2015; Durusu-Ciftci et al., 2020; Islam et al., 2013). Moreover, some researchers find that energy consumption can be affected by the financial development by the savings channel, where the business and household's savings are increased by benefits from the less energy-consuming technologies. The demand for low energy consuming goods can affect the increase of investment in this type of technology and in consequence supports the development of financial markets (Durusu-Ciftci et al., 2020). The summary of comparative analysis of the selected researches for financial development-energy consumption nexus were presented in Table 18.2. The comparison was carried out by author, time range, country/region/group of countries, methodology, and empirical results.

Analyzing presented studies, it can be seen panel data or time series where different econometric approaches such as: autoregressive distributed lags bounds testing (Bekhet et al., 2017; Öztürk and Acaravcı, 2013; Paramati et al., 2016; Shahbaz et al., 2013a, b; Shahbaz and Lean, 2012; Shahzad et al., 2017), generalized method of moments (Coban and Topcu, 2013; Sadorsky, 2010, 2011a), Granger causality (Al-mulali & Lee, 2013; Lee et al., 2008; Ouyang and Li, 2018; Shahbaz and Lean, 2012), and multivariate cointegration, correlation analysis, etc., were used.

Al-mulali and Lee (2013), Islam et al. (2013), Liu et al. (2018), Paramati et al. (2016), Shahbaz et al. (2013a), Shahbaz and Lean (2012), Shahzad et al. (2017), Ziaei (2015) in their empirical results found exists of the bidirectional causality between energy consumption and growth financial development in the examined group or single of countries. Also some of researchers pointed out that financial development and economic growth positively affect energy (Al-mulali & Lee, 2013; Islam et al., 2013; Mukhtarov et al., 2018; Paramati et al., 2016; Sadorsky, 2010, 2011a). Shahbaz et al. (2013a, b), Muhammad Shahbaz et al. (2013), Ouyang and Li (2018) found that no causality exists between energy consumption and growth in case of western region of China.

Other research results highlighted a one-way causal relationship that extends from financial development to energy consumption. For example, in the long-run: Pradhan et al. (2018) for Financial Action Task Force Countries, Paramati et al. (2016) for 20 EMC' countries, Bekhet et al. (2017) in Bahrain, Shahbaz et al. (2013a) in Indonesia, Öztürk and Acaravcı (2013) in case of

TABLE 18.2 Summary of the selected studies for financial development-energy consumption nexus.

Authors	Period	Country	Methodology	Results
Yue et al. (2019)	2006–2015	21 transitional countries	Panel smooth transition regression	Stock markets development led decreased energy consumption in China and Poland
Liu et al. (2018)	1980–2014	China	VECM-GC	Short-run: FD → EC Long-run: EC ↔ FD
Mukhtarov et al. (2018)	1992–2015	Azerbaijan	Johansen cointegration/ Gregory −Hansen cointegration	Cointegrated and the positive effect FD on EC
Ouyang and Li (2018)	1996–2015	Chinese	Panel GC	Eastern and Central region: EC ↔ FD Western region: EC≠FD
Pradhan et al. (2018)	1961–2015	35 FATFC	Panel GC	Short-run: Various results for different proxies. Long-run: FD → EC
Bekhet et al. (2017)	1980–2011	GCC	ARDL bounds test, GC	Oman EC → FD UAE, Bahrain: Long-run: FD → EC
Shahzad et al. (2017)	1971–2011	Pakistan	ARDL bounds test, GC	EC ↔ FD
Paramati et al. (2016)	1991–2012	20 EMC	Panel ARDL Panel FMOLS Panel GC	Positive effect Long-run: FD → EC
Furuoka (2015)	1980–2012	Asian countries	Panel GC	EC → FD
Chang (2015)	1999–2008	53 countries	Panel threshold regression	The effects of FD on EC are different for advanced economies and emerging as well as developing economies.
Ziaei (2015)	1989–2011	13 EC and 12 EAO	Panel VAR Impulse-response	Between energy consumption and economic growth has different relationship in different countries. Asia EC ↔ FD

Continued

TABLE 18.2 Summary of the selected studies for financial development-
energy consumption nexus.—cont'd

Authors	Period	Country	Methodology	Results
Coban and Topcu (2013)	1990–2011	EU	System GMM	No significant impact for EU 27. Various results for old/new members.
Al-mulali and Lee (2013)	1980–2009	GCC	Panel DOLS Panel GC	Positively effect. Long-run: EC ↔ FD
Shahbaz et al. (2013a)	1975–2011	Indonesia	ARDL bounds test/GC	Long-run: FD → EC
Shahbaz et al. (2013b)	1971–2011	China	ARDL bounds test/Johansen cointegration/ GC	Long-run: EC → FD
Shahbaz et al. (2013c)	1971–2011	Malaysia	VECM-GC	Long-run: EC ↔ FD
Islam et al. (2013)	1971–2018	Malaysia	ARDL bounds test, GC	Cointegrated and short-run FD → EC, Long-run EC ↔ FD
Öztürk and Acaravcı (2013)	1960–2007	Turkey	ARDL, VECM -GC	Short-run: EC → FD Long-run: FD → EC
Shahbaz and Lean (2012)	1971–2008	Tunisia	ARDL/ Johansen cointegration/ GC	EC ↔ FD
Sadorsky (2011a)	1996–2006	9 CEEC	System GMM	Positive effect
Sadorsky (2010)	1990–2006	22 EMC	System GMM	Positive effect.

AC, asymmetric causality; ARDL, autoregressive distributed lag model; BHC, Bayer and Hack cointegration; CEEC, Central and Eastern European countries; EAO, East Asia and Oceania; EC, energy consumption; EC, European countries; EC≠FD means that no causality exists between energy consumption and growth; EC→FD means that the causality runs from energy consumption to financial development; EC↔FD means that bidirectional causality exists between energy consumption and growth financial development; EMC, emerging market countries; EU, European Union; FATFC, Financial Action Task Force Countries; FD, financial development; FD→EC means that the causality runs from financial development to energy consumption; FMOLS, fully modified ordinary least square; GC, Granger causality; GCC, Gulf Cooperation Council; GCC, Gulf Cooperation Council; GMM, generalized method of moments; NIC, newly industrialized countries; T&Y, Toda & Yamamoto causality; VECM, vector error correction model.
Source: Own Estimation.

Turkey and in the short-run: Islam et al. (2013) in Malaysia, Liu et al. (2018) for China, and Yue et al. (2019) for the 21 transitional countries. Yue et al. (2019) also pointed out that stock markets development led decreased energy consumption in China and Poland.

The opposite one-way causal relationship running from energy consumption to financial development was noticed by Furuoka (2015) for Asian countries, Bekhet et al. (2017) for UAE, and Shahbaz et al. (2013b) for China.

2.3 Globalization and energy consumption

The impact of the globalization on the energy consumption becomes an important issue concerning the environmental pollution from carbon dioxide (CO_2) emission. As was explained in the previous part, the energy consumption is directly related to economic activity.

Moreover, researchers pointed out the globalization effect on energy consumption by the three channels (Shahbaz et al., 2018a,b):

- Scale effect with other factors constant when the globalization will rise energy consumption as a results of enlarged economic activity (Cole, 2006);
- Technique effect where the globalization is in the form of trade and capital inflows, globalization empowers economies to retrench energy consumption by importing new technology without hindering the economic activity (Antweiler et al., 2001; Dollar and Kraay, 2004);
- Composition effect in which the globalization influence energy consumption when energy demand decrease with the rise of economics activity (Stern, 2007).

Various research has used different indicators of globalization to analyze the relationship between globalization and energy consumption. For example, trade openness (exports/imports) by Narayan and Prasad (2008), Sadorsky (2010, 2011b); Trade liberalization by Cole (2006) were used as an indicator of globalization.

In the studies of the globalization-energy consumption nexus, the autoregressive distributed lags bounds testing, vector autoregressive regression, Granger causality, nonlinear autoregressive distributed lags bounds testing, and generalized method of moments were most often methodology approaches used. Comparison analyze of the selected studies for globalization-energy consumption nexus was presented in Table 18.3.

It should also be noted that the relationship between globalization and energy consumption nexus is often part of the complex analysis with financial development or economic growth nexus. Recently, due to the increased importance of environmental protection, scientists have focused on the impact of renewable energy sources on economic growth (Paramati et al., 2016; Shahzad et al., 2017).

TABLE 18.3 Summary of the selected studies for globalization-energy consumption nexus.

Author(s)	Period	Region/country	Method(s)	Results
Chen et al. (2019)	1980–2016	16 CEEC	Dynamic seemingly unrelated regression and Dumitrescu–Hurlin causality	G → renewable energy consumption, EG → renewable energy consumption, FD → EC
Shahbaz et al. (2018a)	1970–2015	BRICS region	NARDL bounds test	EC is positively and negatively affected by positive and negative G shocks. A positive shock in EG stimulates EC while a negative shock reduces EC
Koengkan (2017)	1991–2012	12 Latin American and Caribbean countries	Panel cointegration (Westerlund)	Globalization positive effect primary energy consumption
Shahbaz et al. (2016)	1971–2012	India	Bayer–Hanck cointegration ARDL bounds cointegration, VECM-GC	Long-run: G decreases EC FD decreases EC Urbanization increases EC EG ↔ EC FD ↔ EC G → EC
Shahbaz et al. (2014)	1980–2010	91 low-, middle- and high-income economies.	Panel cointegration, nonhomogenous and homogenous causality	TO ↔ EC low- and middle-income countries: U-shaped association between TO and EC high-income countries: an inverted U-shaped relationship

Continued

TABLE 18.3 Summary of the selected studies for globalization-energy consumption nexus.—cont'd

Author(s)	Period	Region/country	Method(s)	Results
Sadorsky (2012)	1980–2007	7 South America countries	Panel cointegration, VECM-GC	Cointegrated, short-run EC → IM; EX ↔ EC Long-run causality relationship between EC and TO
Sadorsky (2011b)	1980–2007	8 Middle Eastern countries	Panel GC	Short-run EX → EC Long-run EX → EC; IM → EC
Narayan and Smyth (2009)	1974–2002	Iran, Israel, Kuwait, Oman, Saudi Arabia, and Syria	Panel FMOLS, Panel GC	EX → EC IM ↔ EC
Cole (2006)	1975–1995	32 developing and develop countries	Panel data	Trade openness can influence energy consumption via the scale, technique, and composite effect.

AC, asymmetric causality; *ARDL*, autoregressive distributed lag model; *BRICS*, emerging national economies: Brazil, Russia, India, China, and South Africa; *CEEC*, Central and Eastern European countries; *EC*, energy consumption; *EG*, economic growth; *EMC*, emerging market countries; *EX → EC* means unidirectional causality from exports to energy consumption; *EX*, export; *FD*, financial development; *FMOLS*, fully modified ordinary least square; *G → EC* means unidirectional causality from globalization to energy consumption; *G*, globalization; *GC*, Granger causality; *GMM*, generalized method of moments; *IM ↔ EC* means bidirectional causality between imports and energy consumption; *IM*, import; *TO ↔ EC* means bidirectional causality between trade openness and energy consumption; *TO*, trade openness; *VECM*, vector error correction model.
Source: Own Estimation.

3. The model and data

This study examines the relationship between globalization and energy consumption by incorporating economic growth and financial development into the energy demand function for Poland. The functional forms of the model yield:

$$\ln EC_t = \beta_1 + \beta_2 \ln G_t + \beta_3 \ln EG_t + \beta_4 \ln FD_t + \beta_5 DUM + \varepsilon_t \quad (18.1)$$

$$\ln EC_t = \beta_1 + \beta_2 \ln ECG_t + \beta_3 \ln EG_t + \beta_4 \ln FD_t + \beta_5 DUM + \varepsilon_t \quad (18.2)$$

$$\ln EC_t = \beta_1 + \beta_2 \ln TG_t + \beta_3 \ln EG_t + \beta_4 \ln FD_t + \beta_5 DUM + \varepsilon_t \quad (18.3)$$

$$\ln EC_t = \beta_1 + \beta_2 \ln FG_t + \beta_3 \ln EG_t + \beta_4 \ln FD_t + \beta_5 DUM + \varepsilon_t \quad (18.4)$$

$$\ln EC_t = \beta_1 + \beta_2 \ln SG_t + \beta_3 \ln EG_t + \beta_4 \ln FD_t + \beta_5 DUM + \varepsilon_t \quad (18.5)$$

$$\ln EC_t = \beta_1 + \beta_2 \ln IPG_t + \beta_3 \ln EG_t + \beta_4 \ln FD_t + \beta_5 DUM + \varepsilon_t \quad (18.6)$$

$$\ln EC_t = \beta_1 + \beta_2 \ln IFG_t + \beta_3 \ln EG_t + \beta_4 \ln FD_t + \beta_5 DUM + \varepsilon_t \quad (18.7)$$

$$\ln EC_t = \beta_1 + \beta_2 \ln CG_t + \beta_3 \ln EG_t + \beta_4 \ln FD_t + \beta_5 DUM + \varepsilon_t \quad (18.8)$$

$$\ln EC_t = \beta_1 + \beta_2 \ln PG_t + \beta_3 \ln EG_t + \beta_4 \ln FD_t + \beta_5 DUM + \varepsilon_t \quad (18.9)$$

Annual data between 1980 and 2015 were used for the analyses, with all variables transformed into natural logarithms. In these, EC_t is energy consumption per capita as measured by kg of oil equivalent, G_t denotes the KOF globalization index, ECG_t is the KOF economic globalization index, TG_t is the KOF trade globalization index, FG_t is the KOF financial globalization index, SG_t is the KOF social globalization index, IPG_t is the KOF interpersonal globalization index, IFG_t is the KOF information globalization index, CG_t is the KOF cultural globalization index, PG_t is the KOF political globalization index; EG_t is GDP per capita in constant 2010 US\$, representing economic growth, FD_t is the IMF financial development index, and finally, DUM denotes the dummy variables created on the structural breaks, suggested by Zivot and Andrews, 1992 unit root tests, for the dependent variables in the relevant equations. We also considered Poland as going through a democratic transition for the year 1990 when creating the dummy variables: the value of the dummy variable was 1 if the period was after a structural break time and 0 otherwise. 1990 is the beginning of the transition of Poland to a market-based system. The most significant dummy variables were incorporated into the estimated models. Energy consumption was obtained from the World Bank database. Financial development index data and GDP per capita data were obtained from the IMF database and KOF globalization indices from the Swiss Economic Institute.

4. Methodological framework

4.1 Nonlinearity tests

In the literature, traditional nonlinearity tests are based upon the assumption of variables are stationarity. Harvey and Leybourne (2007) suggested using the following equation, which allows the existence of I (0) and I (1) processes together.

$$y_t = \beta_0 + \beta_1 y_{t-1} + \beta_2 y_{t-1}^2 + \beta_3 y_{t-1}^3 + \beta_4 \Delta y_{t-1} + \beta_5 (\Delta y_{t-1})^2 + \beta_6 (\Delta y_{t-1})^3 + \varepsilon_t$$

The null hypothesis of linearity and alternative hypothesis of nonlinearity can be represented respectively as:

$$H_0: \beta_2, \beta_3, \beta_5, \beta_6 = 0$$

$$H_1: \text{at least one of } \beta_2, \beta_3, \beta_5, \beta_6 \neq 0$$

Harvey and Leybourne (2007) proposed the following test statistics

$$W_T^* = \exp(-b|DF_T|)^{-1} W_T$$

$$W_T = \frac{RSS_1 - RSS_0}{RSS_0 / T}$$

where b is nonzero constant, DF_T is the standard ADF t-statistic obtained from the restricted regression.

Harvey et al. (2008) developed a new linearity test that can be applied when the unit root properties of the data are uncertain. This new test eliminates problems of traditional linearity tests which are based upon the assumption that the variables are $I(0)$ or $I(1)$ properties. When the time series is stationary at the level $I(0)$, the following model is estimated.

$$y_t = \beta_0 + \beta_1 y_{t-1} + \beta_2 y_{t-1}^2 + \beta_3 y_{t-1}^3 + \sum_{j=1}^{p} \beta_{4,j} \Delta y_{t-j} + \varepsilon_t$$

where Δ is the difference operator, p is the number of lags.
The null hypothesis of linearity can be represented respectively as:

$$H_{0,0}: \beta_2 = \beta_3 = 0$$

and the alternative hypothesis of nonlinearity can be represented respectively as:

$$H_{1,0}: \beta_2 \neq 0 \text{ and/or } \beta_3 \neq 0$$

The standard Wald statistic for testing these hypotheses is given by

$$W_0 = T\left(\frac{RSS_0^r}{RSS_0^u} - 1\right)$$

where T is the number of observations, RSS_0^r and RSS_0^u represent the residual sums of squares from the unrestricted and restricted OLS regression.

If the series are nonstationary $I\,(1)$, the regression model is:

$$\Delta y_t = \lambda_1 \Delta y_{t-1} + \lambda_2 (\Delta y_{t-1})^2 + \lambda_3 (\Delta y_{t-1})^3 + \sum_{j=1}^{p} \lambda_{4,j} \Delta y_{t-j} + \varepsilon_t$$

The null hypothesis of linearity can be represented respectively as:

$$H_{0,1} : \lambda_2 = \beta\lambda_3 = 0$$

and the alternative hypothesis of nonlinearity can be represented respectively as:

$$H_{1,1} : \lambda_2 \neq 0 \text{ and/or } \lambda_3 \neq 0$$

$$W_1 = T\left(\frac{RSS_1^r}{RSS_1^u} - 1\right)$$

where T is the number of observations, RSS_1^r and RSS_1^u represent the residual sums of squares from the unrestricted and restricted OLS regression.

Harvey et al. (2008), offer weighted average statistic when y_t is not known whether stationary or unit root

$$W_\lambda = \{1 - \lambda\} W_0 + \lambda W_1$$

where λ is a function that converges in probability to zero when y_t is I(0) and to one when y_t is $I(1)$.

4.2 The nonlinear autoregressive distributed lag bounds testing approach for cointegration

We utilize an asymmetric ARDL (NARDL) model that the general form of asymmetric long-run relationship is represented below:

$$y_t = \beta^+ x_t^+ + \beta^- x_t^- + u_t$$

where β^+ and β^- represent the associated long-run parameters. y_t is the dependent variable. x_t is a $k \times 1$ vector of regressors as:

$$x_t = x_0 + x_t^+ + x_t^-$$

In this method, decomposition of the exogenous variables into their negative and positive partial sums for decreases and increases can be calculated as follows:

$$x_t^+ = \sum_{i=1}^{t} \Delta x_i^+ = \sum_{i=1}^{t} \max(\Delta x_i, \, 0)$$

$$x_t^- = \sum_{i=1}^{t} \Delta x_i^- = \sum_{i=1}^{t} \min(\Delta x_i, \, 0)$$

Shin et al. (2014) extended the ARDL method developed by Pesaran et al. (2001) with the idea of cumulative positive and negative partial sums as follows:

$$y_t = \sum_{j=1}^{p} \varnothing_i y_{t-j} + \sum_{j=0}^{q} \left(\theta_j^{+\prime} x_{t-j}^+ + \theta_j^{-\prime} x_{t-j}^- \right) + \varepsilon_t$$

The NARDL model which is estimated that covers the short and long-run of the positive and negative partial sums in our study takes the following equation form:

$$\Delta LNEC_t = \theta_0 + \rho LNEC_{t-1} + \theta_1^+ \ln G_{t-1}^+ + \theta_1^- LNG_{t-1}^- + \theta_2^+ LNEG_{t-1}^+$$

$$+ \theta_2^- LNEG_{t-1}^- + \theta_3^+ LNFD_{t-1}^+ + \theta_3^- LNFD_{t-1}^- + \sum_{i=1}^{p-1} \alpha_i \Delta LNEC_{t-i}$$

$$+ \sum_{i=0}^{q-1} \pi_i^+ \Delta LNG_{t-i}^+ + \sum_{i=0}^{q-1} \pi_i^- \Delta LNG_{t-i}^- + \sum_{i=0}^{q-1} \pi_i^+ \Delta LNEG_{t-i}^+$$

$$+ \sum_{i=0}^{q-1} \pi_i^- \Delta LNEG_{t-i}^- + \sum_{i=0}^{q-1} \pi_i^+ \Delta LNFD_{t-i}^+$$

$$+ \sum_{i=0}^{q-1} \pi_i^- \Delta LNFD_{t-i+}^- \psi DUM + \varepsilon_t$$

4.3 Asymmetric causality

In the study, causality relationships between variables examined by the Hatemi-J (2012) causality test which enables the examination of the different causality relationship of asymmetric shocks. This test does not require any previous test in terms of a unit root or cointegration test; however, we do need to determine the maximum integration level of variables for d_{max} value in the estimation equation. The positive and negative shocks of each variable can be defined cumulatively as follows (Hatemi-J, 2012):

$$y_{1t}^+ = \sum_{i=1}^{t} \varepsilon_{1i}^+, y_{1t}^- = \sum_{i=1}^{t} \varepsilon_{1i}^-, y_{2t}^+ = \sum_{i=1}^{t} \varepsilon_{2i}^+, y_{2t}^- = \sum_{i=1}^{t} \varepsilon_{2i}^-, \tag{18.14}$$

VAR(p) model that used to examine the causality relationship between these constituents by the context of positive cumulative shocks as follows (Hatemi-J, 2012):

$$y_1^+ = v + A_1 y_{t-1}^+ + \ldots + A_p y_{t-1}^+ + u_1^+, \tag{18.15}$$

y_1^+ is the 2×1 variable vector, v is the 2×1 constant term vector, and u_1^+ is the error term vector. A_r matrix is the 2×2 matrix for r-lag parameter. VAR model estimation by using Hatemi-J criterion (HJC) information criteria is the next step of the causality test. Also, critical value table creates by using bootstrap techniques. HJC information criteria is shown below (Hatemi-J, 2003):

$$HJC = \ln\left(\det\widehat{\Omega}_j\right) + j\left(\frac{n^2 \ln T + 2n^2 \ln(\ln T)}{2T}\right), \quad j = 0, \ldots, p.$$

The VAR (p) model can be defined as the following (Hatemi-J, 2012):

$$Y = DZ + \delta, \tag{18.16}$$

The null hypothesis of Granger causality test is no granger causality relationship between the variables. The null hypothesis is tested by the following test method (Hatemi-J, 2003, 2012):

$$WALD = (C\beta)' \left[C\left((Z'Z)^{-1} \otimes S_U\right) C' \right]^{-1} (C\beta), \tag{18.17}$$

$\beta = vec(D)$ and vec indicate the column-stacking operator, $\otimes$ is the Kronecker multiplier, and C is a $p \times n(n(1 + n(p + d)))$ matrix. $S_U = \frac{\hat{\delta}_U' \hat{\delta}_U}{T - q}$ is the variance-covariance matrix of the unrestricted VAR model, and q is the number of parameters in each equation of the VAR model.

5. Empirical results

In the study, firstly we apply the Harvey et al. (2008) and Harvey and Leybourne (2007) nonlinearity tests to investigate the linearity of variables (Table 18.4). According to nonlinearity tests, our dependent variable energy consumption and financial development independent variable that included in the all models are determined nonlinear.

According to the results of nonlinearity tests, we decided to apply asymmetric techniques which allow to analysis positive and negative relationship between variables. For the cointegration analysis, we use NARDL and for the causality, we use Hatemi-J (2012) asymmetric causality test (Table 18.8 and 18.9).

5.1 Nonlinearity test results

It was found that LNIFG, LNTG, LNSG, and LNCG series is nonlinear according to both linearity tests. According to Harvey et al. (2008) linearity test,

TABLE 18.4 Nonlinearity tests.

Variables	Harvey at al. (2008)	Harvey and Leybourne (2007)		
		10%	5%	1%
LNIFG	5.43*	11.82*	12.26**	13.09
LNPG	1.52	6.43	6.51	6.64
LNTG	15.21***	10.88*	11.25**	11.94
LNSG	7.16**	8.1*	8.21	8.42
LNIPG	2.89	5.3	5.42	5.64
LNG	4.48	10.49*	10.62**	10.86
LNFG	1.03	2.79	2.85	2.97
LNFD	3.81	17.09*	17.2**	17.4***
LNEG	3.75	0.86	0.98	1.24
LNECG	7.27**	2.81	3.02	3.42
LNEC	0.25	36.81*	37.02**	37.39***
LNCG	8.26**	25.6*	25.8**	26.17***

a: (*) Significant at the 10%; (**) Significant at the 5%; (***) Significant at the 1%.
Source: Own Estimation.

LNECG was found nonlinear. It was found that LNG, LNFD, and LNEC series is nonlinear according to Harvey and Leybourne (2007) test. Some of series was found nonlinear according to linearity tests. Linearity tests recommended the nonlinear approach for the empirical analysis (Table 18.4).

5.2 Unit root tests

In the stationarity analysis, (Phillips and Perron, 1988) and (Zivot and Andrews, 1992) unit root tests were used. According to the unit root tests results, series are determined as being stationary at first difference but none of the series is stationary at second difference (Table 18.5 and 18.6).

According to unit root test results, we use the ARDL bounds test approach by the new asymmetric version that enables to investigate cointegration relationship among various stationary levels.

5.3 The long-run and short-run analysis

NARDL bounds test approach was used for cointegration analysis which enables to examine the asymmetric relationship among variables.

TABLE 18.5 Unit root test results.

		With constant		Without constant and trend	
At level		t-Statistic	Prob.	t-Statistic	Prob.
	LNEC	−1.84	0.36	−1.22	0.20
	LNEG	0.94	0.99	2.28	0.99
	LNFD	−1.68	0.43	−0.29	0.58
	LNG	−0.34	0.91	3.17	1.00
	LNECG	0.16	0.97	2.74	1.00
	LNTG	0.42	0.98	2.05	0.99
	LNFG	−0.38	0.90	4.97	1.00
	LNSG	−0.40	0.90	2.98	1.00
	LNIPG	0.03	0.96	3.80	1.00
	LNIFG	−0.28	0.92	2.59	1.00
	LNCG	−0.82	0.80	2.01	0.99
	LNPG	−1.29	0.62	1.12	0.93
At first difference	d(LNEC)	−5.26	0.00*	−5.26	0.00***
	d(LNEG)	−3.31	0.02**	−2.54	0.01***
	d(LNFD)	−4.04	0.00*	−4.10	0.00***
	d(LNG)	−4.14	0.00*	−2.98	0.00***
	d(LNECG)	−7.24	0.00*	−6.16	0.00***
	d(LNTG)	−6.01	0.00*	−5.50	0.00***
	d(LNFG)	−8.10	0.00*	−6.06	0.00***
	d(LNSG)	−3.36	0.02**	−2.26	0.03**
	d(LNIPG)	−4.18	0.00*	−3.23	0.00***
	d(LNIFG)	−6.21	0.00*	−5.38	0.00***
	d(LNCG)	−3.16	0.03**	−2.49	0.01***
	d(LNPG)	−3.67	0.01*	−3.55	0.00***

a: (*) Significant at the 10%; (**) Significant at the 5%; (***) Significant at the 1% and (no) Not Significant, b: Lag Length based on SIC, c: Probability based on MacKinnon (1996) one-sided P-values.
Source: Own Estimation.

TABLE 18.6 Unit root test results.

Variable	Level		First difference	
	Model A	Model C	Model A	Model C
LNEG Test-Stat.	-4.98**	-8.40***		
1%	-5.34	-5.57		
5%	-4.93	-5.08		
10%	-4.58	-4.82		
Break	1989	1988		
LNEC Test-Stat.	-5.15**	-4.79		-6.30***
1%	-5.34	-5.57		-5.57
5%	-4.93	-5.08		-5.08
10%	-4.58	-4.82		-4.82
Break	1990	1990		1992
LNFD Test-Stat.	-4.54	-5.23**	-6.79***	
1%	-5.34	-5.57	-5.34	
5%	-4.93	-5.08	-4.93	
10%	-4.58	-4.82	-4.58	
Break	1987	1988	1991	1991

Variable	Level		First difference	
	Model A	Model C	Model A	Model C
LNFG Test-Stat.	-3.66	-4.68	-7.33***	-7.16***
1%	-5.34	-5.57	-5.34	-5.57
5%	-4.93	-5.08	-4.93	-5.08
10%	-4.58	-4.82	-4.58	-4.82
Break	2002	2002	2005	2005
LNIFG Test-Stat.	-4.61*	-3.12		-4.86*
1%	-5.34	-5.57		-5.57
5%	-4.93	-5.08		-5.08
10%	-4.58	-4.82		-4.82
Break	1995	2004		1994
LNTG Test-Stat.	-6.01***	-5.99***		
1%	-5.34	-5.57		
5%	-4.93	-5.08		
10%	-4.58	-4.82		
Break	1989	1989		

	Statistic	(1)	(2)	(3)	(4)
LNG	Test-Stat.	−3.55	−3.53	−6.27***	−6.28***
	1%	−5.34	−5.57	−5.34	−5.57
	5%	−4.93	−5.08	−4.93	−5.08
	10%	−4.58	−4.82	−4.58	−4.82
	Break	2002	2002	1990	1990
LNCG	Test-Stat.	−6.01***	−5.99***		
	1%	−5.34	−5.57		
	5%	−4.93	−5.08		
	10%	−4.58	−4.82		
	Break	1989	1989		
LNECG	Test-Stat.	−3.87	−3.57	−9.11***	−8.99***
	1%	−5.34	−5.57	−5.34	−5.57
	5%	−4.93	−5.08	−4.93	−5.08
	10%	−4.58	−4.82	−4.58	−4.82
	Break	2002	2002	2005	2005
LNIPG	Test-Stat.	−6.55***	−5.18**	−5.20**	−8.21***
	1%	−5.34	−5.57	−5.34	−5.57
	5%	−4.93	−5.08	−4.93	−5.08
	10%	−4.58	−4.82	−4.58	−4.82
	Break	1992	2005	1989	1993
LNSG	Test-Stat.	−4.5	−3.61		
	1%	−5.34	−5.57		
	5%	−4.93	−5.08		
	10%	−4.58	−4.82		
	Break	1991	1991		
LNPG	Test-Stat.	−5.77***	−5.17***		
	1%	−5.34	−5.57		
	5%	−4.93	−5.08		
	10%	−4.58	−4.82		
	Break	1993	1993		

a: (*) Significant at the 10%; (**) Significant at the 5%; (***) Significant at the 1% and (no) Not Significant, b: Lag Length based on SIC, c: Test Stat. indicates test statistics.
Source: Own Estimation.

454 Energy-Growth Nexus in an Era of Globalization

TABLE 18.7 Summarized nonlinear autoregressive distributed lag bounds test results.

Model	Equation	F statistics (F_{PSS})	Outcome
Model 1	$\ln EC_t = \beta_1 + \beta_2 \ln G_t + \beta_3 \ln EG_t + \beta_4 \ln FD_t + \beta_5 DUM + \varepsilon_t$	3.9748	No cointegration
Model 2	$\ln EC_t = \beta_1 + \beta_2 \ln ECG_t + \beta_3 \ln EG_t + \beta_4 \ln FD_t + \beta_5 DUM + \varepsilon_t$	3.3714	No cointegration
Model 3	$\ln EC_t = \beta_1 + \beta_2 \ln TG_t + \beta_3 \ln EG_t + \beta_4 \ln FD_t + \beta_5 DUM + \varepsilon_t$	3.5862	No cointegration
Model 4	$\ln EC_t = \beta_1 + \beta_2 \ln FG_t + \beta_3 \ln EG_t + \beta_4 \ln FD_t + \beta_5 DUM$	5.1517***	Cointegration
Model 5	$\ln EC_t = \beta_1 + \beta_2 \ln SG_t + \beta_3 \ln EG_t + \beta_4 \ln FD_t + \beta_5 DUM + \varepsilon_t$	2.4293	No cointegration
Model 6	$\ln EC_t = \beta_1 + \beta_2 \ln IPG_t + \beta_3 \ln EG_t + \beta_4 \ln FD_t + \beta_5 DUM + \varepsilon_t$	1.6073	No cointegration
Model 7	$\ln EC_t = \beta_1 + \beta_2 \ln IFG_t + \beta_3 \ln EG_t + \beta_4 \ln FD_t + \beta_5 DUM + \varepsilon_t$	1.3689	No cointegration
Model 8	$\ln EC_t = \beta_1 + \beta_2 \ln CG_t + \beta_3 \ln EG_t + \beta_4 \ln FD_t + \beta_5 DUM + \varepsilon_t$	2.5439	No cointegration
Model 9	$\ln EC_t = \beta_1 + \beta_2 \ln PG_t + \beta_3 \ln EG_t + \beta_4 \ln FD_t + \beta_5 DUM + \varepsilon_t$	3.3279	No cointegration

a: (*) Significant at the 10%; (**) Significant at the 5%; (***) Significant at the 1%.
Source: Own Estimation.

TABLE 18.8 Nonlinear autoregressive distributed lag results.

Panel A: cointegration		
F_PSS	5.1557	

Panel B: long-run coefficients		
	Coefficient	F-stat	Probability
$LNFG^+$	−1.851	9.699	0.009***
$LNFG^-$	−1.231	2.995	0.109
$LNEG^+$	2.316	8.345	0.014**
$LNEG^-$	−0.962	3.701	0.078*
$LNFD^+$	−3.335	9.706	0.009***
$LNFD^-$	1.385	6.178	0.029**

Panel C: short-run coefficients		
	Coefficient	t-stat	Probability
$LNEC_{t-1}$	−0.483	−2.84	0.015**
$LNFG^+_{t-1}$	−0.894	−3.24	0.007***
$LNFG^-_{t-1}$	0.465	1.9	0.082*
$LNEG^+_{t-1}$	−0.595	−2.78	0.017**
$LNEG^-_{t-1}$	1.612	4.83	0.000***
$LNFD^+_{t-1}$	1.119	3.98	0.002***
$LNFD^-_{t-1}$	−0.669	−3.91	0.002***
$\Delta LNEC_{t-1}$	−0.319	−1.51	0.156
$\Delta LNFG^+_t$	−0.830	−3.73	0.003***
$\Delta LNFG^+_{t-1}$	−0.236	−1.09	0.299
$\Delta LNFG^-_t$	0.112	0.48	0.642
$\Delta LNFG^-_{t-1}$	−0.001	−0.01	0.996
$\Delta LNEG^+_t$	0.250	0.48	0.637
$\Delta LNEG^+_{t-1}$	−0.404	−0.94	0.365
$\Delta LNEG^-_t$	1.828	3.46	0.005***
$\Delta LNEG^-_{t-1}$	−0.817	−1.52	0.154
$\Delta LNFD^+_t$	0.564	2.92	0.013**
$\Delta LNFD^+_{t-1}$	−0.327	−1.74	0.108
$\Delta LNFD^-_t$	−0.230	−1.73	0.109

Continued

TABLE 18.8 Nonlinear autoregressive distributed lag results.—cont'd

Panel C: short-run coefficients			
	Coefficient	t-stat	Probability
$\Delta LNFD^-_{t-1}$	0.300	1.58	0.140
DUM	0.020	0.5	0.629
Constant	4.020	2.91	0.013**

Panel D: asymmetry test results		
	F-stat	Probability
W^{FG}_{LR}	9.81	0.009***
W^{EG}_{LR}	6.98	0.021**
W^{FD}_{LR}	8.10	0.015**
W^{FG}_{SR}	2.91	0.113
W^{EG}_{SR}	0.95	0.349
W^{FD}_{SR}	0.14	0.714

Panel E: diagnostic tests		
	Statistics	Probability
Portmanteau autocorrelation	17.48	0.29
Breusch/Pagan heteroscedasticity	0.43	0.51
Ramsey RESET function	0.44	0.72
Jarque—Bera normality	9.74	0.07

a: (*) Significant at the 10%; (**) Significant at the 5%; (***) Significant at the 1%.
Source: Own Estimation.

According to the bounds test results, only Model four identified a cointegration relationship among the variables. In this model, the financial globalization, economic growth, and financial development indices were the explanatory variables (Table 18.7).

According to the asymmetric long-run results shown in Panel B, positive financial globalization shocks significantly reduce energy consumption. Energy consumption is negatively affected by negative shocks in economic growth and positively affected by positive shocks in economic growth. In Poland, positive shocks in financial development reduce energy consumption. On the other hand, energy consumption affected positively by negative shocks in financial development (Table 18.8).

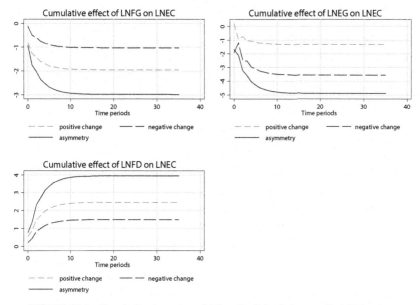

FIGURE 18.4 Cumulative dynamic multipliers for Poland. *Source: Own Estimation.*

According to the asymmetric short-run results shown in Panel C, energy consumption is inversely affected by financial globalization. That means positive shocks in financial globalization reduce energy consumption and negative shocks in financial globalization increase energy consumption. Positive shocks in economic growth reduce energy consumption and negative shocks in economic growth increase energy consumption. Positive shocks in financial development increase energy consumption and negative shocks in financial development reduce energy consumption (Table 18.8).

A dynamic multiplier graph for NARDL was plotted as shown in Fig. 18.4 to evaluate the adjustment of the asymmetry in the current long-term equilibrium after switching to a new long-erm equilibrium due to negative and positive shocks.

Fig. 18.4 indicates the adjustment pattern of energy consumption to negative and positive shocks in financial globalization, financial development, and economic growth for the Poland economy. The asymmetric response in energy consumption due to positive and negative financial globalization shocks is significant that confirms that financial globalization overall positively impacts energy consumption. After about 10 years, the asymmetrical relationship between energy consumption and financial globalization, economic growth and financial development has been disappeared. Reduces in financial globalization have a greater impact than increases in energy consumption.

5.4 Causality test results

Asymmetric causality testing can figure out interesting results about the natural dynamics of the period of investigation. Asymmetric causality test enables to investigate positive and negative shocks permanent effects on the underlying variable. Classical causality test does not allow to investigate whether a positive or negative shock has significant predictive value.

In this section, we investigate the asymmetric causality relationship between energy consumption and globalization types, economic growth and financial development variables.

The results of the causality tests are summarized in Table 18.9. The asymmetric causality tests results indicate that there is a unidirectional causality between globalization and energy consumption in Poland. However,

TABLE 18.9 Hatemi-J (2012) asymmetric causality test.

Direction of causality	Test statistics	Critical values		
		1%	5%	10%
$LNG^+ -> LNEC^+$	0.630	11.261	5.301	3.413
$LNG^+ -> LNEC^-$	6.806**	12.847	6.331	4.090
$LNG^- -> LNEC^-$	0.102	24.834	6.940	3.863
$LNG^- -> LNEC^+$	6.747**	9.657	5.034	3.322
$LNECG^+ -> LNEC^+$	2.610	10.234	5.081	3.348
$LNECG^+ -> LNEC^-$	0.102	9.977	4.999	3.384
$LNECG^- -> LNEC^-$	0.296	10.312	5.136	3.413
$LNECG^- -> LNEC^+$	0.126	10.239	5.042	3.359
$LNTG^+ -> LNEC^+$	3.923*	10.495	5.086	3.395
$LNTG^+ -> LNEC^-$	1.499	10.045	5.237	3.475
$LNTG^- -> LNEC^-$	3.033	11.238	4.896	3.180
$LNTG^- -> LNEC^+$	0.356	10.938	5.173	3.469
$LNFG^+ -> LNEC^+$	0.018	9.982	4.990	3.337
$LNFG^+ -> LNEC^-$	1.181	10.471	5.142	3.485
$LNFG^- -> LNEC^-$	0.118	12.523	5.277	3.430
$LNFG^- -> LNEC^+$	0.581	10.786	4.987	3.277
$LNSG^+ -> LNEC^+$	4.435	22.213	10.452	7.028
$LNSG^+ -> LNEC^-$	0.119	8.292	4.474	2.916

Continued

TABLE 18.9 Hatemi-J (2012) asymmetric causality test.—cont'd

Direction of causality	Test statistics	Critical values		
		1%	5%	10%
$LNSG^- - > LNEC^-$	0.025	20.379	5.279	3.083
$LNSG^+ - > LNEC^-$	0.213	9.589	4.597	3.115
$LNIPG^+ - > LNEC^+$	0.048	10.445	5.128	3.400
$LNIPG^+ - > LNEC^-$	4.341	20.468	10.465	7.325
$LNIPG^- - > LNEC^-$	0.274	23.589	10.559	6.972
$LNIPG^- - > LNEC^+$	17.772**	21.627	10.308	7.211
$LNIFG^+ - > LNEC^+$	5.931	20.978	10.143	6.996
$LNIFG^+ - > LNEC^-$	0.150	20.285	9.782	6.575
$LNIFG^- - > LNEC^-$	3.139	18.229	8.721	6.038
$LNIFG^- - > LNEC^+$	3.023	22.238	10.398	7.262
$LNCG^+ - > LNEC^-$	0.983	22.910	11.146	7.523
$LNPG^+ - > LNEC^+$	0.692	10.601	5.121	3.400
$LNPG^+ - > LNEC^-$	3.017	23.571	12.158	8.498
$LNPG^- - > LNEC^-$	0.012	18.314	6.701	4.008
$LNPG^- - > LNEC^+$	7.066	23.046	10.713	7.333
$LNFD^+ - > LNEC^+$	0.969	11.407	5.154	3.392
$LNFD^+ - > LNEC^-$	17.405**	23.632	11.626	8.137
$LNFD^- - > LNEC^-$	0.057	18.163	7.835	4.735
$LNFD^- - > LNEC^+$	1.795	18.208	8.676	6.106
$LNEG^+ - > LNEC^+$	1.000	10.495	5.214	3.367
$LNEG^+ - > LNEC^-$	1.123	22.662	10.468	7.039
$LNEG^- - > LNEC^-$	0.049	24.245	11.403	7.408
$LNEG^- - > LNEC^+$	0.517	21.771	10.473	7.330
$LNEC^+ - > LNG^+$	0.937	10.592	5.234	3.438
$LNEC^+ - > LNG^-$	0.333	18.887	9.568	6.524
$LNEC^- - > LNG^-$	0.251	21.356	6.622	3.380
$LNEC^- - > LNG^+$	5.902	27.800	12.954	8.686
$LNEC^+ - > LNFD^+$	0.145	9.966	5.167	3.548

Continued

TABLE 18.9 Hatemi-J (2012) asymmetric causality test.—cont'd

Direction of causality	Test statistics	Critical values		
		1%	5%	10%
$LNEC^+ - > LNFD^-$	0.357	11.109	5.146	3.284
$LNEC^- - > LNFD^-$	0.149	11.659	5.031	3.252
$LNEC^- - > LNFD^+$	0.193	13.531	5.924	3.579
$LNEC^+ - > LNPG^+$	0.563	9.966	5.140	3.502
$LNEC^+ - > LNPG^-$	0.005	11.841	5.326	3.373
$LNEC^- - > LNPG^-$	0.008	11.331	5.417	3.474
$LNEC^- - > LNPG^+$	1.866	15.606	6.063	3.274
$LNEC^+ - > LNEG^+$	0.196	10.082	4.945	3.455
$LNEC^+ - > LNEG^-$	0.462	21.309	10.400	7.091
$LNEC^- - > LNEG^-$	3.566	31.591	13.095	8.385
$LNEC^- - > LNEG^+$	0.866	23.120	11.397	7.593
$LNEC^+ - > LNCG^+$	1.452	21.336	10.466	7.256
$LNEC^+ - > LNCG^-$	15.907**	27.776	12.674	8.216
$LNEC^+ - > LNIFG^+$	2.451	21.119	10.016	6.802
$LNEC^+ - > LNIFG^-$	0.778	22.072	9.759	6.508
$LNEC^- - > LNIFG^-$	50.565***	27.057	12.286	7.958
$LNEC^- - > LNIFG^+$	7.992	28.208	12.766	8.322
$LNEC^+ - > LNIPG^+$	2.171	10.797	5.376	3.454
$LNEC^+ - > LNIPG^-$	19.327**	22.935	10.389	6.985
$LNEC^- - > LNIPG^-$	0.256	33.180	13.779	8.548
$LNEC^- - > LNIPG^+$	3.486	22.761	10.757	7.253
$LNEC^+ - > LNSG^+$	3.510	24.026	11.823	8.053
$LNEC^+ - > LNSG^-$	4.020*	12.190	5.223	3.268
$LNEC^- - > LNSG^-$	0.134	25.942	6.023	3.706
$LNEC^- - > LNSG^+$	7.237**	14.125	5.967	3.786
$LNEC^+ - > LNFG^+$	0.120	11.549	5.465	3.632
$LNEC^+ - > LNFG^-$	7.650	25.109	11.649	7.731
$LNEC^- - > LNFG^-$	1.292	23.931	7.291	4.161

Continued

TABLE 18.9 Hatemi-J (2012) asymmetric causality test.—cont'd

Direction of causality	Test statistics	Critical values		
		1%	5%	10%
$LNEC^- -> LNFG^+$	0.643	23.270	10.645	7.397
$LNEC^+ -> LNTG^+$	0.575	10.158	5.232	3.495
$LNEC^+ -> LNTG^-$	0.241	12.403	5.452	3.423
$LNEC^- -> LNTG^-$	85.239***	26.419	8.243	4.475
$LNEC^- -> LNTG^+$	0.167	24.970	6.641	3.740
$LNEC^+ -> LNECG^+$	0.310	10.648	5.236	3.430
$LNEC^+ -> LNECG^-$	0.275	9.522	5.086	3.414
$LNEC^- -> LNECG^-$	5.158*	28.994	8.172	4.455
$LNEC^- -> LNECG^+$	0.008	12.510	5.876	3.834

a: (*) Significant at the 10%; (**) Significant at the 5%; (***) Significant at the 1% and (no) Not Significant, b: Lag Length based on HJC.
Source: Own Estimation.

this causality is asymmetric as positive shocks in globalization causality of negative shocks in energy consumption and negative shocks in globalization causality of negative shocks in energy consumption. From trade globalization and interpersonal globalization to energy consumption is determined causality. Also, there is a causality relationship from negative shocks in energy consumption to negative shocks in trade globalization and positive shocks in energy consumption to negative shocks in interpersonal globalization. It means that, there is a bidirectional causality relationship between trade globalization and energy consumption. Positive shocks in energy consumption causality of negative shocks in cultural globalization and social globalization. Between cultural globalization and energy consumption, and social globalization and energy consumption have determined unidirectional causality relationship. Positive shocks in financial development causality of negative shocks in energy consumption.

6. Conclusion

Analyzing the relation of GDP and electricity consumption in Poland in recent years can be distinguished several periods.

The first until the summer of 1996—2003, i.e., the period of "zero-energy" growth. At that time, GDP growth amounted to approx. 27% without increasing electricity consumption—due to the use of large energy efficiency reserves throughout the economy. Another period to 2004—2008, when taking into account the increase in electricity consumption by about 2% per year in conditions of growth (over 5%). The period from the third to the summer of 2009—2015. It is a time of a slow increase in electricity consumption (on average about 0.8% a year) with a moderate GDP growth rate (on average about 3% a year). In the last 2 years (2015 and 2016), electricity consumption has grown on average by almost 2% annually with an average GDP growth of approx. 3% annually. This is a much larger increase in energy consumption in relation to GDP growth than in several previous years (Ouyang and Li, 2018).

Poland economy is one of the rising economies in the EU. However, for the EU, energy efficiency is also important and our results support economic growth has a positive impact on energy consumption. From this result, can figure out the energy efficiency is in Poland is low. Nevertheless, they only rely on simple indicators, e.g., energy consumption based on GDP, it is easy to formulate incorrect opinions—such as those about the poor energy efficiency of the Polish economy. Our empirical results that positive shocks in economic growth increase energy consumption support to growth hypothesis which indicate a relationship between economic growth to energy consumption. To sum up, the empirical results of the study show that the economic growth of Poland is energy-dependent, so one can state that electricity consumption is a limiting factor in the economic growth of Poland. It means that the energy policy may have a strong negative impact on the economic growth and development of the Polish economy.

According to the determination of the World WorldBank (2020), financial development is the process of decreasing financial cost as acquiring information and enforcing contracts. Financial development impact on energy consumption is asymmetric that means improvements in financial development reduce energy consumption. Financial development can create an opportunity for small and medium-sized enterprises (SMEs) to access finance. Thus, SMEs can grow by using more efficient production techniques. Financial development can lead to a reduction in energy consumption by providing easily accessible and cost-effective capital to access energy-friendly technologies (Chang, 2015; Durusu-Ciftci et al., 2020; Islam et al., 2013). Our NARDL and causality test results support this finding for the case of Poland.

Financial globalization which indicates to an individual country's linkages to international capital markets has determined a positive impact on energy consumption. That means increases in financial globalization decrease energy consumption. This result also supports by causality test that determined a unidirectional causality from globalization to energy consumption.

References

Akarca, A.T., Long, T.V., 1980. On the relationship between energy and GNP: a reexamination. J. Energy Dev. 326—331.

Al-mulali, U., Lee, J.Y.M., 2013. Estimating the impact of the financial development on energy consumption: evidence from the GCC (Gulf Cooperation Council) countries. Energy 60, 215−221.

Altınay, G., Karagöl, E., 2004. Structural break, unit root, and the causality between energy consumption and GDP in Turkey. Energy Econ. 26 (6), 985−994.

Ang, J.B., 2007. CO_2 emissions, energy consumption, and output in France. Energy Policy 35 (10), 4772−4778.

Antweiler, W., Copeland, B.R., Taylor, M.S., 2001. Is free trade good for the environment? Am. Econ. Rev. 91 (4), 877−908.

Bekhet, H.A., Matar, A., Yasmin, T., 2017. CO_2 emissions, energy consumption, economic growth, and financial development in GCC countries: dynamic simultaneous equation models. Renew. Sustain. Energy Rev. 70, 117−132.

Belke, A., Dobnik, F., Dreger, C., 2011. Energy consumption and economic growth: new insights into the cointegration relationship. Energy Econ. 33 (5), 782−789.

Belloumi, M., 2009. Energy consumption and GDP in Tunisia: cointegration and causality analysis. Energy Policy 37 (7), 2745−2753.

Chang, S.C., 2015. Effects of financial developments and income on energy consumption. Int. Rev. Econ. Finance 35, 28−44.

Chen, S., Saud, S., Bano, S., Haseeb, A., 2019. The nexus between financial development, globalization, and environmental degradation: fresh evidence from Central and Eastern European Countries. Environ. Sci. Pollut. Res. 26 (24), 24733−24747.

Coban, S., Topcu, M., 2013. The nexus between financial development and energy consumption in the EU: a dynamic panel data analysis. Energy Econ. 39, 81−88.

Cole, M.A., 2006. Does trade liberalization increase national energy use? Econ. Lett. 92 (1), 108−112.

Dollar, D., Kraay, A., 2004. Trade, growth, and poverty. Econ. J. 114 (493), F22−F49.

Durusu-Ciftci, D., Soytas, U., Nazlioglu, S., 2020. Financial development and energy consumption in emerging markets: smooth structural shifts and causal linkages. Energy Econ. 104729.

Erdal, G., Erdal, H., Esengün, K., 2008. The causality between energy consumption and economic growth in Turkey. Energy Policy 36 (10), 3838−3842.

Erol, U., Yu, E.S., 1987. On the causal relationship between energy and income for industrialized countries. J. Energy Dev. 113−122.

EU, 2019. 2018 Assessment of the Progress Made by Member States towards the National Energy Efficiency Targets for 2020 and towards the Implementation of the Energy Efficiency Directive as Required by Article 24(3) of the Energy Efficiency Directive 2012/27/EU. Retrieved from Brussels. https://ec.europa.eu/commission/sites/beta-political/files/report-2018-assessment-progress-energy-efficiency-targets-april2019_en.pdf.

Fatai, K., Oxley, L., Scrimgeour, F., 2002. Energy consumption and employment in New Zealand: searching for causality. In: Paper Presented at the NZAE Conference, Wellington.

Furuoka, F., 2015. Financial development and energy consumption: evidence from a heterogeneous panel of Asian countries. Renew. Sustain. Energy Rev. 52, 430−444.

Ghali, K.H., El-Sakka, M.I., 2004. Energy use and output growth in Canada: a multivariate cointegration analysis. Energy Econ. 26 (2), 225−238.

Gorynia, M., Wolniak, R., 2002. The participation of transitional economy in globalisation−the case of Poland. J. Euro-Asian Manage. 6 (2), 57−75.

Gygli, S., Haelg, F., Potrafke, N., Sturm, J.-E., 2019. The KOF globalisation index−revisited. Rev. Int. Organ. 14 (3), 543−574.

Halıcıoğlu, F., 2009. An econometric study of CO_2 emissions, energy consumption, income and foreign trade in Turkey. Energy Policy 37 (3), 1156−1164.

Harvey, D.I., Leybourne, S.J., 2007. Testing for time series linearity. Econom. J. 10 (1), 149−165.

Harvey, D.I., Leybourne, S.J., Xiao, B., 2008. A powerful test for linearity when the order of integration is unknown. Stud. Nonlinear Dynam. Econom. 12 (3).

Hatemi-J, A., 2003. A new method to choose optimal lag order in stable and unstable VAR models. Appl. Econ. Lett. 10 (3), 135−137.

Hatemi-J, A., 2012. Asymmetric causality tests with an application. Empir. Econ. 43 (1), 447−456.

Hondroyiannis, G., Lolos, S., Papapetrou, E., 2002. Energy consumption and economic growth: assessing the evidence from Greece. Energy Econ. 24 (4), 319−336.

Huang, B.-N., Hwang, M.J., Yang, C.W., 2008. Causal relationship between energy consumption and GDP growth revisited: a dynamic panel data approach. Ecol. Econ. 67 (1), 41−54.

IMF, 2019. World Economic Outlook (WEO). October 2019. Retrieved from. https://www.imf.org/external/pubs/ft/weo/2020/01/weodata/index.aspx.

Islam, F., Shahbaz, M., Ahmed, A.U., Alam, M.M., 2013. Financial development and energy consumption nexus in Malaysia: a multivariate time series analysis. Econ. Modell. 30, 435−441.

Jakovac, P., 2018. Causality between energy consumption and economic growth: literature review. In: Paper Presented at the INTCESS2018- 5th International Conference on Education and Social Sciences 5-7 February 2018- Istanbul, Turkey, Istanbul, Turkey.

Kasperowicz, R., 2014. Economic growth and energy consumption in 12 European countries: a panel data approach. J. Int. Stud. 7 (3), 112−122.

Koengkan, M., 2017. Is the globalization influencing the primary energy consumption? The case of Latin America and Caribbean countries. Cadernos UniFOA 12 (33), 57−67.

Komal, R., Abbas, F., 2015. Linking financial development, economic growth and energy consumption in Pakistan. Renew. Sustain. Energy Rev. 44, 211−220.

Kraft, J., Kraft, A., 1978. On the relationship between energy and GNP. J. Energy Dev. 3, 401−403.

Lee, C.-C., Chang, C.-P., 2005. Structural breaks, energy consumption, and economic growth revisited: evidence from Taiwan. Energy Econ. 27 (6), 857−872.

Lee, C.-C., Chang, C.-P., 2008. Energy consumption and economic growth in Asian economies: a more comprehensive analysis using panel data. Resour. Energy Econ. 30 (1), 50−65.

Lee, C.-C., Chang, C.-P., Chen, P.-F., 2008. Energy-income causality in OECD countries revisited: the key role of capital stock. Energy Econ. 30 (5), 2359−2373.

Levine, R., 1999. Financial Development and Economic Growth: Views and Agenda.

Liu, L., Zhou, C., Huang, J., Hao, Y., Trade, 2018. The impact of financial development on energy demand: evidence from China. Emerging Markets Finance 54 (2), 269−287.

MacKinnon, J., 1996. Numerical distribution functions for unit root and cointegration tests. J. Appl. Econ. 11, 601−618.

Mukhtarov, S., Mikayilov, J.I., Mammadov, J., Mammadov, E., 2018. The impact of financial development on energy consumption: evidence from an oil-rich economy. Energies 11 (6).

Narayan, P.K., Prasad, A., 2008. Electricity consumption−real GDP causality nexus: evidence from a bootstrapped causality test for 30 OECD countries. Energy Policy 36 (2), 910−918.

Narayan, P.K., Smyth, R., 2009. Multivariate Granger causality between electricity consumption, exports and GDP: evidence from a panel of Middle Eastern countries. Energy Policy 37 (1), 229−236.

Ouyang, Y.F., Li, P., 2018. On the nexus of financial development, economic growth, and energy consumption in China: new perspective from a GMM panel VAR approach. Energy Econ. 71, 238−252.

Öztürk, İ., 2010. A literature survey on energy−growth nexus. Energy Policy 38, 340−349.

Öztürk, İ., Acaravcı, A., 2013. The long-run and causal analysis of energy, growth, openness and financial development on carbon emissions in Turkey. Energy Econ. 36, 262—267.

Öztürk, İ., Aslan, A., Kalyoncu, H., 2010. Energy consumption and economic growth relationship: evidence from panel data for low and middle income countries. Energy Policy 38 (8), 4422—4428.

Paramati, S.R., Ummalla, M., Apergis, N., 2016. The effect of foreign direct investment and stock market growth on clean energy use across a panel of emerging market economies. Energy Econ. 56, 29—41.

Payne, J.E., 2009. On the dynamics of energy consumption and output in the US. Appl. Energy 86 (4), 575—577.

Pesaran, M.H., Shin, Y.C., Smith, R.J., 2001. Bounds testing approaches to the analysis of level relationships. J. Appl. Econom. 16 (3), 289—326. https://doi.org/10.1002/jae.616.

Phillips, P.C., Perron, P., 1988. Testing for a unit root in time series regression. Biometrika 75 (2), 335—346.

Pradhan, R.P., Arvin, M.B., Nair, M., Bennett, S.E., Hall, J.H., 2018. The dynamics between energy consumption patterns, financial sector development and economic growth in Financial Action Task Force (FATF) countries. Energy 159, 42—53.

Saboori, B., Sapri, M., bin Baba, M., 2014. Economic growth, energy consumption and CO_2 emissions in OECD (Organization for Economic Co-operation and Development)'s transport sector: a fully modified bi-directional relationship approach. Energy 66, 150—161.

Sadorsky, P., 2010. The impact of financial development on energy consumption in emerging economies. Energy Policy 38 (5), 2528—2535.

Sadorsky, P., 2011a. Financial development and energy consumption in Central and Eastern European frontier economies. Energy Policy 39 (2), 999—1006.

Sadorsky, P., 2011b. Trade and energy consumption in the Middle East. Energy Econ. 33 (5), 739—749.

Sadorsky, P., 2012. Energy consumption, output and trade in South America. Energy Econ. 34 (2), 476—488.

Sbia, R., Shahbaz, M., Hamdi, H., 2014. A contribution of foreign direct investment, clean energy, trade openness, carbon emissions and economic growth to energy demand in UAE. Econ. Modell. 36, 191—197.

Shahbaz, M., Hoang, T.H.V., Mahalik, M.K., Roubaud, D., 2017. Energy consumption, financial development and economic growth in India: new evidence from a nonlinear and asymmetric analysis. Energy Econ. 63, 199—212.

Shahbaz, M., Hye, Q.M.A., Tiwari, A.K., Leitao, N.C., 2013. Economic growth, energy consumption, financial development, international trade and CO_2 emissions in Indonesia. Renew. Sustain. Energy Rev. 25, 109—121.

Shahbaz, M., Khan, S., Tahir, M.I., 2013. The dynamic links between energy consumption, economic growth, financial development and trade in China: fresh evidence from multivariate framework analysis. Energy Econ. 40, 8—21.

Shahbaz, M., Lean, H.H., 2012. Does financial development increase energy consumption? The role of industrialization and urbanization in Tunisia. Energy Policy 40, 473—479.

Shahbaz, M., Mallick, H., Mahalik, M.K., Sadorsky, P., 2016. The role of globalization on the recent evolution of energy demand in India: implications for sustainable development. Energy Econ. 55, 52—68.

Shahbaz, M., Nasreen, S., Ling, C.H., Sbia, R., 2014. Causality between trade openness and energy consumption: what causes what in high, middle and low income countries. Energy Policy 70, 126—143.

Shahbaz, M., Shahzad, S.J.H., Alam, S., Apergis, N., 2018a. Globalisation, economic growth and energy consumption in the BRICS region: the importance of asymmetries. J. Int. Trade Econ. Dev. 27 (8), 985−1009.

Shahbaz, M., Shahzad, S.J.H., Mahalik, M.K., Sadorsky, P., 2018b. How strong is the causal relationship between globalization and energy consumption in developed economies? A country-specific time-series and panel analysis. Appl. Econ. 50 (13), 1479−1494.

Shahbaz, M., Solarin, S.A., Mahmood, H., Arouri, M., 2013. Does financial development reduce CO_2 emissions in Malaysian economy? A time series analysis. Econ. Modell. 35, 145−152.

Shahzad, S.J.H., Kumar, R.R., Zakaria, M., Hurr, M., 2017. Carbon emission, energy consumption, trade openness and financial development in Pakistan: a revisit. Renew. Sustain. Energy Rev. 70, 185−192.

Shin, Y., Yu, B., Greenwood-Nimmo, M., 2014. Modelling asymmetric cointegration and dynamic multipliers in a nonlinear ARDL framework. In: Festschrift in Honor of Peter Schmidt. Springer, pp. 281−314.

Smiech, S., Papiez, M., 2014. Energy consumption and economic growth in the light of meeting the targets of energy policy in the EU: the bootstrap panel Granger causality approach. Energy Policy 71, 118−129.

Soytaş, U., Sarı, R., 2003. Energy consumption and GDP: causality relationship in G-7 countries and emerging markets. Energy Econ. 25 (1), 33−37.

Soytaş, U., Sarı, R., Özdemir, Ö., 2001. Energy consumption and GDP relation in Turkey: a cointegration and vector error correction analysis. Econ. Bus. Trans. Facilitat. Compet. Change Global Environ. Proc. 1, 838−844.

Stern, D.I., 2007. The effect of NAFTA on energy and environmental efficiency in Mexico. Pol. Stud. J. 35 (2), 291−322.

Streimikiene, D., Kasperowicz, R., 2016. Economic growth and energy consumption: comparative analysis of V4 and the "old" EU countries. Renew. Sustain. Energy Rev. 59, 1545−1549.

WorldBank, 2020. Global Financial Development Report. Retrieved from. https://www.worldbank.org/en/publication/gfdr/gfdr-2016/background/financial-development.

Yu, E.S., Choi, J.-Y., 1985. The causal relationship between energy and GNP: an international comparison. J. Energy Dev. 249−272.

Yue, S., Lu, R., Shen, Y., Chen, H., 2019. How does financial development affect energy consumption? Evidence from 21 transitional countries. Energy Policy 130, 253−262.

Zhang, X.-P., Cheng, X.-M., 2009. Energy consumption, carbon emissions, and economic growth in China. Ecol. Econ. 68 (10), 2706−2712.

Ziaei, S.M., 2015. Effects of financial development indicators on energy consumption and CO2 emission of European, East Asian and Oceania countries. Renew. Sustain. Energy Rev. 42, 752−759.

Zivot, E., Andrews, D.W.K., 1992. Further evidence on the great crash, the oil-price shock, and the unit-root hypothesis. J. Bus. Econ. Stat. 10 (3), 251−270.

Index

Note: 'Page numbers followed by "b" indicate boxes, those followed by "f" indicate figures and those followed by "t" indicate tables.'

Printed in the United States
by Baker & Taylor Publisher Services